T0361855

# Applied Hydrodynamics

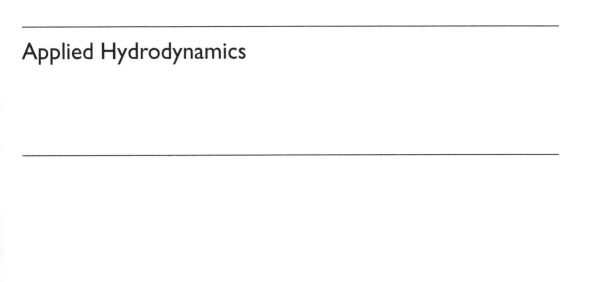

# Applied Hydrodynamics

## An Introduction

## Hubert Chanson
*The University of Queensland, Brisbane, Australia*

**CRC Press**
Taylor & Francis Group
Boca Raton  London  New York

CRC Press is an imprint of the
Taylor & Francis Group, an **informa** business

A BALKEMA BOOK

CRC Press
Taylor & Francis Group
6000 Broken Sound Parkway NW, Suite 300
Boca Raton, FL 33487-2742

First issued in hardback 2019

ISBN-13: 978-1-138-00093-3 (hbk)

Typeset by MPS Limited, Chennai, India

*Library of Congress Cataloging-in-Publication Data*

Chanson, Hubert.
   Applied hydrodynamics : an introduction / Hubert Chanson, The University of Queensland, Brisbane, Australia.
      pages cm
   Includes bibliographical references and index.
   ISBN 978-1-138-00093-3 (hardback)
   1. Hydrodynamics.   I. Title.
   TC171.C47   2014
   532'.5–dc23
                                                                        2013021204

## Dedication/Dédicace

*To Ya Hui.*
*Pour Bernard, Nicole et André.*

# Table of Contents

## Assignments

# List of symbols

A     cross-section area $(m^2)$;

C     celerity (m/s);

$C_D$     drag coefficient; $C_D = \dfrac{\text{Drag}}{\frac{1}{2} \times \rho \times V_O^2 \times \text{chord}}$ for a two-dimensional object;

$C_L$     lift coefficient; $C_L = \dfrac{\text{Lift}}{\frac{1}{2} \times \rho \times V_O^2 \times \text{chord}}$ for a two-dimensional object;

$C_c$     contraction coefficient;

$C_d$     discharge coefficient;

$C_v$     energy loss coefficient;

$D_H$     hydraulic diameter (m):

$$D_H = 4 \times \frac{\text{cross-sectional area}}{\text{wetted perimeter}} = \frac{4 \times A}{P_w}$$

d     flow depth (m);

e     internal energy per unit mass (J/kg);

$F_p$     pressure force (N);

$F_v$     volume force (N);

f     Darcy coefficient (or head loss coefficient, friction factor);

$f_{visc}$     viscous force (N);

g     gravity constant $(m/s^2)$: $g = 9.80 \, m/s^2$ (in Brisbane);

H     1. total head (m) defined as: $H = \dfrac{P}{\rho \times g} + z + \dfrac{V^2}{2 \times g}$

       2. piezometric head (m) defined as: $H = \dfrac{P}{\rho \times g} + z$

h     height (m);

K     1. vortex strength $(m^2/s)$ or circulation;
       2. hydraulic conductivity (m/s);

k     permeability $(m^2)$;

k     constant of proportionality;

$k_s$     equivalent sand roughness height (m);

L     length (m);

P     absolute pressure (Pa);

$P_d$     dynamic pressure (Pa);

$P_s$     static pressure (Pa);

Q    discharge ($m^3/s$);

q    discharge per meter width ($m^2/s$);

R    1. circle radius (m);

     2. cylinder radius (m);

$R_1$   radius (m);

Ro   gas constant: Ro = 8.3143 J/Kmole;

r    polar radial coordinate (m);

T    thermodynamic (or absolute) temperature (K);

U    volume force potential;

V    velocity (m/s);

v    specific volume ($m^3/kg$):

$$v = \frac{1}{\rho}$$

W    complex potential: $W = \phi + i \times \psi$;

w    complex velocity: $w = -V_x + i \times V_y$;

x    Cartesian coordinate (m);

y    Cartesian coordinate (m);

z    1. altitude (m);

     2. complex number (Chapters I-5 & I-6);

$\phi$    velocity potential ($m^2/s$);

$\Gamma$    circulation ($m^2/s$);

$\gamma$    specific heat ratio:

$$\gamma = \frac{C_p}{C_v}$$

$\mu$    1. strength of doublet ($m^3/s$);

     2. dynamic viscosity ($N.s/m^2$ or Pa.s);

$\nu$    kinematic viscosity ($m^2/s$):

$$\nu = \frac{\mu}{\rho}$$

$\pi$    $\pi$ = 3.141592653589793238462643;

$\theta$    polar coordinate (radian);

$\rho$    density ($kg/m^3$);

$\sigma$    surface tension (N/m);

$\tau$    shear stress (Pa);

$\tau_o$    average shear stress (Pa);

$\omega$    1. speed of rotation (rad/s);

     2. hydrodynamic frequency (Hz) of vortex shedding;

$\vec{\Psi}$    stream function vector; for a two-dimensional flow in the {x, y} plane:

$$\vec{\Psi} = (0, 0, \psi);$$

$\psi$    two-dimensional flow stream function ($m^2/s$);

## Subscript

n    normal component;

o    reference conditions: e.g., free-stream flow conditions;

r   radial component;
s   streamwise component;
x   x-component;
y   y-component;
z   z-component;
θ   ortho-radial component;

## Notes

1. Water at atmospheric pressure and 20.2 Celsius has a kinematic viscosity of exactly $10^{-6}$ m²/s.
2. Water in contact with air has a surface tension of about 0.073 N/m.

## Dimensionless numbers

Ca   Cauchy number (Henderson 1966):

$$Ca = \frac{\rho \times V^2}{E_{co}}$$ where $E_{co}$ is the fluid compressibility;

$C_D$   drag coefficient for a structural shape:

$$C_D = \frac{\tau_o}{\dfrac{1}{2} \times \rho \times V^2} = \frac{\text{shear stress}}{\text{dynamic pressure}}$$

where $\tau_o$ is the shear stress (Pa);
Note: other notations include $C_d$ and $C_f$;

Fr   Reech-Froude number:

$$Fr = \frac{V}{\sqrt{g \times d_{charac}}}$$

Note: some authors use the notation:

$$Fr = \frac{V^2}{g \times d_{charac}} = \frac{\rho \times V^2 \times A}{\rho \times g \times A \times d_{charac}} = \frac{\text{inertial force}}{\text{weight}}$$

M   Sarrau-Mach number:

$$M = \frac{V}{C}$$

Nu   Nusselt number:

$$Nu = \frac{h_t \times d_{charac}}{\lambda} = \frac{\text{heat transfer by convection}}{\text{heat transfer by conduction}}$$

where $h_t$ is the heat transfer coefficient (W/m²/K) and $\lambda$ is the thermal conductivity

Re    Reynolds number:

$$Re = \frac{V \times d_{charac}}{\nu} = \frac{\text{inertial forces}}{\text{viscous forces}}$$

Re*    shear Reynolds number:

$$Re^* = \frac{V^* \times k_s}{\nu}$$

St    Strouhal number:

$$St = \frac{\omega \times d_{charac}}{V_o}$$

We    Weber number:

$$We = \frac{V^2}{\dfrac{\sigma}{\rho \times d_{charac}}} = \frac{\text{inertial forces}}{\text{surface tension forces}}$$

Note: some authors use the notation: $We = \dfrac{V}{\sqrt{\dfrac{\sigma}{\rho \times d_{charac}}}}$

## Comments

The variable $d_{charac}$ characterises the geometric characteristic length (e.g. pipe diameter, flow depth, sphere diameter, …).

# Acknowledgements

The author wants to thank especially Professor Colin J. Apelt, University of Queensland, for his help, support and assistance all along the academic career of the writer. He thanks also Dr Sergio Montes, University of Tasmania for his positive feedback and advice on the course material.

He expresses his gratitude to all the people who provided photographs and illustrations of interest, including:

Mr Jacques-Henri Bordes (France);
Mr and Mrs J. Chanson (Paris, France);
Mr A. Chanson (Brisbane, Australia);
Ms Y.H. Chou (Brisbane, Australia);
Mr Francis Fruchard (Lyon, France)
Prof. C. Letchford, University of Tasmania;
Dassault Aviation;
Dryden Aircraft Photo Collection (NASA);
Equipe Cousteau, France;
NASA Earth Observatory;
Northrop Grumman Corporation;
Officine Maccaferri, Italy;
Rafale International;
Stéphane Saissi (France);
Sequana-Normandie;
VF communication, La Grande Arche;
Vought Retiree Club;
Washington State Department of Transport;

At last, but not the least, the author thanks all the people including students and former students, professionals, and colleagues who gave him information, feedback and comments on his lecture material. In particular, he acknowledges: Mr and Mrs J. Chanson (Paris, France); Ms Y.H. Chou (Brisbane, Australia); Mr G. Illidge (University of Queensland); Mrs N. Lemiere (France); Professor N. Rajaratnam (University of Alberta, Canada); Mr R. Stonard (University of Queensland).

# About the author

*Hubert Chanson, Professor in Hydraulic Engineering and Applied Fluid Mechanics*

Hubert Chanson received a degree of 'Ingénieur Hydraulicien' from the Ecole Nationale Supérieure d'Hydraulique et de Mécanique de Grenoble (France) in 1983 and a degree of 'Ingénieur Génie Atomique' from the 'Institut National des Sciences et Techniques Nucléaires' in 1984. He worked for the industry in France as a R&D engineer at the Atomic Energy Commission from 1984 to 1986, and as a computer professional in fluid mechanics for Thomson-CSF between 1989 and 1990. From 1986 to 1988, he studied at the University of Canterbury (New Zealand) as part of a Ph.D. project.

Hubert Chanson is a Professor in hydraulic engineering and applied fluid mechanics at the University of Queensland since 1990. His research interests include design of hydraulic engineering and structures, experimental investigations of two-phase flows, coastal hydrodynamics, water quality modelling, environmental management and natural resources. He authored single-handedly eight books including: "Hydraulic Design of Stepped Cascades, Channels, Weirs and Spillways" (*Pergamon*, 1995), "Air Bubble Entrainment in Free-Surface Turbulent Shear Flows" (*Academic Press*, 1997), "The Hydraulics of Open Channel Flow: An Introduction" (*Butterworth-Heinemann*, 1999 & 2004), "The Hydraulics of Stepped Chutes and Spillways" (*Balkema*, 2001), "Environmental Hydraulics of Open Channel Flows" (*Elsevier*, 2004), "Applied Hydrodynamics: an Introduction to Ideal and Real Fluid Flows" (*CRC Press*, 2009) and "Tidal Bores, Aegir, Eagre, Mascaret, Pororoca: Theory and Observations" (*World Scientific*, 2011). He co-authored the book "Fluid Mechanics for Ecologists" (*IPC Press*, 2002) and he edited several books (*Balkema* 2004, *IEaust* 2004, *The University of Queensland* 2006, 2008, 2010, *Engineers Australia* 2011). His textbook "The Hydraulics of Open Channel Flow: An Introduction" has already been translated into Chinese (*Hydrology Bureau of Yellow River Conservancy Committee*) and Spanish (*McGraw Hill Interamericana*), and it was re-edited in 2004. His publication record includes nearly 650 international refereed papers, and his work was cited over 4,000 times since 1990. Hubert Chanson has been active also as consultant for both governmental agencies and private organisations. He chaired the Organisation of the 34th IAHR World Congress held in Brisbane, Australia between 26 June and 1 July 2011.

The International Association for Hydraulic engineering and Research (IAHR) presented Hubert Chanson the 13th Arthur Ippen award for outstanding achievements in hydraulic engineering. This award is regarded as the highest achievement in hydraulic research. The American Society of Civil Engineers, Environmental and Water Resources Institute (ASCE-EWRI) presented him with the 2004 award for the best practice paper in the Journal of Irrigation and Drainage Engineering ("Energy Dissipation and Air Entrainment in Stepped Storm Waterway" by Chanson and Toombes 2002). In 1999 he was awarded a Doctor of Engineering from the University of Queensland for outstanding research achievements in gas-liquid bubbly flows.

He has been awarded eight fellowships from the Australian Academy of Science. In 1995 he was a Visiting Associate Professor at National Cheng Kung University (Taiwan R.O.C.) and he was Visiting Research Fellow at Toyohashi University of Technology (Japan) in 1999 and 2001. In 2008, he was an invited Professor at the University of Bordeaux (France). In 2004 and 2008, he was a visiting Research Fellow at Laboratoire Central des Ponts et Chaussées (France). In 2008 and 2010, he was an invited Professor at the Université de Bordeaux, I2M, Laboratoire des Transferts, Ecoulements, Fluides, et Energetique, where he is an adjunct research fellow.

Hubert Chanson was invited to deliver keynote lectures at the 1998 ASME Fluids Engineering Symposium on Flow Aeration (Washington DC), at the Workshop on Flow Characteristics around Hydraulic Structures (Nihon University, Japan 1998), at the first International Conference of the International Federation for Environmental Management System IFEMS'01 (Tsurugi, Japan 2001), at the 6th International Conference on Civil Engineering (Isfahan, Iran 2003), at the 2003 IAHR Biennial Congress (Thessaloniki, Greece), at the International Conference on Hydraulic Design of Dams and River Structures HDRS'04 (Tehran, Iran 2004), at the 9th International Symposium on River Sedimentation ISRS04 (Yichang, China 2004), at the International Junior Researcher and Engineer Workshop on Hydraulic Structures IJREW'06 (Montemor-o-Novo, Portugal 2006), at the 2nd International Conference on Estuaries & Coasts ICEC-2006 (Guangzhou, China 2006), at the 16th Australasian Fluid Mechanics Conference 16AFMC (Gold Coast, Australia 2007), at the 2008 ASCE-EWRI World Environmental and Water Resources Congress (Hawaii, USA 2008), the 2nd International Junior Researcher and Engineer Workshop on Hydraulic Structures IJREW'08 (Pisa, Italy 2008), the 11th Congrès Francophone des Techniques Laser CTFL 2008 (Poitiers, France 2008), International Workshop on Environmental Hydraulics IWEH09 (Valencia, Spain 2009), 17th Congress of IAHR Asia and Pacific Division (Auckland, New Zealand 2010), International Symposium on Water and City in Kanazawa – Tradition, Culture and Climate (Japan 2010), 2nd International Conference on Coastal Zone Engineering and Management (Arabian Coast 2010) (Oman 2010), NSF Partnerships for International Research and Education (PIRE) Workshop on "Modelling of Flood Hazards and Geomorphic Impacts of Levee Breach and Dam Failure" (Auckland 2012). He gave invited lectures at the International Workshop on Hydraulics of Stepped Spillways (ETH-Zürich, 2000), at the 2001 IAHR Biennial Congress (Beijing, China), at the International Workshop on State-of-the-Art in Hydraulic Engineering (Bari, Italy 2004), at the Australian Partnership for Sustainable Repositories Open Access Forum (Brisbane, Australia 2008), at the 4th International Symposium on Hydraulic Structures (Porto, Portugal 2012). He lectured several short courses in Australia and overseas (e.g. Taiwan, Japan, Italy).

His Internet home page is {http://www.uq.edu.au/~e2hchans}. He also developed a gallery of photographs website {http://www.uq.edu.au/~e2hchans/photo.html} that received more than 2,000 hits per month since inception. His open access publication webpage is the most downloaded publication record at the University of Queensland open access repository: {http://espace.library.uq.edu.au/list.php?browse=author&author_id=193}.

| | |
|---|---|
| {http://www.uq.edu.au/~e2hchans/photo.html} | Gallery of photographs in water engineering and environmental fluid mechanics |
| {http://www.uq.edu.au/~e2hchans/url_menu.html} | Internet technical resources in water engineering and environmental fluid mechanics |
| {http://www.uq.edu.au/~e2hchans/reprints.html} | Reprints of research papers in water engineering and environmental fluid mechanics |
| {http://espace.library.uq.edu.au/list.php?browse=author&author_id=193} | Open access publications at UQeSpace |

# Preface

Fluid dynamics is the engineering science dealing with forces generated by fluids in motion. Fluid dynamics and hydrodynamics play a vital role in our everyday life, from the ventilation of our home, the air flow around cars and aircrafts, wind loads on building. When driving a car, the air flow around the vehicle body induces some drag which increases with the square of the car speed, contributing to excess fuel consumption, as well as some downforce used in motor car racing. This advanced undergraduate and post-graduate textbook is designed especially assist senior undergraduate and postgraduate students in Aeronautical, Civil, Environmental, Hydraulic and Mechanical Engineering.

The textbook derives from a series of lecture notes developed by the author for the past twenty two years. The notes were enhanced based upon some extensive feedback from his students as well as colleagues. The present work draws upon the strength of the acclaimed text "Applied Hydrodynamics: An Introduction to Ideal and Real Fluid Flows" (Chanson 2009), with the inclusion of a number of updates, revisions, corrections, the addition of more exercises, some Internet-based resources and a series of new digital movies (see Appendix F).

## Reviews of "Applied Hydrodynamics: An Introduction to Ideal and Real Fluid Flows"

*"The book contains a lot of applications and exercises. It handles some aspects in more detail than other books in hydrodynamics. [...] A great number of the chosen applications comes from phenomena in nature and from technical applications."* (Prof. B. Platzer, in Z. Angew Math. Mech., 2011, Vol. 91, No. 5, p. 399.)

*"This book merits being read and even studied by a very large spectrum of people who should be able to find it on the shelves of the professional and university libraries that respect themselves. [...] There is an abyss of ignorance concerning Hydrodynamics bases. [...] This population of modellers [the users of commercial hydraulics simulation software] receive now, with Hubert Chanson's book, a tool for such understanding as well as the material for individual catching up with desired knowledge profile."* (Dr. J. Cunge, in Journal of Hydraulic Research, 2013, Vol. 51, No. 1, pp. 109–110.)

*"Professor Chanson's book will be an important addition to the field of hydrodynamics. I am glad to recommend it to instructors, students, and researchers who are in need of a clear and updated presentation of the fundamentals of fluid mechanics and their applications to engineering practice."* (Dr. O. Castro-Orgaz, in Journal of Hydraulic Engineering, 2013, Vol. 139, No. 4, p. 460.)

# Chapter 1

# Presentation

**SUMMARY**

The thrust of the textbook is presented and discussed. Then, after a short paragraph on fluid properties, the fundamental equations of real fluid flows are detailed. The particular case of ideal fluid is presented in the next chapter (Chap. I-1).

## 1 PRESENTATION

Fluid dynamics is the engineering science dealing with forces and energies generated by fluids in motion. The study of hydrodynamics involves the application of the fundamental principles of mechanics and thermodynamics to understand the dynamics of fluid flow motion. Fluid dynamics and hydrodynamics play a vital role in everyday lives. Practical examples include the flow motion in the kitchen sink, the exhaust fan above the stove, and the air conditioning system in our home. When we drive a car, the air flow around the vehicle body induces some drag which increases with the square of the car speed and contributes to fuel consumption. Engineering applications encompass fluid transport in pipes and canals, energy generation, environmental processes and transportation (cars, ships, aircrafts). Other applications includes coastal structures, wind flow around buildings, fluid circulations in lakes, oceans and atmosphere (Fig. 1), even fluid motion in the human body. Further illustrations are presented in Appendix F in the form of movies.

Civil, environmental and mechanical engineers require basic expertise in hydrodynamics, turbulence, multiphase flows and water chemistry. The education of these fluid dynamic engineers is a challenge for present and future generations. Although some introduction course is offered at undergraduate levels, most hydrodynamic subjects are offered at postgraduate levels only, and they rarely develop the complex interactions between air and water. Too many professionals and government administrators do not fully appreciate the complexity of fluid flow motion, nor the needs for further higher education of quality. Today's engineering problems require engineers with hydrodynamic expertise for a broad range of situations spanning from design and evaluation to maintenance and decision-making. These challenges imply a sound understanding of the physical processes and a solid grasp of the physical laws governing fluid flow motion.

### 1.1 Structure of the book

This text deals with the topic of applied hydrodynamics. A particular problem may be analysed for a system of constant mass. The description of the flow motion is called a system approach and the basic equations are the integral forms of the continuity, momentum and energy principles. The technique yields global results without entering into the details of the flow field at the small scale. Another technique is called the field approach. It gives a description of the flow field (pressure,

*Figure 1*   Geophysical vortical flow: swirling sediment in Gulf of Alaska on 13 March 2008 (NASA image by Norman Kuring, MODIS Ocean Color Teams, Courtesy of NASA Earth Observatory) – The soft shades of turquoise highlight some intense sediment mixing caused likely by strong winds and high waves in the shallow waters over the continental shelf – The deeper waters beyond the shelf edge in the lower right corner of the image are dark blue, not clouded by sediment.

velocity) at each point in the coordinate system. It is based upon the differential forms of the basic principles: conservation of mass, of momentum and of energy. In this text, we will use primarily a field approach to gain a complete solution of the two- or three-dimensional flow properties.

The textbook material is regrouped into two complementary sections: ideal fluid flow and real fluid flow. The former deals with two- and possibly three-dimensional fluid motions that are not subjected to boundary friction effects, while the latter considers the flow regions affected by boundary friction and turbulent shear.

Section I develops the basic theory of fluid mechanics of ideal fluid with irrotational flow motion. Under an appropriate set of conditions, the continuity and motion equations may be solved analytically. This technique is well-suited to two-dimensional flows in regions where the effects of boundary friction are negligible: e.g., outside of boundary layers. The outcomes include the entire flow properties (velocity magnitude and direction, pressure) at any point. Although no ideal fluid actually exists, many real fluids have small viscosity and the effects of compressibility may be negligible. For fluids of low viscosity the viscosity effects are appreciable only in a narrow region surroundings the fluid boundaries. For incompressible flow where the boundary layer remains thin, non-viscous fluid results may be applied to real fluid to a satisfactory degree of approximation. Applications include the motion of a solid through an ideal fluid which is applicable with slight modification to the motion of an aircraft through the air, of a submarine through the oceans, flow through the passages of a pump or compressor, or over the crest of a dam, and some geophysical flows. While the complex notation is introduced in the chapters I-5 to I-7, it is not central to the lecture material and could be omitted if the reader is not familiar with complex variables.

*Figure 2*  Paraglider above Dune du Pilat (France) on 7 Sept. 2008.

In Nature, three types of shear flows are encountered commonly: (1) jets and wakes, (2) developing boundary layers, and (3) fully-developed open channel flows. Section II presents the basic boundary layer flows and shear flows. The fundamentals of boundary layers are reviewed. The results are applied to both laminar and turbulent boundary layers. Basic shear flow and jet applications are developed. The text material aims to emphasise the inter-relation between ideal and real-fluid flows. For example, the calculations of an ideal flow around a circular cylinder are presented in Chapter I-4 and compared with real-fluid flow results. Similarly, the ideal fluid flow equations provide the boundary conditions for the developing boundary layer flows (Chap. II-3 and II-4). The calculations of lift force on air foil, developed for ideal-fluid flows (Chap. I-6), give good results for real-fluid flow past a wing at small to moderate angles of incidence (Fig. 2).

The lecture material is supported by a series of appendices (A to F), while some major home-work assignments are developed before the bibliographic references. The appendices include some basic fluid properties, unit conversion tables, mathematical aids, an introduction to an ideal-fluid flow software, and some presentation of relevant video movies.

Computational fluid dynamics (CFD) is largely ignored in the book. It is a subject in itself and its inclusion would yield a too large material for an intermediate textbook. In many universities, computational fluid dynamics is taught as an advanced postgraduate subject for students with solid expertise and experience in fluid mechanics and hydraulics. In line with the approach of Liggett (1994), this book aims to provide a background for studying and applying CFD.

The lecture material is designed as an intermediate course in fluid dynamics for senior undergraduate and postgraduate students in Civil, Environmental, Hydraulic and Mechanical Engineering. Basic references on the topics of Section I include Streeter (1948) and Vallentine (1969). The first four chapters of the latter reference provides some very pedagogical lecture material for simple flow patterns and flow net analysis. References on the topics of real fluid flows include Schlichting (1979) and Liggett (1994). Relevant illustrations of flow motion comprise Van Dyke (1982), JSME (1988) and Homsy (2000, 2004), as well as Appendix F of this book.

### Warning

Sign conventions differ between various textbooks. In the present manuscript, the sign convention may differ sometimes from the above references.

## 2   FLUID PROPERTIES

The density $\rho$ of a fluid is defined as its mass per unit volume. All real fluids resist any force tending to cause one layer to move over another but this resistance is offered only while movement is taking place. The resistance to the movement of one layer of fluid over an adjoining one is referred to the viscosity of the fluid. Newton's law of viscosity postulates that, for the straight parallel motion of a given fluid, the tangential stress between two adjacent layers is proportional to the velocity gradient in a direction perpendicular to the layers (Fig. 3):

$$\tau = \mu \times \frac{\partial V}{\partial y} \tag{1}$$

where $\mu$ is the dynamic viscosity of the fluid.

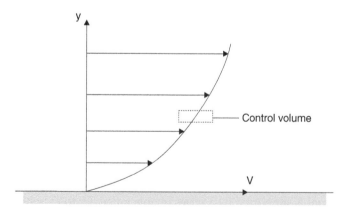

*Figure 3*  Sketch of a two-dimensional flow past a solid boundary.

## Notes

– Isaac Newton (1642–1727) was an English mathematician (see Glossary).
– The kinematic viscosity is the ratio of viscosity to mass density:

$$\nu = \frac{\mu}{\rho}$$

– A Newtonian fluid is one in which the shear stress, in one-directional flow, is proportional to the rate of deformation as measured by the velocity gradient across the flow (i.e. Equation (1)). The common fluids such as air, water and light petroleum oils, are Newtonian fluids. Non-Newtonian fluids will not be considered any further.
– A glossary of technical terms is provided in Appendix A.
– Basic fluid properties including density and viscosity of air and water, including freshwater and seawater, are reported in Appendix B. Tables for unit conversion are presented in Appendix C.

## 3   REAL FLUID FLOWS

### 3.1   Presentation

All fluid flow situations are subjected to the following relationships: the first and second laws of thermodynamics, the law of conservation of mass, Newton's law of motion and the boundary conditions. Other relations (e.g. state equation, Newton's law of viscosity) may apply.

### 3.2   The continuity equation

The law of conservation of mass states that the mass within a system remains constant with time, disregarding relativity effects:

$$\frac{DM}{Dt} = \frac{D}{Dt} \int_x \int_y \int_z \rho \times dx \times dy \times dz = 0 \tag{2}$$

where M is the total mass, t is the time, and x, y and z are the Cartesian co-ordinates. For an infinitesimal small control volume the continuity equation is:

$$\frac{\partial \rho}{\partial t} + \text{div}(\rho \times \vec{V}) = 0 \tag{3a}$$

where $\vec{V}$ is the velocity vector and div is the divergent vector operator. In Cartesian coordinates, it yields:

$$\frac{\partial \rho}{\partial t} + \sum_{i=x,y,z} \frac{\partial(\rho \times V_i)}{\partial x_i} = 0 \tag{3b}$$

where $V_x$, $V_y$ and $V_z$ are the velocity components in the x-, y- and z-directions respectively.
    For an incompressible flow (i.e. $\rho$ = constant) the continuity equation becomes:

$$\text{div}\,\vec{V} = 0 \tag{4a}$$

and in Cartesian coordinates:

$$\sum_{i=x,y,z} \frac{\partial V_i}{\partial x_i} = 0 \qquad (4b)$$

## Notes

– The word *Cartesian* is named after the Frenchman Descartes (App. A). It is spelled with a capital C. René Descartes (1596–1650) was a French mathematician, scientist, and philosopher who is recognised as the father of modern philosophy.
– In Cartesian coordinates, the velocity components are $V_x$, $V_y$, $V_z$:

$$\vec{V} = (V_x, V_y, V_z)$$

– The vector notation is used herein to lighten the mathematical writings. The reader will find some relevant mathematical aids in Appendix C.
– For a two-dimensional and incompressible flow the continuity equations is:

$$\frac{\partial V_x}{\partial x} + \frac{\partial V_y}{\partial y} = 0$$

– Considering an incompressible fluid flowing in a pipe the continuity equation may be integrated between two cross sections of areas $A_1$ and $A_2$. Denoting $V_1$ and $V_2$ the mean velocity across the sections, we obtain:

$$Q = V_1 \times A_1 = V_2 \times A_2$$

The above relationship is the integral form of the continuity equation.

## 3.3   The motion equation

### 3.3.1   Equation of motion

Newton's second law of motion is expressed for a system as:

$$\frac{D}{Dt}(M \times \vec{V}) = \sum \vec{F} \qquad (5)$$

where $\sum \vec{F}$ refers to the resultant of all external forces acting on the system, including body forces such as gravity, and $\vec{V}$ is the velocity of the centre of mass of the system. The forces acting on the control volume are (a) the surface forces (i.e. shear forces) and (b) the volume force (i.e. gravity). For an infinitesimal small volume, the momentum equation is applied to the i-component of the vector equation:

$$\frac{D(\rho \times V_i)}{Dt} = \left( \frac{\partial(\rho \times V_i)}{\partial t} + \sum_{j=x,y,z} V_j \times \frac{\partial(\rho \times V_i)}{\partial x_j} \right) = \rho F_{V_i} + \sum_{j=x,y,z} \frac{\partial \sigma_{ij}}{\partial x_j} \qquad (6)$$

where $F_V$ is the resultant of the volume forces, $\sigma$ is the stress tensor (see Notes below) and i, j = x, y, z.

If the volume forces $\vec{F}_V$ are derived from a potential U (U $= -g \times z$ for the gravity force), they can be rewritten as:

$$\vec{F}_V = -\overrightarrow{\text{grad}\,U} \tag{7}$$

where grad is the gradient vector operator. In Cartesian coordinates, it yields:

$$F_{V_x} = -\frac{\partial U}{\partial x}$$

$$F_{V_y} = -\frac{\partial U}{\partial y}$$

$$F_{V_z} = -\frac{\partial U}{\partial z}$$

For a Newtonian fluid, the shear forces are (1) the pressure forces and (2) the resultant of the viscous forces on the control volume. Hence, for a Newtonian fluid, the momentum equation becomes:

$$\frac{D(\rho \times \vec{V})}{Dt} = \rho \times \vec{F}_V - \overrightarrow{\text{grad}\,P} + \vec{f}_{visc} \tag{8a}$$

where $\vec{f}_{visc}$ is the resultant of the viscous forces on the control volume and P is the pressure. In Cartesian coordinates, it yields:

$$\frac{D(\rho \times \vec{V}_i)}{Dt} = \rho \times F_{V_i} - \frac{\partial P}{\partial x_i} + \vec{f}_{visc_i} \tag{8b}$$

Assuming a constant viscosity over the control volume and using the expressions of shear and normal stresses in terms of the viscosity and velocity gradients, the equation of motion becomes

$$\frac{D(\rho \times V_i)}{Dt} = \rho \times F_{V_i} - \frac{\partial P}{\partial x_i} - \frac{2}{3} \times \mu \times \sum_{j=x,y,z} \frac{\partial^2 V_j}{\partial x_i\, \partial x_j} + \mu \times \sum_{j=x,y,z} \left( \frac{\partial^2 V_i}{\partial x_j\, \partial x_j} + \frac{\partial^2 V_j}{\partial x_j\, \partial x_i} \right) \tag{8c}$$

## Notes

– The gravity force equals:

$$\vec{F}_V = -\overrightarrow{\text{grad}}(g \times z)$$

$$U = \vec{g} \times \vec{x}$$

where the z-axis is positive upward (i.e. U $= g \times z$) and g is the gravity acceleration (Appendix B).
– The i-component of the vector of viscous forces is:

$$f_i = \text{div}\,\tau_i = \sum_{j=x,y,z} \frac{\partial \tau_{ij}}{\partial x_j}$$

– For a Newtonian fluid the stress tensor is (Streeter 1948, p. 22):

$$\sigma_{ij} = -P \times \delta_{ij} + \tau_{ij}$$

$$\tau_{ij} = -\frac{2 \times \mu}{3} \times e \times \delta_{ij} + 2 \times \mu \times e_{ij}$$

where P is the static pressure, $\tau_{ij}$ is the shear stress tensor, $\delta_{ij}$ is the identity matrix element: $\delta_{ii} = 1$ and $\delta_{ij} = 0$ (for i different of j), $e_{ij} = \frac{1}{2} \times \left( \frac{\partial V_i}{\partial x_j} + \frac{\partial V_j}{\partial x_i} \right)$ and $e = \text{div } \vec{V} = \sum_{j=x,y,z} \frac{\partial V_i}{\partial x_i}$

Note that for an incompressible flow the continuity equation gives: $e = \text{div } \vec{V} = 0$
– The equations of motion were first rigorously developed by Leonhard Euler and are usually referred as Euler's equations of motion.
– Leonhard Euler (1707–1783) was a Swiss mathematician and a close friend of Daniel Bernoulli (Swiss mathematician and hydrodynamist, 1700–1782).

### 3.3.2 Navier-Stokes equation

For an incompressible flow (i.e. $\rho = $ constant) the derivation of the equations of motion yields to the Navier-Stokes equation:

$$\rho \times \frac{D\vec{V}}{Dt} = \rho \times \vec{F_V} - \overrightarrow{\text{grad } p} + \mu \times \Delta \vec{V} \tag{9a}$$

In Cartesian coordinates, it becomes:

$$\rho \times \left( \frac{\partial V_i}{\partial t} + V_j \times \sum_{j=x,y,z} \frac{\partial V_i}{\partial x_j} \right) = \rho \times FV_i - \frac{\partial P}{\partial x_i} + \left( \mu \times \sum_{j=x,y,z} \frac{\partial^2 V_i}{\partial x_j \, \partial x_j} \right) \tag{9b}$$

Dividing by the density, the Navier-Stokes equation becomes:

$$\frac{DV_i}{Dt} = F_{v_i} - \frac{1}{\rho} \times \frac{\partial P}{\partial x_i} + \frac{\mu}{\rho} \times \Delta V_i \tag{9c}$$

Note that $\nu = \mu / \rho$ is the kinematic viscosity.
For a two-dimensional flow and gravity forces, the Navier-Stokes equation is:

$$\rho \left( \frac{\partial V_x}{\partial t} + V_x \times \frac{\partial V_x}{\partial x} + V_y \times \frac{\partial V_x}{\partial y} \right) = -\frac{\partial}{\partial x}(P + \rho \times g \times z) + \mu \times \left( \frac{\partial^2 V_x}{\partial x \, \partial x} + \frac{\partial^2 V_x}{\partial y \, \partial y} \right) \tag{10a}$$

$$\rho \left( \frac{\partial V_y}{\partial t} + V_x \times \frac{\partial V_y}{\partial x} + V_y \times \frac{\partial V_y}{\partial y} \right) = -\frac{\partial}{\partial y}(P + \rho \times g \times z) + \mu \times \left( \frac{\partial^2 V_y}{\partial x \, \partial x} + \frac{\partial^2 V_y}{\partial y \, \partial y} \right) \tag{10b}$$

where z is taken as a coordinate which is positive vertically upward. Then (dz/dy) is the cosine of the angle between the x-axis and the z-axis, and similarly (dz/dy) for the y-axis and z-axis.

## Notes

− The viscous force term is a Laplacian:

$$\Delta V_i = \frac{\partial^2 V_i}{\partial x^2} + \frac{\partial^2 V_i}{\partial y^2} + \frac{\partial^2 V_i}{\partial z^2} = \nabla \times \nabla V_i = \text{div } \overrightarrow{\text{grad}} \ V_i$$

− The equations were first derived by Navier in 1822 and Poisson in 1829 by an entirely different method. They were derived in a manner similar as above by Saint-Venant in 1843 and Stokes in 1845.
− Henri Navier (1785−1835) was a French engineer who primarily designed bridge but also extended Euler's equations of motion. Siméon Denis Poisson (1781−1840) was a French mathematician and scientist. He developed the theory of elasticity, a theory of electricity and a theory of magnetism. The Frenchman Adhémar Jean Claude Barré De Saint-Venant (1797−1886) developed the equations of motion of a fluid particle in terms of the shear and normal forces exerted on it. George Gabriel Stokes (1819−1903), British mathematician and physicist, is known for his research in hydrodynamics and a study of elasticity (see Glossary, App. A).

# Irrotational flow motion of ideal fluid

Cyclonic Clouds over the South Atlantic Oceanon 29 April 2009 (Image acquired by MODIS Rapid Response System, Courtesy of NASA Earth Observatory) – The S-shaped swirl is in fact two cyclones that seem to be feeding on each other.

# Introduction to ideal fluid flows

**SUMMARY**

After a short paragraph on the ideal fluid properties, the basic equations are simplified for the case of an ideal fluid flow.

## 1 PRESENTATION

Although no ideal fluid exists, many real fluid flows have small viscosity and the effects of compressibility are negligible. When the boundary layer regions remain thin, the non-viscous fluid results may be applied to real fluid flow motion to a satisfactory degree of approximation. Applications include the motion of a solid through an ideal fluid which is applicable with slight modification to the motion of an aircraft through the air, of a submarine through the oceans, flow through the passages of a pump or compressor, or over the crest of a dam, and some geophysical flows.

## 2 DEFINITION OF AN IDEAL FLUID

An ideal fluid is defined as a non-viscous and incompressible fluid. That is, the fluid has zero viscosity and a constant density:

$$\rho = \text{constant} \tag{1.1}$$

$$\mu = 0 \tag{1.2}$$

An ideal fluid flow must satisfy: (1) the continuity equation, (2) the equations of motion at every point at every instant and (3) neither penetration of fluid into nor gaps between fluid and boundary at any solid boundary.

> **Remember**
>
> An ideal fluid has zero viscosity ($\mu = 0$) and a constant density ($\rho = $ constant). Since it has a nil viscosity, an ideal fluid cannot sustain shear stress. It is also termed frictionless.

## 3 IDEAL FLUID FLOWS

### 3.1 Presentation

For fluids of low viscosity the effects of viscosity are appreciable only in a narrow region surroundings the fluid boundaries, and ideal fluid flow results may be applied. Converging or accelerating flow situations generally have thin boundary layers but decelerating flows may have flow separation and development of large wake that is difficult to predict with non-viscous fluid equations.

Figure 1.1 illustrates a number of engineering flow applications where both ideal-fluid and real-fluid flow regions are observed. Figure 1.1A presents the supercritical flow downstream of a sluice gate. A boundary layer develops from about the vena contracta downstream of the gate. Figure 1.1B shows a turbulent flow past a smooth flat plate. At the plate, the velocity is zero and the flow region affected by the presence of the plate is called the boundary layer, while ideal fluid flow calculations may applied to the flow outside of the boundary layer. Figure 1.1C presents the flow region around an immersed body (e.g. a torpedo), while Figure 1.1D illustrates the effect of

(A) Flow downstream of a sluice gate

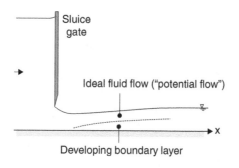

(B) Uniform flow past a flat plate

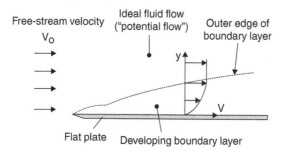

(C) Flow regions around an immersed body

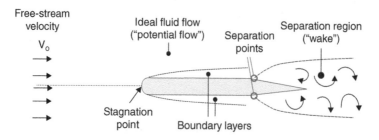

*Figure 1.1*   Examples of ideal and real fluid flow situations in engineering applications.

incidence angle on flow separation behind an aerofoil. Figures 1.1E shows the flow around high-speed catamaran and the wake region behind it, while Figure 1.1F illustrates the wake behind a ferry ship. In each example, the assumption of ideal-fluid flow is reasonable outside of the boundary layers and outside of the wake regions.

(D) Separation caused by increased angle of incidence of an aerofoil

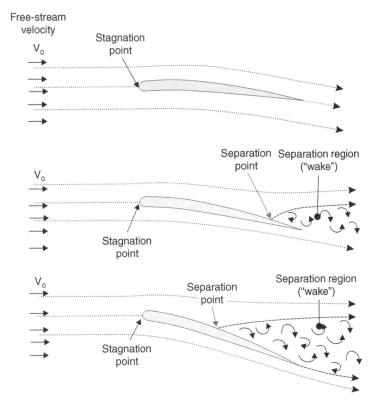

(E) Flow past a catamaran ferry on the Brisbane river (Australia) on 5 May 2002 – Note the white waters behind the catamaran highlighting the ship wake

*Figure 1.1* Continued.

(F) Wake behind the ferry to Moreton Island with the Port of Brisbane in background on 6 July 2011

*Figure 1.1* Continued.

## Notes

– A boundary layer is the flow region next to a solid boundary where the flow field is affected by the presence of the boundary and where friction plays an essential part (Chap. II-3 and II-4). A boundary layer flow is characterised by a range of velocities across the boundary layer region from zero at the boundary to the free-stream velocity at the outer edge of the boundary layer.
– A wake is the separation region downstream of the streamline that separates from a boundary. It is also called a wake region.
– In a boundary layer, a deceleration of fluid particles leading to a reversed flow within the boundary layer is called a separation. The decelerated fluid particles are forced outwards and the boundary layer is separated from the wall. At the point of separation, the velocity gradient normal to the wall is zero:

$$\left(\frac{\partial V_x}{\partial y}\right)_{y=0} = 0$$

In a boundary layer, the separation point is the intersection of the solid boundary with the streamline dividing the separation zone and the deflected outer flow. The separation point is a stagnation point.
– A stagnation point is defined as the point where the velocity is zero. When a streamline intersects itself, the intersection is a stagnation point. For irrotational flow a streamline intersects itself at right-angle at a stagnation point.

---

### Application

For an ideal fluid flow (i.e. neglecting separation), the lift force on a plane wing may be analytically estimated as:

$$\text{Lift} = \frac{1}{2} \times \rho \times V^2 \times 2 \times \pi \times L \times W \times \sin \alpha \qquad (1.3)$$

where L is the wing chord, W is the wing span and α is the angle of incidence. This relationship is derived from the Kutta-Joukowski law for two-dimensional cylinders and airfoils without camber (see Chapter I-6). Two practical applications are detailed in Table 1.1.

At take off, the lift on a commercial jumbo jet is generated by the large wing surface area. In a military fighter, the lift is caused by the large angle of incidence and high take-off speed. Figure 1.2 presents two aircrafts with a relatively high incidence angle to increase their lift force at take-off and landing. The movie Para10.mov illustrates the landing of a para-glider who increased his sail's angle of incidence to reduce his velocity without stalling, shortly before touch down.

Note that these ideal fluid flow calculations neglect the effect of separation. Figure 1.1D illustrates the effect of the angle of incidence on flow separation.

Table 1.1 Lift force for two types of aircraft.

| Aircraft | Aircraft mass kg | Chord L m | Wing span W m | Angle of incidence α deg. | Take-off speed V m/s | Lift force N |
|---|---|---|---|---|---|---|
| Jet fighter | 16 E+3 | 5 | 9 | 20 | 98 | 5.6 E+5 |
| Boeing 747 | 300 E+3 | 9 | 60 | 10 | 83 | 2.4 E+6 |

(A) C-17 Globemaster III at take-off with extended flaps (Courtesy of NASA Dryden Flight Research Center Photo Collection, Photo Number EC01-0325-8 by Tony Landis)

(B) Airbus A300-600ST Beluga at landing at Nantes airport on 10 August 2008   Note the extended flaps to increase the lift and reduce the stall speed

Figure 1.2 Aircrafts at take off and landing.

**Notes**

- The chord or chord length is the straight line distance joining the leading and trailing edges of an airfoil.
- The span is the maximum lateral distance (from tip to tip) of an airplane.
- The angle of incidence or angle of attack is the angle between the approaching flow velocity vector and the chordline.

## 3.2 Fundamental equations for ideal fluid flows

The fundamental equations developed in Chapter 2 may be simplified for an ideal fluid flow. The continuity and Navier-Stokes equations are presented below.

### 3.2.1 Continuity equation

For an incompressible flow, the continuity equation is:

$$\text{div} \, \vec{V} = 0 \tag{1.4a}$$

and in Cartesian coordinates:

$$\sum_{i=x,y,z} \frac{\partial V_i}{\partial x_i} = 0 \tag{1.4b}$$

**Notes**

- This expression is valid for viscous and non-viscous fluids, and steady and unsteady flows.
- For a two-dimensional flow the continuity equation in Cartesian coordinates becomes:

$$\frac{\partial V_x}{\partial x} + \frac{\partial V_y}{\partial y} = 0$$

### 3.2.2 Navier-Stokes equation

For an ideal fluid flow the Navier-Stokes equation becomes:

$$\rho \times \frac{D \vec{V}}{Dt} = \rho \times \vec{F_v} - \overrightarrow{\text{grad}} \, P \tag{1.5a}$$

and in Cartesian coordinates:

$$\rho \times \frac{D V_i}{Dt} = \rho \times F_{v_i} - \frac{\partial P}{\partial x_i} \tag{1.5b}$$

For a two-dimensional flow and gravity forces, the Navier-Stokes equations written in Cartesian coordinates are:

$$\rho \left( \frac{\partial V_x}{\partial t} + V_x \times \frac{\partial V_x}{\partial x} + V_y \times \frac{\partial V_x}{\partial y} \right) = -\frac{\partial}{\partial x}(P + \rho \times g \times z) \tag{1.6a}$$

$$\rho \left( \frac{\partial V_y}{\partial t} + V_x \times \frac{\partial V_y}{\partial x} + V_y \times \frac{\partial V_y}{\partial y} \right) = -\frac{\partial}{\partial y}(P + \rho \times g \times z) \tag{1.6b}$$

where z is taken as a coordinate which is positive vertically upward.

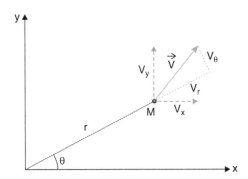

*Figure 1.3* Definition sketch of radial and ortho-radial velocity components.

In polar system of coordinates, the Navier-Stokes equations are:

$$\rho\left(\frac{\partial V_r}{\partial t} + V_r \times \frac{\partial V_r}{\partial r} + \frac{V_\theta}{r} \times \frac{\partial V_r}{\partial \theta} - \frac{V_\theta^2}{r}\right) = -\frac{\partial}{\partial r}(P + \rho \times g \times z) \tag{1.7a}$$

$$\rho\left(\frac{\partial V_\theta}{\partial t} + V_r \times \frac{\partial V_\theta}{\partial r} + \frac{V_\theta}{r} \times \frac{\partial V_\theta}{\partial \theta} + \frac{V_r \times V_\theta}{r}\right) = -\frac{1}{r} \times \frac{\partial}{\partial \theta}(P + \rho \times g \times z) \tag{1.7b}$$

where $V_r$ and $V_\theta$ are the radial and ortho-radial velocity components (Fig. 1.3).

## Note

This transformation from Cartesian to polar coordinates is discussed in Appendix D.

# Exercises

## 1.1 Physical properties of fluids

Give the following fluid and physical properties (at 20 Celsius and standard pressure) with a 4-digit accuracy.

|  | Value | Units |
|---|---|---|
| Air density: | | |
| Water density: | | |
| Air dynamic viscosity: | | |
| Water dynamic viscosity: | | |
| Gravity constant (in Brisbane): | | |
| Surface tension (air and water): | | |

### Solution

See Appendix B.

## 1.2 Basic equations (1)

What is the definition of an ideal fluid?

What is the dynamic viscosity of an ideal fluid?

Can an ideal fluid flow be supersonic? Explain.

### Solution

No, because an ideal fluid is incompressible.

## 1.3 Basic equations (2)

From what fundamental equation does the Navier-Stokes equations derive from: (a) continuity, (b) momentum equation, (c) energy equation?

Do the Navier-Stokes equations apply for any type of flow? If not for what type of flow does it apply for?

In a fluid mechanics textbook, find the Navier-Stokes equations for a two-dimensional flow in polar coordinates.

### Solution

For an incompressible flow the continuity equation is:

$$\mathrm{div}\,\vec{V} = \frac{\partial V_x}{\partial x} + \frac{\partial V_x}{\partial y} = 0$$

The Navier-Stokes equation for a two-dimensional flow is:

$$\rho\left(\frac{\partial V_x}{\partial t} + V_x \times \frac{\partial V_x}{\partial x} + V_y \times \frac{\partial V_x}{\partial y}\right) = -\frac{\partial}{\partial x}(P + \rho \times g \times z) + \mu \times \left(\frac{\partial^2 V_x}{\partial x\,\partial x} + \frac{\partial^2 V_x}{\partial y\,\partial y}\right)$$

$$\rho\left(\frac{\partial V_y}{\partial t} + V_x \times \frac{\partial V_y}{\partial x} + V_y \times \frac{\partial V_y}{\partial y}\right) = -\frac{\partial}{\partial y}(P + \rho \times g \times z) + \mu \times \left(\frac{\partial^2 V_y}{\partial x\,\partial x} + \frac{\partial^2 V_y}{\partial y\,\partial y}\right)$$

The relationships between Cartesian coordinates and polar coordinates are:

$$x = r \times \cos\theta \qquad\qquad y = r \times \sin\theta$$

$$r^2 = x^2 + y^2 \qquad\qquad \theta = \tan^{-1}\left(\frac{y}{x}\right)$$

The velocity components in polar coordinates are deduced from the Cartesian coordinates by a rotation of angle $\theta$:

$$V_x = V_r \times \cos\theta - V_\theta \times \sin\theta \qquad V_y = V_r \times \sin\theta + V_\theta \times \cos\theta$$

$$V_r = V_x \times \cos\theta + V_y \times \sin\theta \qquad V_\theta = V_y \times \cos\theta - V_x \times \sin\theta$$

The continuity and Navier-Stokes equations are transformed using the following formula:

$$\frac{dr}{dx} = \cos\theta \qquad\qquad \frac{dr}{dy} = \sin\theta$$

$$\frac{d\theta}{dx} = -\frac{\sin\theta}{r} \qquad\qquad \frac{d\theta}{dy} = \frac{\cos\theta}{r}$$

and

$$dx = dr \times \cos\theta - r \times d\theta \times \sin\theta \qquad dy = dr \times \sin\theta + r \times d\theta \times \cos\theta$$

$$dr = dx \times \cos\theta + dy \times \sin\theta \qquad d\theta = \frac{1}{r} \times (dy \times \cos\theta - dx \times \sin\theta)$$

In polar coordinates the continuity equation is:

$$\frac{1}{r} \times \frac{\partial(r \times V_r)}{\partial r} + \frac{1}{r} \times \frac{\partial V_\theta}{\partial \theta} = 0$$

and the Navier-Stokes equations are:

$$\rho\left(\frac{\partial V_r}{\partial t} + V_r \times \frac{\partial V_r}{\partial r} + \frac{V_\theta}{r} \times \frac{\partial V_r}{\partial \theta} - \frac{V_\theta^2}{r}\right) = -\frac{\partial}{\partial r}(P + \rho \times g \times z) + \mu \times \left(\Delta V_r - \frac{V_r}{r^2} - \frac{2}{r^2} \times \frac{\partial V_\theta}{\partial \theta}\right)$$

$$\rho\left(\frac{\partial V_\theta}{\partial t} + V_r \times \frac{\partial V_\theta}{\partial r} + \frac{V_\theta}{r} \times \frac{\partial V_\theta}{\partial \theta} + \frac{V_r \times V_\theta}{r}\right) = -\frac{1}{r} \times \frac{\partial}{\partial \theta}(P + \rho \times g \times z) + \mu \times \left(\Delta V_\theta + \frac{2}{r^2} \times \frac{\partial V_r}{\partial \theta} - \frac{V_\theta}{r}\right)$$

where:

$$\Delta V_r = \frac{1}{r} \times \frac{\partial}{\partial r}\left(r \times \frac{\partial V_r}{\partial r}\right) + \frac{1}{r^2} \times \frac{\partial^2 V_r}{\partial \theta^2} + \frac{\partial^2 V_r}{\partial z^2}$$

$$\Delta V_\theta = \frac{1}{r} \times \frac{\partial}{\partial r}\left(r \times \frac{\partial V_\theta}{\partial r}\right) + \frac{1}{r^2} \times \frac{\partial^2 V_\theta}{\partial \theta^2} + \frac{\partial^2 V_\theta}{\partial z^2}$$

## 1.4 Basic equations (3)

For a two-dimensional flow, write the Continuity equation and the Momentum equation *for a real fluid* in the Cartesian system of coordinates.

What is the name of this equation?

$$\frac{\partial V_x}{\partial x} + \frac{\partial V_y}{\partial y} = 0$$

Does the above equation apply to any flow situation? If not for what type of flow field does the equation apply for?

Write the above equation in a polar system of coordinates.

For a two-dimensional *ideal fluid* flow, write the continuity equation and the Navier-Stokes equations (in the two components) in Cartesian coordinates and in polar coordinates.

### Solution

For an incompressible flow the continuity equation is:

$$\operatorname{div} \vec{V} = \frac{\partial V_x}{\partial x} + \frac{\partial V_y}{\partial y} = 0$$

The Navier-Stokes equation for a two-dimensional flow of ideal fluid is:

$$\rho \left( \frac{\partial V_x}{\partial t} + V_x \times \frac{\partial V_x}{\partial x} + V_y \times \frac{\partial V_x}{\partial y} \right) = -\frac{\partial}{\partial x} (P + \rho \times g \times z)$$

$$\rho \left( \frac{\partial V_y}{\partial t} + V_x \times \frac{\partial V_y}{\partial x} + V_y \times \frac{\partial V_y}{\partial y} \right) = -\frac{\partial}{\partial y} (P + \rho \times g \times z)$$

In polar coordinates the continuity equation is:

$$\frac{1}{r} \times \frac{\partial (r \times V_r)}{\partial r} + \frac{1}{r} \times \frac{\partial V_\theta}{\partial \theta} = 0$$

and the Navier-Stokes equations are:

$$\rho \left( \frac{\partial V_r}{\partial t} + V_r \times \frac{\partial V_r}{\partial r} + \frac{V_\theta}{r} \times \frac{\partial V_r}{\partial \theta} - \frac{V_\theta^2}{r} \right) = -\frac{\partial}{\partial r} (P + \rho \times g \times z)$$

$$\rho \left( \frac{\partial V_\theta}{\partial t} + V_r \times \frac{\partial V_\theta}{\partial r} + \frac{V_\theta}{r} \times \frac{\partial V_\theta}{\partial \theta} + \frac{V_r \times V_\theta}{r} \right) = -\frac{1}{r} \times \frac{\partial}{\partial \theta} (P + \rho \times g \times z)$$

## 1.5 Mathematics

If f is a scalar function, and F a vector, verify in a Cartesian system of coordinates that:

$$\overrightarrow{\operatorname{curl}} (\overrightarrow{\operatorname{grad}} f) = 0$$

$$\operatorname{div} (\overrightarrow{\operatorname{curl}} \vec{F}) = 0$$

## 1.6    Flow situations

Sketch the streamlines of the following two-dimensional flow situations:

A.    A laminar flow past a circular cylinder,
B.    A turbulent flow past a circular cylinder,
C.    A laminar flow past a flat plate normal to the flow direction,
D.    A turbulent flow past a flat plate normal to the flow direction.

In each case, show the possible extent of the wake (if any). Indicate clearly in which regions the ideal fluid flow assumptions are valid, and in which areas they are not.

# Ideal fluid flows and irrotational flow motion

**SUMMARY**

The concept of ideal fluid flow is defined and the main equations are presented. Then the concepts of irrotational flows, stream function, and velocity potential are developed. At last energy considerations are discussed.

## 1 PRESENTATION

### 1.1 Ideal fluid

An **ideal fluid** is one that is *frictionless* and *incompressible*. It has zero viscosity and it cannot sustain a shear stress at any point.

An understanding of two-dimensional and three-dimensional flow of ideal fluid provides the engineer with a much broader approach to real-fluid flow situations. Although no ideal fluid actually exists, many real fluids have small viscosity and the effects of compressibility may be small. Applications include the motion of a solid through an ideal fluid are applicable with slight modification to the motion of an aircraft through the air, of a submarine through the oceans, flow through the passages of a pump or compressor, or over the crest of a dam.

Prandtl's hypothesis states that, for fluids of low viscosity, the effects of viscosity are appreciable only in a narrow region surrounding the fluid boundaries. For incompressible flow situations in which the boundary layer remains thin, the ideal fluid flow results may be applied to flow of a real fluid to a satisfactory degree of approximation *outside the boundary layer*.

An additional assumption of **irrotational flow** will be introduced and developed.

#### Notes

– A non-viscous (or inviscid) fluid cannot sustain shear stress. It is called friction less fluid.
– Ludwig Prandtl (1875–1953) was a German physicist and aerodynamist. He introduced the concept of boundary layer in "On Fluid Motion with Very Small Friction" (1904).
   Prandtl, L. (1904). "Uber Flussigkeitsbewegung bei sehr kleiner Reibung." *Verh. III Int. Math. Kongr.*, Heidelberg, Germany.

### 1.2 Fundamental equations

An ideal fluid must satisfy the following basic principles and associated boundary conditions.

(a)   The continuity equation:

$$\operatorname{div} \vec{V} = 0 \tag{2.1a}$$

where $\vec{V}$ is the velocity vector. In Cartesian coordinates, it yields:

$$\frac{\partial V_x}{\partial x} + \frac{\partial V_y}{\partial y} + \frac{\partial V_z}{\partial z} = 0 \qquad (2.1b)$$

where x, y and z are the Cartesian coordinates.

(b)   The equations of motion at every point at every instant:

$$\rho \times \frac{D\vec{V}}{Dt} = \rho \times \vec{F_v} - \overrightarrow{grad\ P} \qquad (2.2)$$

where $\rho$ is the fluid density, $F_v$ is the resultant of the volume forces and P is the pressure. In Cartesian coordinates, it yields:

$$\rho \times \frac{DV_x}{Dt} = \rho \times F_{v_x} - \frac{\partial P}{\partial x} \qquad (2.3a)$$

$$\rho \times \frac{DV_y}{Dt} = \rho \times F_{v_y} - \frac{\partial P}{\partial y} \qquad (2.3b)$$

$$\rho \times \frac{DV_z}{Dt} = \rho \times F_{v_z} - \frac{\partial P}{\partial z} \qquad (2.3c)$$

(c)   The solid boundary condition: that is, neither penetration of fluid into nor gap between fluid and boundary.

The unknowns in an ideal fluid flow situation with given boundaries are the velocity vector and the pressure at every point at every instant.

# 2   IRROTATIONAL FLOWS

## 2.1   Introduction

The assumption that the flow is irrotational provides a means to integrate the motion equations if the volume forces are derivable from a potential (i.e. gravity force). This will be developed in the "Velocity potential" paragraph. The concept of rotation and vorticity are introduced before the irrotational flow condition is defined.

## 2.2   Rotation and vorticity

The rotation component of a fluid particle about an axis (e.g. the z axis) is defined as the average angular velocity of any two infinitesimal linear elements in the particle that are perpendicular to each other and to the axis (e.g. the z axis).

The rotation vector is related to the velocity vector as:

$$\vec{\omega} = \frac{1}{2} \times \overrightarrow{curl}\ \vec{V}$$

$$= \frac{1}{2} \times \left(\frac{\partial V_z}{\partial y} - \frac{\partial V_y}{\partial z}\right)\vec{i} + \frac{1}{2} \times \left(\frac{\partial V_x}{\partial z} - \frac{\partial V_z}{\partial x}\right)\vec{j} + \frac{1}{2} \times \left(\frac{\partial V_y}{\partial x} - \frac{\partial V_x}{\partial y}\right)\vec{k} \qquad (2.4)$$

where curl is the curl vector operator (Appendix D).

The **vorticity** vector is defined as twice the rotation vector:

$$\overrightarrow{\text{Vort}} = \overrightarrow{\text{curl}} \; \overrightarrow{V} = \left(\frac{\partial V_z}{\partial y} - \frac{\partial V_y}{\partial z}\right)\overrightarrow{i} + \left(\frac{\partial V_x}{\partial z} - \frac{\partial V_z}{\partial x}\right)\overrightarrow{j} + \left(\frac{\partial V_y}{\partial x} - \frac{\partial V_x}{\partial y}\right)\overrightarrow{k} \tag{2.5}$$

**Note**

For a two-dimensional flow in the {x, y} plane, the rotation vector and the vorticity vector are perpendicular to the {x, y} plane. That is, their components in the x- and y-directions are zero.

## 2.3  Irrotational flow

An **irrotational flow** is defined as a flow motion in which the rotation and hence vorticity are zero everywhere:

$$\overrightarrow{\text{curl}} \, \overrightarrow{V} = 0 \tag{2.6}$$

In Cartesian coordinates, the condition of irrotational flow becomes:

$$\left(\frac{\partial V_z}{\partial y} - \frac{\partial V_y}{\partial z}\right) = 0 \tag{2.7a}$$

$$\left(\frac{\partial V_x}{\partial z} - \frac{\partial V_z}{\partial x}\right) = 0 \tag{2.7b}$$

$$\left(\frac{\partial V_y}{\partial x} - \frac{\partial V_x}{\partial y}\right) = 0 \tag{2.7c}$$

The individual particles of a frictionless incompressible fluid initially at rest cannot be caused to rotate. Considering a small free body of fluid in the shape of a sphere (Fig. 2.1), the surfaces forces must act normal to its surface since the fluid is frictionless: i.e., there is no shear stress. Hence they act through the centre of the sphere. The volume force also acts at the centre of mass. Simply no torque can be exerted on the sphere and the spherical volume of fluid remains without rotation if it was initially at rest.

Conversely, once an ideal fluid has rotation, there is no way of altering it as no torque can be exerted on an elementary sphere of the fluid. Rotation, or lack of rotation of the fluid particles, is a property of the fluid itself and not of its position in space.

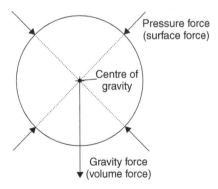

*Figure 2.1*  Forces acting on a spherical control volume of an ideal fluid.

**Notes**

– For a two-dimensional flow, the condition of irrotationality is:

$$\frac{\partial V_y}{\partial x} = \frac{\partial V_x}{\partial y}$$

– There may be isolated points or lines in an irrotational flow where the above condition (Eq. (2.6)) is not satisfied. Such points or lines are known as **singularities**. They are points or lines where the velocity is zero or theoretically infinite.
– The French mathematician Joseph-Louis Lagrange showed that Euler's equations of motion (Chapter 2, paragraph 3.3.1) could be solved analytically for irrotational flow motion of ideal fluid. He argued on the permanence of irrotational fluid motion by introducing the velocity potential (paragraph 4, in this chapter).
– Leonhard Euler (1707–1783) was a Swiss mathematician.

## 3 STREAM FUNCTION – STREAMLINES

### 3.1 Stream function

For an incompressible flow the continuity equation is:

$$\operatorname{div} \vec{V} = 0 \tag{2.1}$$

This relationship is equivalent to:

$$\vec{V} = -\overrightarrow{\operatorname{curl}} \vec{\Psi} \tag{2.8}$$

because: $\operatorname{div}(\overrightarrow{\operatorname{curl}} \vec{F}) = 0$ (see Appendix D).

For a two-dimensional flow, the vector $\vec{\Psi}$ becomes:

$$\vec{\Psi} = (0, 0, \psi) \tag{2.9}$$

where $\psi(x, y, t)$ is called the **stream function**. The use of the continuity equation implies that:

$$D\psi = -V_x \times dy + V_y \times dx$$

is an absolute differential which satisfies:

$$V_x = -\frac{\partial \psi}{\partial y} \tag{2.10a}$$

$$V_y = \frac{\partial \psi}{\partial x} \tag{2.10b}$$

for a two-dimensional flow.

**Since the stream function $\psi$ satisfies the continuity equation, it does exist.**

**Notes**

– The stream function was introduced by the French mathematician Joseph-Louis Lagrange (Lagrange 1781, pp. 718–720, Chanson 2007).

- Joseph-Louis Lagrange (1736–1813) was a French mathematician and astronomer. During the 1789 Revolution, he worked on the committee to reform the metric system. He was Professor of mathematics at the École Polytechnique and École Normale from the start in 1794–1795.
- The stream function is defined for steady and unsteady *incompressible* flows because it does satisfy the continuity principle.
- Stream functions may also be defined for three-dimensional incompressible flows. The stream function is then a vector:

$$\vec{\Psi} = (\psi_x, \psi_y, \psi_z)$$

- The stream function has the dimensions $L^2 T^{-1}$ (i.e. m²/s).
- For a two-dimensional flow, the relationship between velocity and stream function in polar coordinates is:

$$V_r = -\frac{1}{r} \times \frac{\partial \psi}{\partial \theta}$$

$$V_\theta = -\frac{\partial \psi}{\partial r}$$

## 3.2 Streamlines

A **streamline** is the line drawn so that the velocity vector is always tangential to it (i.e. no flow across a streamline). For a two-dimensional flow the streamline equation is:

$$\frac{dy}{dx} = \tan \theta = \frac{V_y}{V_x} \tag{2.11a}$$

which may be rewritten as:

$$\frac{dx}{V_x} = \frac{dy}{V_y} \tag{2.11b}$$

If the values of $V_x$ and $V_y$ are substituted a function of $\psi$ in the streamline equation (i.e. $-V_x \times dy + V_y \times dx = 0$), it follows that, along a streamline, the total differential of the stream function is:

$$D\psi = \frac{\partial \psi}{\partial x} \times dx + \frac{\partial \psi}{\partial y} \times dy = 0 \tag{2.12}$$

Simply, along a streamline, the stream function $\psi$ is constant.

For a two-dimensional flow, the volumetric flow rate $\delta q$ (m²/s) between two streamlines is:

$$\delta q = \delta \psi = V \times \delta n \tag{2.13}$$

where V is the velocity magnitude between the streamlines and $\delta n$ is the distance between two adjacent streamlines (Fig. 2.2).

Some important characteristics of streamlines are (Vallentine 1969, p. 14):

1   There can be no flow across a streamline.
2   Streamlines converging in the direction of the flow indicate a fluid acceleration.
3   Streamlines do not cross.
4   In steady flow the pattern of streamlines does not change with time.
5   Solid stationary boundaries are streamlines provided that separation of the flow from the boundary does not occur.

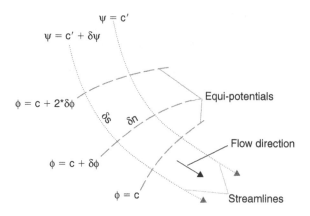

Figure 2.2 Sketch of streamlines and equipotentials.

Figure 2.3 Jean-Charles de Borda (1733–1799).

## Notes

- The concept of streamline was first introduced by the Frenchman J.C. de Borda (Fig. 2.3).
- Jean-Charles de Borda (1733–1799) was a French mathematician and military engineer. He achieved the rank of Capitaine de Vaisseau and participated to the American War of Independence with the French Navy. He investigated the flow through orifices and developed the Borda mouthpiece. During the French Revolution, he worked with Joseph-Louis Lagrange and Pierre-Simon Laplace (1749–1827) on the metric system.
- The definitions of stream function and streamline do not rely on the irrotational flow assumption.
- The concept of stream function is valid for any incompressible flow because "*since the stream function $\psi$ satisfies the continuity equation, it does exist*".

### On the stagnation point

A **stagnation point** is defined as a point where the velocity is zero. The stagnation point is a flow singularity. For a two-dimensional flow the stream function at a stagnation point is such as:

$$V_x = -\frac{\partial \psi}{\partial y} = 0$$

$$V_y = \frac{\partial \psi}{\partial x} = 0$$

(2.14)

It can be shown that, at a stagnation point, the streamline ($\psi = 0$) crosses itself: i.e., a double point. For irrotational flow, the two branches of the streamline cut at right angles.

## 3.3   Stream function of irrotational flow

For two-dimensional flows, the irrotational flow condition (Eq. (2.7)) applies:

$$\frac{\partial V_y}{\partial x} = \frac{\partial V_x}{\partial y}$$

Substituting the stream function $\psi$, it gives:

$$\Delta \psi = \frac{\partial^2 \psi}{\partial x^2} + \frac{\partial^2 \psi}{\partial y^2} = 0$$

(2.15)

where $\Delta$ is the Laplacian differential operator (Appendix D). Only one $\psi$ function satisfies both the Laplace equation (i.e. $\Delta \psi = 0$) and the boundary conditions for a particular flow pattern.

---

**Application**

1   The stream function of an uniform flow parallel to the x-axis is:

$$\psi = V \times y$$

2   The stream function of a radial flow to a point outlet at the origin (flow rate q) is:

$$\psi = -\frac{q}{2 \times \pi} \times \theta$$

where $\theta$ is the angular coordinate.

3   The stream function of a flow past a cylinder (radius R) at the origin in a fluid of infinite extent, and whose undisturbed velocity is $V_o$, is:

$$\psi = V_o \times y \times \left(1 - \frac{R^2}{x^2 + y^2}\right)$$

4   Equation (2.15) is a Laplace equation in terms of the stream function $\psi$.
5   The solution to the Laplace equation is uniquely determined if (a) the value of the function is specified on all boundaries (i.e. Dirichlet boundary conditions) or (b) the normal derivative of the function is specified on all boundaries (Neumann boundary conditions).

---

*Figure 2.4* Joseph-Louis Lagrange (1736–1813).

## 4 VELOCITY POTENTIAL

### 4.1 Definition

The **velocity potential** is defined as a scalar function of space and time $\phi(x, y, z, t)$ such that its negative derivative with respect to any direction is the fluid velocity in that direction:

$$V_x = -\frac{\partial \phi}{\partial x} \tag{2.16a1}$$

$$V_y = -\frac{\partial \phi}{\partial y} \tag{2.16a2}$$

$$V_z = -\frac{\partial \phi}{\partial z} \tag{2.16a3}$$

In vector notation, it becomes:

$$\vec{V} = -\overrightarrow{\text{grad}}\, \phi \tag{2.16b}$$

### Notes

– The velocity potential was first introduced by the French mathematician Joseph-Louis Lagrange in 1781 in his paper "Mémoire sur la théorie du mouvement des fluides" (Lagrange 1781, Liggett 1994) (Fig. 2.4).
– The velocity potential function only exists for irrotational flows (see next paragraph).

- Velocity potential functions are defined for two-dimensional and three-dimensional, steady and unsteady flows. The velocity potential $\phi$ is always a scalar.
- The velocity potential function $\phi$ has the dimension $L^2T^{-1}$ (i.e. $m^2/s$) that is the same as for the two-dimensional flow stream function $\psi$.
- In polar coordinates, the relationship between the velocity vector and the velocity potential is:

$$V_r = -\frac{\partial \phi}{\partial r}$$

$$V_\theta = \frac{1}{r} \times \frac{\partial \phi}{\partial \theta}$$

$$V_z = -\frac{\partial \phi}{\partial z}$$

- Although many applications of the velocity potential only exists for irrotational flows of ideal fluids, J.L. Lagrange demonstrated that the velocity potential exists for irrotational flow motion of both ideal- and real-fluids (Chanson 2007).

## 4.2  Velocity potential and irrotational flow

The existence of a velocity potential implies irrotational flow motion. Indeed, if the velocity potential exists, the vorticity may be rewritten as:

$$\overrightarrow{Vort} = \overrightarrow{curl}\ \overrightarrow{V} = -\overrightarrow{curl}\ \overrightarrow{grad}\ \phi = 0 \qquad (2.17)$$

since: $\overrightarrow{curl}\ \overrightarrow{grad}\ \phi = 0$ (see Appendix D). The assumption of irrotational flow and the assumption that a velocity potential exists are one and the same thing. In conclusion, the velocity potential function exists if and only if flow is irrotational. Corollary a flow is irrotational if and only if a velocity potential function exists.

For an incompressible flow, the continuity equation may then be rewritten in terms of the velocity potential:

$$div(-\overrightarrow{grad}\ \phi) = -\Delta\phi = 0 \qquad \text{Continuity equation} \qquad (2.18)$$

The continuity equation (Eq. (2.18)) is a Laplace equation in terms of the velocity potential which must be satisfied at every point throughout the fluid.

In summary, considering the irrotational flow motion of an ideal fluid, it was shown that:

(a)   the stream function $\psi$ exists since it satisfies the continuity equation;
(b)   the velocity potential $\phi$ exists;
(c)   the condition of irrotational flow written in term of the stream function yields to a Laplace equation: $\Delta\psi = 0$; and
(d)   the continuity equation can be rewritten in term of the velocity potential as another Laplace equation: $\Delta\phi = 0$.

A velocity potential can be found for each stream function. If the stream function satisfies the Laplace equation (Eq. (2.15)), the velocity potential also satisfies it (Eq. (2.18)) but with different boundary conditions. Hence the velocity potential may be considered as the stream function for another flow case. The velocity potential $\phi$ and the stream function $\psi$ are called conjugate functions.

## Properties of the Laplace equation

A interesting property of any linear homogenous differential equation, including the Laplace equation, is the principle of superposition. If any two functions are solutions, their sum, or any linear combination, is also a solution. Hence, if $\phi_1$ and $\phi_2$ satisfy the Laplace equation subject to the respective boundary conditions $B_1(\phi_1)$ and $B_2(\phi_2)$, any linear combination of these solutions: $(a \times \phi_1 + b \times \phi_2)$ satisfies the Laplace equation and the boundary conditions $a \times B_1(\phi_1) + b \times B_2(\phi_2)$. This is the principle of superposition.

Another property of the Laplace equation is the unicity of its solution for a well-defined set of boundary conditions.

### Notes

– An important difference between velocity potential and stream function lies in the fact that the velocity potential exists for irrotational flows only, while the stream function is not restricted to irrotational flows but to incompressible fluids.
– Pierre-Simon Laplace (1749–1827) was a French mathematician, astronomer and physicist. He is best known for his investigations into the stability of the solar system.

## 4.3  Application to the equations of motion: the Bernoulli equation

The equations of motion are:

$$\frac{\partial V_i}{\partial t} + \sum_{j=x,y,z} V_j \times \frac{\partial V_i}{\partial x_j} = -\frac{\partial}{\partial x_i}\left(\frac{P}{\rho} + U\right) \tag{2.2}$$

where U is the volume force potential: $\overrightarrow{F_v} = -\overrightarrow{\text{grad}}\,U$.

Substituting the irrotational flow conditions (Eq. (2.6)):

$$\frac{\partial V_k}{\partial x_j} = \frac{\partial V_j}{\partial x_k} \quad \text{for } j \neq k$$

and the velocity potential (Eq. (2.16)):

$$V_i = -\frac{\partial \phi}{\partial x_i}$$

the motion equations may be expressed as (Streeter 1948, p. 24):

$$\frac{\partial}{\partial x_i}\left(-\frac{\partial \phi}{\partial t} + \frac{V_j^2}{2} + \sum_{j=x,y,z}\frac{P}{\rho} + U = 0\right) \tag{2.19}$$

### Notes

– The gravity force is a volume force:

$$\overrightarrow{F_v} = -\overrightarrow{\text{grad}}(g \times z)$$

where the z axis is positive upward: i.e., $U = g \times z$.
– If the only volume force is the gravity force, the integrated form of the Bernoulli equation is:

$$-\frac{d\phi}{dt} + \frac{V^2}{2} + \frac{P}{\rho} + g \times z = F(t)$$

Introducing the gravity force, the integration with respect to $x_i$ yields:

$$-\frac{\partial \phi}{\partial t} + \frac{V^2}{2} + \frac{P}{\rho} + g \times z = F_i(x_j, x_k, t) \tag{2.20}$$

where i, j, k = x, y, z and V is defined as the magnitude of the velocity defined as: $V^2 = V_x^2 + V_y^2 + V_z^2$. The integration of the three motion equations are identical and the left-hand sides of the equations are the same:

$$F_x(y, z, t) = F_y(x, z, t) = F_z(x, y, t)$$

The final integrated form of the three equations of motion is the **Bernoulli equation for unsteady flow**:

$$-\frac{d\phi}{dt} + \frac{V^2}{2} + \frac{P}{\rho} + g \times z = F(t) \tag{2.21}$$

containing an arbitrary function of time F(t) that is independent of space (x, y, z).

## Notes

- Daniel Bernoulli (1700–1782) was a Swiss mathematician and hydrodynamist, who developed the equation in his 'Hydrodynamica' textbook (1st draft in 1733) (Carvill 1981, p. 3).
- In steady flow the Bernoulli equation reduces to:

$$\frac{V^2}{2} + \frac{P}{\rho} + g \times z = \text{Constant}$$

---

## Application

Usually, it is convenient to write the pressure P as:

$$P = P_s + P_d \tag{2.22}$$

where $P_s$ is the static (ambient) pressure and $P_d$ the dynamic pressure (i.e. pressure due to changes in velocity). Assuming that the gravity is the only volume force acting, Equation (2.21) becomes:

$$-\frac{\partial\phi}{\partial t} + \frac{V^2}{2} + \frac{P_d}{\rho} + \frac{P_s}{\rho} + g \times z = F(t)$$

If the pressure distribution is hydrostatic:

$$\left(\frac{P_s}{\rho} + g \times z\right) = \text{constant}$$

then the Bernoulli equation becomes:

$$-\frac{\partial\phi}{\partial t} + \frac{V^2}{2} + \frac{P_d}{\rho} = F(t) \tag{2.22}$$

For steady flows with hydrostatic pressure distribution, the Bernoulli equation becomes:

$$\frac{V^2}{2} + \frac{P_d}{\rho} = \text{constant} \tag{2.23a}$$

If the dynamic pressure and velocity are known at one point ($P_{d_o}$, $V_o$), the variation in pressure may be determined if the change in velocity magnitude is known:

$$P_d = P_{d_o} + \rho \times \frac{V_o^2}{2} \times \left(1 - \left(\frac{V}{V_o}\right)^2\right) \tag{2.23b}$$

## 4.4   Discussion

In the irrotational flow motion of ideal fluid, the problem resolves itself into a geometrical problem: to find the velocity potential $\phi$, or the stream function $\psi$, that satisfies the continuity principle and the boundary conditions. As the continuity equation and boundary equations are both kinematical (i.e. contain no density term), the magnitude and directions of the velocity at all points are independent of the particular fluid. Basically the solution of the velocity field is independent of the fluid properties.

Once the velocity is known, the pressure may be determined from the Bernoulli equation (Eq. (2.23)).

---

**Remarks**

In ideal fluid flows with irrotational motion, the basic equations may be solved analytically (Chapters I-3 and I-4). They yield the velocity field. In turn the pressure field may be calculated using the Bernoulli equation. The technique is particularly useful to predict hydrodynamic forces on structures in converging or accelerating flow conditions. In real fluid flows, it is usual to combined ideal-fluid and real-fluid flow calculations to assess fluid-structure interactions (see Chapters II-3 and II-4). An example is wind loading on buildings and man-made structures. Figure 2.5 illustrates severe damage during wind storm episodes.

---

(A) Damage in La Rochelle marina after a strong wind storm on 26–28 December 1999 (Courtesy of J.H. Bordes) – During the storm event, wind speeds exceeded 100 knots (51 m/s), combined with high tidal range and storm surge effects

*Figure 2.5* Wind storm damage.

(B) Remnants of the original Pont Roche-Bernard across the Vilaine River (France) in 2009 – Completed in 1838, the 349 m long suspension bridge failed in 1862 during a wind storm because of wind-structure interactions leading to resonance

*Figure 2.5*  Continued.

## 5  ENERGY CONSIDERATIONS

### 5.1  Energy equation for non-viscous fluids

Considering a control volume [CV], the total kinetic energy $E_k$ in the fluid is:

$$E_k = \int_{CV} \rho \times \frac{V^2}{2} \times dVolume = \frac{1}{2} \times \iiint \rho \times V^2 \times dx \times dy \times dz \qquad (2.24)$$

The total potential energy $E_p$ of the fluid region is:

$$E_p = \int_{CV} \rho \times U \times dVolume = \frac{1}{2} \times \iiint \rho \times U \times dx \times dy \times dz \qquad (2.25)$$

where U is the potential of the volume forces (i.e. potential energy per unit mass).

For incompressible fluids the energy equation becomes (Streeter 1948, p. 31; Vallentine 1969, p. 46):

$$\frac{D}{Dt}(E_k + E_p) = \int_{CS} P \times V_n \times dArea \tag{2.26}$$

where $V_n$ is the velocity of the boundary normal to itself in the direction of the fluid and [CS] is the control surface. Hence for an incompressible frictionless fluid, the total increase in kinetic plus potential energy equals the work done by the pressures on its surface.

**Note**

This result is valid for both rotational and irrotational flows.

## 5.2 Kinetic energy in irrotational flows

When the flow motion is irrotational (i.e. $V = \nabla\phi$), the use of the Green's theorem permits the derivation of an expression of the kinetic energy term:

$$E_k = \frac{\rho}{2} \times \iiint (\nabla\phi) \times (\nabla\phi) \times dx \times dy \times dz = -\frac{\rho}{2} \times \iint \phi \times \frac{d\phi}{dn} \times dS \tag{2.27}$$

where $\phi$ is the velocity potential. The velocity of the fluid along the inward normal, n, to an element of surface dS is: $V_n = -d\phi/dn$.

The surface integral (i.e. right term of the above equation) represents the work done by an impulsive pressure ($\rho \times \phi$) in starting the motion from rest.

**Notes**

– George Green (1793–1841) was an English mathematician.
– Green's theorem states that:

$$\iint \vec{A} \times \vec{dS} = -\iiint \operatorname{div} \vec{A} \times dx \times dy \times dz$$

where A is a vector whose components are any finite, single-valued, differentiable functions of space in a connected region completely bounded by one or more closed surfaces S of which dS is an element and the direction normal to the surface element dS is directed into the region (Vallentine 1969, p. 250). Green's theorem can be regarded as a transformation of a surface integral into a volume integral.

## 5.3 Discussion

The following theorems are proved as a consequence of the kinetic energy considerations listed above and they are limited to irrotational flows of ideal fluid (Streeter 1948, p. 37; Vallentine 1969, p. 49):

(a) Irrotational motion is impossible if all of the boundaries are fixed.
(b) Irrotational motion of a fluid will cease when the boundaries come at rest.
(c) The pattern of irrotational flow which satisfies the Laplace equation and prescribed boundary conditions is unique and is determined by the motion of the boundaries.

(d)   Irrotational motion of a fluid at rest at infinity is impossible if the interior boundaries are at rest.

(e)   Irrotational motion of a fluid at rest at infinity is unique and determined by the motion of the interior solid boundaries.

## 6  CONCLUSION

For an ideal fluid (i.e. incompressible and frictionless) and for an irrotational flow motion (i.e. a velocity potential exists), the solution of the flow problem is a solution of the Laplace equation (Eq. (2.6) or (2.16)) in term of $\phi$ or $\psi$ that also satisfies the boundary conditions.

The velocity field may be obtained from the definition of the velocity potential (or stream function) and the pressure distribution throughout the fluid is deduced from the Bernoulli equation.

# Exercises

## 2.1  Quiz

- What is the definition of the velocity potential?
- Is the velocity potential a scalar or a vector?
- Units of the velocity potential?
- What is definition of the stream function? Is it a scalar or a vector? Units of the stream function?

For an ideal fluid with irrotational flow motion:

- Write the condition of irrotationality as a function of the velocity potential.
- Does the velocity potential exist for (a) an irrotational flow and (b) for a real fluid?
- Write the continuity equation as a function of the velocity potential.

Further, answer the following questions:

- What is a stagnation point?
- For a two-dimensional flow, write the stream function conditions.
- How are the streamlines at the stagnation point?

### Solution

- $\overrightarrow{V} = -\overrightarrow{grad}\ \phi$
- A scalar.
- The stream function $\psi$ and the velocity potential $\phi$ are in $m^2/s$.
- The stream function is a vector. It does exist for incompressible flow because it satisfies the continuity equation.
- The existence of a velocity potential implies irrotational flow. Indeed the vorticity becomes:

$$\overrightarrow{Vort} = \overrightarrow{curl}\ \overrightarrow{V} = -\overrightarrow{curl}\ \overrightarrow{grad}\ \phi = 0$$

as: $\overrightarrow{curl}\ \overrightarrow{grad}\ \phi = 0$ (see Appendix D). $\phi$ exist for both ideal and real-fluids with irrotational flow motion,
- The assumption of irrotational flow and the assumption that a velocity potential exists are one and the same thing (Streeter 1948, p. 23).

Conclusion: Velocity potential functions exist if and only if flow is irrotational. Corollary flow is irrotational if and only if velocity potential functions exist.

- $div(-\overrightarrow{grad}\ \phi) = -\Delta\phi = 0$
- A stagnation point is defined as a point where the velocity is zero.
- For a two-dimensional flow the stream function at a stagnation point is such as:

$$V_x = -\frac{\partial\psi}{\partial y} = 0 \quad V_y = \frac{\partial\psi}{\partial x} = 0$$

- It can be shown that at a stagnation point the streamline $\psi = 0$ crosses itself (i.e. a double point).

## 2.2  Basic applications

1.  Considering the following velocity field:

$$V_x = y \times z \times t$$

$$V_y = z \times x \times t$$

$$V_z = x \times y \times t$$

   -  Is the flow a possible flow of an incompressible fluid?
   -  Is the motion irrotational? If yes: what is the velocity potential?

2.  Draw the streamline pattern of the following stream functions:

$$\psi = 50 \times x \tag{1}$$

$$\psi = -20 \times y \tag{2}$$

$$\psi = -40 \times x - 30 \times y \tag{3}$$

$$\psi = -10 \times x^2 \quad \text{(from } x = 0 \text{ to } x = 5) \tag{4}$$

### Solution (1)

The equations satisfy the equation of continuity for:

$$\frac{\partial V_x}{\partial x} = \frac{\partial V_y}{\partial y} = \frac{\partial V_z}{\partial z} = 0$$

so that:

$$\frac{\partial V_x}{\partial x} + \frac{\partial V_y}{\partial y} + \frac{\partial V_z}{\partial z} = 0$$

The components of the vorticity are:

$$\left( \frac{\partial V_z}{\partial y} - \frac{\partial V_y}{\partial z} \right) = (x \times t - x \times t) = 0$$

$$\left( \frac{\partial V_x}{\partial z} - \frac{\partial V_z}{\partial x} \right) = (y \times t - y \times t) = 0$$

$$\left( \frac{\partial V_y}{\partial x} - \frac{\partial V_x}{\partial y} \right) = (z \times t - z \times t) = 0$$

hence vorticity is zero and the field could represent irrotational flow. The velocity potential would then be the solution of:

$$\frac{\partial \phi}{\partial x} = -V_x = -y \times z \times t \phi = -x \times y \times z \times t + f_1(y, z, t)$$

$$\frac{\partial \phi}{\partial y} = -V_y = -x \times z \times t \phi = -x \times y \times z \times t + f_2(x, z, t)$$

$$\frac{\partial \phi}{\partial z} = -V_z = -x \times y \times t \phi = -x \times y \times z \times t + f_3(x, y, t)$$

and hence:

$$\phi = -x \times y \times z \times t + f(t)$$

is a possible velocity potential.

### Solution (2)

1. Vertical uniform flow: $V_o = +50$ m/s
2. Horizontal uniform flow: $V_o = +20$ m/s
3. Uniform flow: $V_o = 50$ m/s and $\alpha = 127$ degrees
4. Non uniform vertical flow

## 2.3  Two-dimensional flow

Considering a two-dimensional flow, find the velocity potential and the stream function for a flow having components:

$$V_x = -\frac{2 \times x \times y}{(x^2 + y^2)^2}$$

$$V_y = \frac{x^2 - y^2}{(x^2 + y^2)^2}$$

### Solution

In polar coordinate the velocity components are:

$$V_x = -\frac{2 \times \cos\theta \times \sin\theta}{r^2} \quad V_y = \frac{\cos^2\theta - \sin^2\theta}{r^2}$$

Using:

$$V_r = V_x \times \cos\theta + V_y \times \sin\theta \quad V_\theta = V_y \times \cos\theta - V_x \times \sin\theta$$

we deduce:

$$V_r = -\frac{\sin\theta}{r^2} \quad V_\theta = \frac{\cos\theta}{r^2}$$

In polar coordinates the velocity potential and stream function are:

$$V_r = -\frac{\partial\phi}{\partial r} \qquad V_\theta = -\frac{1}{r} \times \frac{\partial\phi}{\partial\theta}$$

$$V_r = -\frac{1}{r} \times \frac{\partial\psi}{\partial\theta} \qquad V_\theta = -\frac{\partial\psi}{\partial r}$$

Hence:

$$\phi = -\frac{\sin\theta}{r} + \text{constant}$$

## 2.4   Applications

(a)   Using the software 2DFlowPlus (Appendix E), investigate the flow field of a vortex (at origin, strength 2) superposed to a sink (at origin, strength 1). Visualise the streamlines, the contour of equal velocity ad the contour of constant pressure.

Repeat the same process for a vortex (at origin, strength 2) superposed to a sink (at $x = -5$, $y = 0$, strength 1). How would you describe the flow region surrounding the vortex.

(b)   Investigate the superposition of a source (at origin, strength 1) and an uniform velocity field (horizontal direction, $V = 1$). How many stagnation point do you observe? What is the pressure at the stagnation point? What is the "half-Rankine" body thickness at $x = +1$? (You may do the calculations directly or use 2DFlowPlus to solve the flow field.)

(c)   Using 2DFlowPlus, investigate the flow past a circular building (for an ideal fluid with irrotational flow motion). How many stagnation points is there? Compare the resulting flow pattern with real-fluid flow pattern behind a circular bluff body (search Reference text in the library).

(d)   Investigate the seepage flow to a sink (well) located close to a lake. What flow pattern would you use?

Note: the software 2DFlowPlus is described in Appendix E and a demonstration version is available at www.dynaflow-inc.com/2DFlow/.

## 2.5   Laplace equation

–   What is the Laplacian of a function? Write the Laplacian of the scalar function $\phi$ in Cartesian and polar coordinates.
–   Rewrite the definition of the Laplacian of a scalar function as a function of vector operators (e.g. grad, div, curl).

### Solution

$$\Delta\phi(x,y,z) = \nabla \times \nabla\phi(x,y,z) = \text{div } \overrightarrow{\text{grad}} \ \phi(x,y,z) = \frac{\partial^2\phi}{\partial x^2} + \frac{\partial^2\phi}{\partial y^2} + \frac{\partial^2\phi}{\partial z^2} \qquad \text{Laplacian of scalar}$$

$$\Delta\overrightarrow{F}(x,y,z) = \nabla \times \nabla\overrightarrow{F}(x,y,z) = \overrightarrow{i}\ \Delta F_x + \overrightarrow{j}\ \Delta F_y + \overrightarrow{k}\ \Delta F_z \qquad \text{Laplacian of vector}$$

$$\Delta\phi(r,\theta,z) = \frac{1}{r} \times \frac{\partial}{\partial r}\left(r \times \frac{\partial\phi}{\partial r}\right) + \frac{1}{r^2} \times \frac{\partial^2\phi}{\partial\theta^2} + \frac{\partial^2\phi}{\partial z^2} \qquad \text{Polar coordinates}$$

It yields:

$$\Delta f = \text{div } \overrightarrow{\text{grad}} \ f$$
$$\Delta\overrightarrow{F} = \overrightarrow{\text{grad}} \ \text{div}\,\overrightarrow{F} - \overrightarrow{\text{curl}}(\overrightarrow{\text{curl}}\ \overrightarrow{F})$$

where f is a scalar.

Note the following operations:

$$\Delta(f + g) = \Delta f + \Delta g$$
$$\Delta(\overrightarrow{F} + \overrightarrow{G}) = \Delta\overrightarrow{F} + \Delta\overrightarrow{G}$$
$$\Delta(f \times g) = g \times \Delta f + f \times \Delta g + 2 \times \overrightarrow{\text{grad}} \ f \times \overrightarrow{\text{grad}} \ g$$

where f and g are scalars.

## 2.6   Basic equations (I)

For a two-dimensional ideal fluid flow, write:

(a)   the continuity equation,
(b)   the streamline equation,

(c)   the velocity potential and stream function,
(d)   the condition of irrotationality and
(e)   the Laplace equation

in polar coordinates.

## Solution

Continuity equation

$$\frac{\partial V_x}{\partial x} + \frac{\partial V_y}{\partial y} = 0 \quad \frac{1}{r} \times \frac{\partial (r \times V_r)}{\partial r} + \frac{1}{r} \times \frac{\partial V_\theta}{\partial \theta} = 0$$

Momentum equation

$$\frac{\partial V_x}{\partial t} + V_x \times \frac{\partial V_x}{\partial x} + V_y \times \frac{\partial V_x}{\partial y} = -\frac{\partial}{\partial x}\left(\frac{P}{\rho} + g \times z\right)$$

$$\frac{\partial V_y}{\partial t} + V_x \times \frac{\partial V_y}{\partial x} + V_y \times \frac{\partial V_y}{\partial y} = -\frac{\partial}{\partial y}\left(\frac{P}{\rho} + g \times z\right)$$

$$\frac{\partial V_r}{\partial t} + V_r \times \frac{\partial V_r}{\partial r} + \frac{V_\theta}{r} \times \frac{\partial V_r}{\partial \theta} - \frac{V_\theta^2}{r} = -\frac{\partial}{\partial r}\left(\frac{P}{\rho} + g \times z\right)$$

$$\frac{\partial V_\theta}{\partial t} + V_r \times \frac{\partial V_\theta}{\partial r} + \frac{V_\theta}{r} \times \frac{\partial V_\theta}{\partial \theta} + \frac{V_r \times V_\theta}{r} = -\frac{1}{r} \times \frac{\partial}{\partial \theta}\left(\frac{P}{\rho} + g \times z\right)$$

Streamline equation

$$V_x \times dy - V_y \times dx = V_r \times r \times d\theta - V_\theta \times dr = 0$$

Velocity potential and stream function

$$V_x = -\frac{\partial \phi}{\partial x} = -\frac{\partial \psi}{\partial y} \quad V_r = -\frac{\partial \phi}{\partial r} = -\frac{1}{r} \times \frac{\partial \psi}{\partial \theta}$$

$$V_y = -\frac{\partial \phi}{\partial y} = +\frac{\partial \psi}{\partial x} \quad V_\theta = -\frac{1}{r} \times \frac{\partial \phi}{\partial \theta} = +\frac{\partial \psi}{\partial r}$$

$$\delta Q = \delta \psi \delta Q = \delta \psi$$

Condition of irrotationality

$$\frac{\partial V_y}{\partial x} - \frac{\partial V_x}{\partial y} = 0 \quad \frac{\partial V_\theta}{\partial r} - \frac{1}{r} \times \frac{\partial V_r}{\partial \theta} = 0$$

Laplace equation

$$\frac{\partial^2 \phi}{\partial x^2} + \frac{\partial^2 \phi}{\partial y^2} = 0 \quad \frac{\partial^2 \phi}{\partial r^2} + \frac{1}{r^2} \times \frac{\partial^2 \phi}{\partial \theta^2} = 0$$

$$\frac{\partial^2 \psi}{\partial x^2} + \frac{\partial^2 \psi}{\partial y^2} = 0 \quad \frac{\partial^2 \psi}{\partial r^2} + \frac{1}{r^2} \times \frac{\partial^2 \psi}{\partial \theta^2} = 0$$

## 2.7   Basic equations (2)

Considering an two-dimensional irrotational flow of ideal fluid:

- write the Navier-Stokes equation (assuming gravity forces),
- substitute the irrotational flow condition and the velocity potential,
- integrate each equation with respect to x and y,
- what is the final integrated form of the three equations of motion?

    This equation is called the Bernoulli equation for unsteady flow.

- For a steady flow write the Bernoulli equation. When the velocity is known, how do you determine the pressure?

### Solution

1.  $$\frac{\partial V_x}{\partial t} + V_x \times \frac{\partial V_x}{\partial x} + V_y \times \frac{\partial V_x}{\partial y} = -\frac{\partial}{\partial x}\left(\frac{P}{\rho} + g \times z\right)$$

    $$\frac{\partial V_y}{\partial t} + V_x \times \frac{\partial V_y}{\partial x} + V_y \times \frac{\partial V_y}{\partial y} = -\frac{\partial}{\partial y}\left(\frac{P}{\rho} + g \times z\right)$$

2.  Substituting the irrotational flow conditions:

    $$\frac{\partial V_x}{\partial y} = \frac{\partial V_y}{\partial x}$$

    and the velocity potential:

    $$V_x = -\frac{\partial \phi}{\partial x} \quad V_y = -\frac{\partial \phi}{\partial y}$$

    the equations may be expressed as (Streeter 1948, p. 24):

    $$\frac{\partial}{\partial x}\left(-\frac{\partial \phi}{\partial t} + \frac{V_x^2}{2} + \frac{V_y^2}{2} + \frac{P}{\rho} + g \times z\right) = 0$$

    $$\frac{\partial}{\partial y}\left(\frac{\partial \phi}{\partial t}\ |\ \frac{V_x^2}{2}\ |\ \frac{V_y^2}{2} + \frac{P}{\rho} + g \times z\right) - 0$$

3.  Integrating with respect to x and y:

    $$-\frac{\partial \phi}{\partial t} + \frac{V^2}{2} + \frac{P}{\rho} + g \times z = F_x(y, t)$$

    $$-\frac{\partial \phi}{\partial t} + \frac{V^2}{2} + \frac{P}{\rho} + g \times z = F_y(x, t)$$

    where V is defined as the magnitude of the velocity: $V^2 = V_x^2 + V_x^2$. The integration of the three motion equations are identical and the left-hand sides of the equations are the same:

    $$F_x(y, t) = F_y(x, t)$$

4.   The final integrated form of the three equations of motion is the Bernoulli equation for unsteady flow:

$$-\frac{\partial \phi}{\partial t} + \frac{V^2}{2} + \frac{P}{\rho} + g \times z = F(t)$$

In the general case of a volume force potential U (i.e. $\overrightarrow{F_v} = -\overrightarrow{grad}\ U$):

$$-\frac{\partial \phi}{\partial t} + \frac{V^2}{2} + \frac{P}{\rho} + U = F(t)$$

# Two-dimensional flows (1) basic equations and flow analogies

## SUMMARY

The main equations for a two-dimensional irrotational flow of ideal fluid are summarised in the first paragraph. Then the construction of flow nets is described. While the technique is restricted to two-dimensional flows, it provides means to solve rapidly some complicated flow situations. Analogies with electrical current, viscous flow and groundwater flow are presented later.

## 1   TWO-DIMENSIONAL FLOWS OF IDEAL FLUID

### 1.1   Equations

For two dimensional irrotational flows of ideal fluid, the basic equations are:

*Irrotational flows*

$$\overrightarrow{\text{curl}}\ \vec{V} = 0 \tag{3.1a}$$

For a two-dimensional flow, the condition of irrotational flow motion becomes:

$$\frac{\partial V_y}{\partial x} - \frac{\partial V_x}{\partial y} = 0 \tag{3.1b}$$

*Velocity potential*

$$\vec{V} = -\overrightarrow{\text{grad}}\ \phi \tag{3.2a}$$

*Continuity equation*

$$\text{div}\ \vec{V} = 0 \tag{3.3a}$$

which leads to the existence of the stream function vector $\vec{\Psi}$:

$$\vec{V} = -\overrightarrow{\text{curl}}\ \vec{\Psi} \tag{3.4a}$$

For a two-dimensional flow, the stream function $\vec{\Psi}$ becomes:

$$\vec{\Psi} = (0, 0, \psi)$$

where $\psi(x, y, t)$ is a scalar function, called commonly the stream function.

$$V_x = -\frac{\partial \psi}{\partial y} \quad V_y = \frac{\partial \psi}{\partial x}$$

*Bernoulli equation*

$$-\frac{\partial\phi}{\partial t} + \frac{V^2}{2} + \frac{P}{\rho} + U = F(t) \tag{3.5}$$

where the volume force potential is $U = g \times z$ for the gravity force, the z-axis is positive upward and F is a scalar function of time only.

## 1.2 Discussion

### 1.2.1 Stream function

For incompressible flows, the velocity vector and the stream function are related by:

$$V_x = -\frac{\partial\psi}{\partial y} \qquad \text{Cartesian coordinates} \tag{3.4b1}$$

$$V_y = \frac{\partial\psi}{\partial x} \tag{3.4b2}$$

In polar coordinates, it becomes:

$$V_r = -\frac{1}{r} \times \frac{\partial\psi}{\partial\theta} \qquad \text{Polar coordinates} \tag{3.4c1}$$

$$V_\theta = \frac{\partial\psi}{\partial r} \tag{3.4c2}$$

In the $\{x, y\}$ plane, the lines of constant stream function $\psi$ are called **streamlines**. The velocity vector is tangential to the streamlines everywhere.

### 1.2.2 Velocity potential

For irrotational flows, the relationship between the velocity vector and velocity potential is:

$$V_x = -\frac{\partial\phi}{\partial x} \qquad \text{Cartesian coordinates} \tag{3.2b1}$$

$$V_y = -\frac{\partial\phi}{\partial y} \tag{3.2b2}$$

In polar coordinates, it becomes:

$$V_r = -\frac{\partial\phi}{\partial r} \qquad \text{Polar coordinates} \tag{3.2c1}$$

$$V_\theta = -\frac{1}{r} \times \frac{\partial\phi}{\partial\theta} \tag{3.2c2}$$

In the $\{x, y\}$ plane, the lines given by $\phi = $ constant are called **equipotential lines**.

### 1.2.3 Continuity equation

For irrotational flow motion, the condition of irrotationality becomes a Laplace equation in terms of the stream function $\psi$:

$$\Delta\psi = \frac{\partial^2\psi}{\partial x^2} + \frac{\partial^2\psi}{\partial y^2} = 0 \tag{3.6}$$

The continuity equation may be rewritten in term of the velocity potential $\phi$:

$$\Delta\phi = \frac{\partial^2\phi}{\partial x^2} + \frac{\partial^2\phi}{\partial y^2} = 0 \tag{3.3b}$$

Both Equations (3.6) and (3.3b) are Laplace differential equations.

One method of obtaining the velocity field for steady irrotational motion of two dimensional flows consists simply in solving the Laplace equation in term of $\phi$, or $\psi$, for the appropriate boundary conditions. The relationship between stream function and velocity potential is found by equating the expressions for velocity components:

$$\frac{\partial\phi}{\partial x} = \frac{\partial\psi}{\partial y} \tag{3.7a}$$

$$\frac{\partial\phi}{\partial y} = -\frac{\partial\psi}{\partial x} \tag{3.7b}$$

Equations (3.7a) and (3.7b) are called the **Cauchy-Riemann equations**.

**Notes**

-  Augustin Louis Cauchy (1789–1857) was a French engineer from the 'Corps des Ponts-et-Chaussées'. He devoted himself later to mathematics and he taught at Ecole Polytechnique, Paris, and at the Collège de France. He was a military engineer from 1810 to 1813 before working with Pierre-Simon Laplace (French mathematician, astronomer and physicist, 1749–1827) and Joseph-Louis Lagrange (French mathematician and astronomer, 1736–1813).
-  Bernhard Georg Friedrich Riemann (1826–1866) was a German mathematician.
-  If $\phi$ and $\psi$ are twice differentiable, they are harmonic functions since they satisfy a Laplace equation.

## 2   FLOW NET

### 2.1   Presentation

Two-dimensional flow patterns can be represented visually by drawing a family of streamlines and the corresponding family of equipotential lines with the constants varying in arithmetical progression. The resulting network of lines is called a **flow net** (Fig. 3.1). Figure 3.1 presents two examples of flow nets: a two-dimensional orifice flow and the seepage flow beneath a concrete dam. Note that Figure 3.1A is an incomplete flow net since not all streamlines are shown for clarity.

In a flow net, it is usual to select the equipotential lines ($\phi$-lines) and streamlines ($\psi$-lines) so that: $\delta\psi = \delta\phi = \delta c$ as illustrated in Figure 2.2. If the distance between two adjacent streamlines is $\delta n$ and the distance between adjacent equipotential lines is $\delta s$, an estimate of the velocity magnitude V is then:

$$V \sim -\frac{\delta\phi}{\delta s} = \frac{\delta c}{\delta s} \tag{3.8a}$$

$$V \sim -\frac{\delta\psi}{\delta n} = \frac{\delta c}{\delta n} \tag{3.8b}$$

where $\delta n$ is the distance between two adjacent streamlines and $\delta s$ is the distance between two adjacent equipotentials (Fig. 2.2). As both velocity magnitude are the same, Equation (3.8) shows that: $\delta s = \delta n$. That is, the flow net consists of an orthogonal grid that reduces to perfect squares when the grid size approaches zero. In uniform flow region, the squares are of equal size. In

(A) Incomplete flow net of the flow through a two-dimensional orifice discharging vertically (after Chanson et al. 2002)

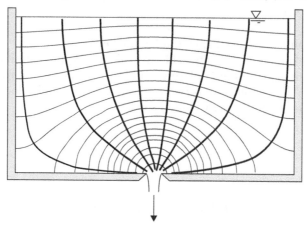

(B) Complete flow net of seepage under a concrete dam with cutoff wall and apron

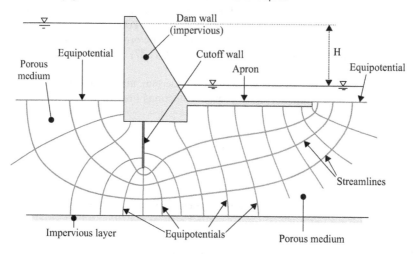

*Figure 3.1*  Examples of flow net.

diverging flows their size increases, and in converging flows they decrease in size, in the direction of flow.

## Discussion

Figure 3.1 illustrates some examples of flow nets. Figure 3.1A presents streamlines and equipotentials of a two-dimensional flow through a vertical orifice when the ratio head above orifice to orifice width is 10. Close to the orifice, the flow is subjected to a strong acceleration, and several streamlines and equipotential lines were not drawn for clarity (Chanson et al. 2002). Figure 3.1B shows the complete flow net analysis of seepage beneath an impervious dam with an apron and a subsurface cutoff wall (see paragraph 3.3 in this chapter).

Another engineering application is the design of a minimum energy loss spillway inlet. The concept was developed to pass large floods with minimum energy loss. The flow in the approach channel is contracted through a streamlined chute and the channel width is minimum at the chute toe. Both the inlet and chute are streamlined to avoid significant form losses, and their design is typically based upon a flow net analysis. The movie IMGP2463.avi illustrates the operation of such a minimum energy loss spillway inlet during a small flood (App. F).

## Notes

– Streamlines are also called $\psi$-lines.
– Equipotentials, or equipotential lines, are sometimes called $\phi$-lines.

## 2.2   Basic properties

Vallentine (1969, p. 33) discussed some basic characteristics of flow nets. These features are critical to the solution of the flow field because the solution is unique:

(a)   Flow nets are based upon the assumption of irrotational flow of ideal fluid, but it is not necessary that the flow is steady.
(b)   For a given set of boundary conditions there is only one possible flow pattern. This is a property of the Laplace equation.
(c)   Streamlines are everywhere tangent to the velocity vector. Fixed boundaries are streamlines. The volumetric flow rate between two streamlines is: $\delta q = \delta \psi$ where $\delta \psi$ is the difference in stream function value between adjacent streamlines.
(d)   There is no velocity component tangent to an equipotential line and hence the velocity vector must be everywhere normal to an equipotential line. Equipotential lines intersect fixed boundaries normally.
(e)   Streamlines and equipotential lines are orthogonal (i.e. intersect at right angles).

---

**Remember**

A property of the Laplace equation is the unicity of the solution. Once an exact solution is found using the flow net, it is the only solution. In other words, both the graphical method (flow net solution), analytical method and numerical method must converge to the same, unique solution.

---

**Application**

Figure 3.1B illustrates the seepage flow beneath an impervious dam. The construction of the flow net starts by drawing obvious streamlines and equipotentials. The dam wall, the cutoff wall and the apron are streamlines, because solid boundaries are streamlines. Further there are two obvious equipotential lines: the reservoir floor upstream of the dam and the river bed downstream of the apron. The flow net is drawn with 4 stream tubes. The graphical analysis yields 15 equipotential tubes.

Since the water level difference between the upstream and downstream reservoirs is H, and the hydraulic conductivity is K, then: $\delta \phi = K \times H/15$ (see paragraph 3.3). As $\delta \phi = \delta \psi$, the seepage flow rate per unit width is: $q = 4 \times \delta \psi = 4/15 \times K \times H$.

Numerical application: if $H = 30\,m$, $K = 1\,E-5\,m/s$, $W = 60\,m$ (dam length across the valley), calculate the seepage flow beneath the dam.

The seepage flow rate is: $Q = q \times W = 0.0048\,m^3/s$. That is, $17.3\,m^3$ per hour. In absence of the cutoff wall and apron, the streamline paths would be much shorter and the seepage flow would be significantly larger.

**Notes**

– A stream tube is a fillet of fluid bounded by two adjacent streamlines.
– An equipotential tube is the region limited by two adjacent equipotentials.
– In the above application, the selection of 4 stream tubes was arbitrary. A greater number of stream tubes enhances the accuracy of the results but increases the workload and the time for completion of the flow net.

## 2.3 Graphical solution

The construction of a flow net requires a sequence of basic steps, as outlined by Vallentine (1969, p. 78):

(a) Construction of the streamlines in the regions where the velocity distributions are evident: e.g., parallel or radial flow, fixed boundaries.
(b) Sketch the remaining portions of streamlines with smooth curves.
(c) Construction of the equipotential lines: (c1) normal to all streamlines and fixed boundaries, and (c2) forming squares with the streamlines. This is a process of trial-and-error. The diagonal of the 'squares' should form smooth curves. At stagnation points the occurrence of five-sided 'square' results from the impossibility to obtain infinite spacing of streamlines in the region.
(d) Location of free surfaces by trial.
(e) Tendency to separate is indicated where the velocity at a boundary reaches a maximum and thereafter decreases in the direction of flow.

The problem of finding the flow net to satisfy given fixed boundaries is purely a **graphical** **exercise**: i.e., the construction of an orthogonal system of lines that compose boundaries and reduce to perfect squares in the limit as the number of lines increases.

**Note**

Free-surfaces are streamlines because there is no flow across a free-surface. The free surface is characterised by:

$$\psi = \text{constant}$$

and

$$\frac{\partial \psi}{\partial n} = 0$$

where n is the direction normal to the free surface. The second condition means that the velocity normal to the streamline is zero: $V_n = 0$. That is, there is no flow across a streamline.

The free surfaces are located by trial and error to satisfy the condition that equipotential lines and streamlines are equally spaced.

**Application No. 1**

Considering an uniform flow past a cambered Joukowski airfoil with zero angle of attack (Fig. 3.2), draw the flow net. Deduce the pressure field on the airfoil surface and total lift force if the uniform velocity away from the airfoil is 8 m/s and the foil chord is 0.5 m. *Assume a wind flow at standard conditions.*

**Solution**

(a)    First the obvious streamlines are drawn. These are the streamlines away from the airfoil and the boundary (perimeter) of the airfoil. Another streamline is that through the stagnation points (i.e. $\psi = 0$) (Fig. 3.2).

(b)    Then the remaining portions of streamlines are sketched with smooth curves.

(c)    Equipotential lines are drawn to be normal to all streamlines and fixed boundaries, and forming squares with the streamlines. This is a trial-and-error method.

Four basic steps of the flow net construction are illustrated in Figure 3.2.

At completion of the flow net, the velocity next to the airfoil is deduced from the continuity equation. In each stream tube, the flow rate per unit width $\delta q$ is constant:

$$\delta q = V_o \times \delta n_o = V \times \delta n \qquad \text{Continuity equation} \qquad (3.9)$$

where $V_o$ is the uniform velocity away from the airfoil, $\delta n_o$ is the distance between two adjacent streamlines in the uniform flow, $V$ is the velocity in a flow net element, and $\delta n$ is the distance between two adjacent streamlines (Fig. 2.2). In this application, $V_o = 8$ m/s and $\delta n_o = 0.032$ m for $\delta \psi = 0.25$. Note that $\delta n_o$ is scaled from the flow net knowing the airfoil chord length (i.e. 0.5 m).

The pressure at the boundary is calculated from the Bernoulli equation. In steady flow, it yields:

$$P_d = P_{d_o} + \rho \times \frac{V_O^2}{2} \times \left(1 + \left(\frac{V}{V_O}\right)^2\right) \qquad (2.23b)$$

where $P_d$ is the dynamic pressure (Chapter 2, paragraph 4.3).

The lift force is calculated by integrating the pressure force component perpendicular to the approach flow direction along the airfoil perimeter:

$$\text{Lift} = - \int_0^{2\times\pi} P \times r \times \sin\theta \times d\theta \qquad (3.10)$$

where $r$ and $\theta$ are the radial coordinates of the foil surface. The integration of Equation (3.10) gives: Lift $\sim +5$ N/m, where the lift is positive upwards.

**Notes**

–  The selection of $\delta\psi = 0.25$ was made arbitrarily to obtain several flow net elements next to the airfoil. The accuracy of the calculations increases with increasingly smaller flow net elements (i.e. $\delta\psi \to 0$).

–  The lift force is the pressure force component in the direction perpendicular to the flow direction.

–  In Equation (3.10), the sign $(-)$ is linked with the selection of the positive flow direction. Herein the lift force is positive along the positive y-direction.

**Remarks**

1. The Joukowski airfoils are a family of two-dimensional airfoils developed by N.E. Joukowski and for which he solved analytically the flow field.
2. Nikolai Egorovich Joukowski (1847–1921) was a Russian mathematician who did some extensive research in aerodynamics. His name is also spelled Zhukovsky or Zhukoskii. In 1906 he published two papers in which he gave a mathematical expression for the lift on an airfoil. Today the theorem is known as the Kutta-Joukowski theorem.
3. For a cambered airfoil (e.g. Fig. 3.2), the Kutta-Joukowski theorem (Chap. I-5) gives the lift per unit width:

$$\frac{\text{Lift}}{\frac{1}{2} \times \rho \times V_o^2} = 2 \times \pi \times \text{chord} \times \sin(\alpha + \beta) \qquad (3.11)$$

where $\alpha$ is the angle of attack and $\beta$ is the airfoil camber.

For the airfoil sketched in Figure 3.2, the angle of attack is zero and the camber is: $\sin \beta = 0.05$. Equation (3.11) predicts Lift $= +6\,\text{N/m}$ which is close to the graphical solution result.

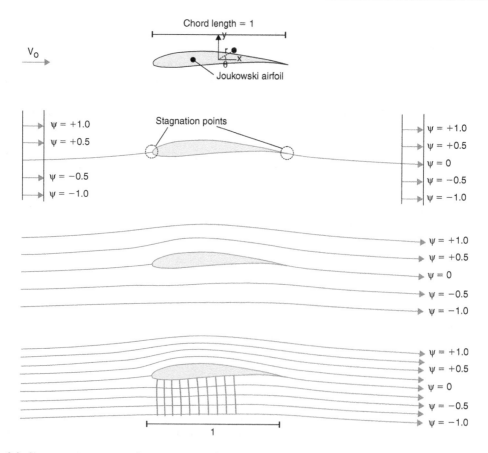

*Figure 3.2* Flow net construction: flow past an aerofoil – Joukowski airfoil with 5% camber, zero angle of attack and 12% thickness ($V_o = 8\,\text{m/s}$, chord: 0.5 m).

## Discussion

Joukowski proposed in 1906 an analytical expression of the lift on an airfoil. His work, called the Kutta-Joukowski theorem, was developed for irrotational flow motion of ideal fluid. Figure 3.3 presents a comparison of lift and drag coefficients for two-dimensional flow past a Joukowski airfoil. The lift and drag coefficients are defined respectively as:

$$C_D = \frac{\text{Drag}}{\frac{1}{2} \times \rho \times V_o^2 \times \text{chord}} \qquad \text{drag coefficient}$$

$$C_L = \frac{\text{Lift}}{\frac{1}{2} \times \rho \times V_o^2 \times \text{chord}} \qquad \text{lift coefficient}$$

where $V_o$ is the uniform velocity away from the air foil. While ideal fluid flow calculations predict zero drag, the lift force calculations are in close agreement with experimental observations. Note that the chord times the span is the largest possible projected area of the foil.

For the airfoil characteristics shown in Figure 3.3, the experiment lift coefficient would be $C_L = 0.7$ for zero angle of attack. For a foil thickness of 12% of the chord (as in Fig. 3.2), the lift force would be: Lift $= +13$ N/m. The result is of the same order of magnitude as the theoretical calculations, but differences must be accounted for a different airfoil profile. Figure 3.3 was not obtained specifically for 5% camber.

In wing design, three-dimensional phenomena associated with wing tips introduce additional complexities that are not considered in the above two-dimensional approach.

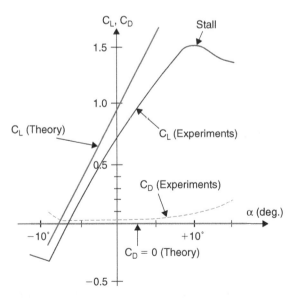

Figure 3.3 Comparison between experimental and theoretical lift and drag predictions for two-dimensional flows past a Joukowski airfoil as functions of the angle of incidence $\alpha$.

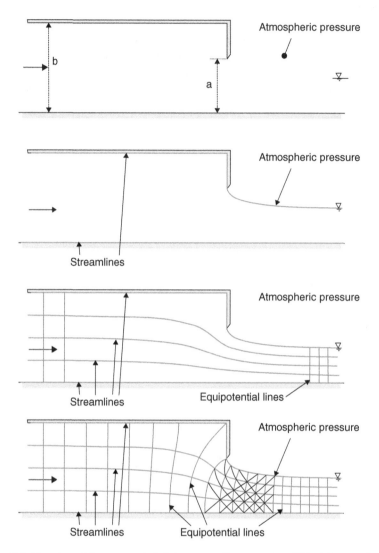

*Figure 3.4*   Flow issued from a vertical sluice gate – Gate opening: 60% of the pipe height.

### Application No. 2

Considering the flow issued from a vertical sluice gate into a horizontal open channel (Fig. 3.4), draw a flow net. Deduce the contraction ratio. *Assume a water flow at standard conditions. Note the sharp edge orifice gate.*

### *Solution*

The problem involves a free-surface downstream of the sluice gate. The free-surface is a streamline and its position is unknown beforehand.

(a)  First the obvious streamlines are drawn. These are the streamlines far upstream and far downstream of the sluice gate, the invert and the pipe boundaries.
(b)  Sketch the remaining portions of streamlines with smooth curves. In particular, the free-surface profile downstream of the sluice gate.
(c)  Construction of the equipotential lines: (c1) normal to all streamlines and fixed boundaries, and (c2) forming squares with the streamlines. This is a process of trial-and-error. The diagonal of the 'squares' should form smooth curves. At stagnation points the occurrence of five-sided 'square' results from the impossibility to obtain infinite spacing of streamlines in the region.
(d)  The location of the free surface is by trial and error.

The coefficient of contraction is defined as the ratio of the downstream water depth to the gate opening. Graphically, the contraction coefficient is about 2/3.

## Notes

The edge of the sluice gate is a singularity because there is a discontinuity of pressure.
   At the free-surface the pressure is atmospheric.
   The diagonal of the "squares" are shown in the lowest figure immediately downstream of the sluice gate. These diagonals should form smooth curves if the flow net is exact.

## Discussion

The analytical solution of a vertical orifice flow, in absence of gravity force, is presented in Chapter I-7 (section 3.2). Further discussion is included in Chapter I-7, section 3.3.
   For a two-dimensional water jet issued from a vertical sluice gate (Fig. 3.4), the contraction coefficient is a function of the relative gate opening a/b where b is the pipe height and a is the gate opening. Ideal-fluid flow calculations are summarised below.

| Relative gate opening a/b (1) | Coefficient of contraction (2) | Remarks (3) |
|---|---|---|
| 1 | 1.0 | Two dimensional pipe flow discharging into air. |
| 0.9 | 0.787 | |
| 0.8 | 0.722 | |
| 0.7 | 0.687 | |
| 0.6 | 0.662 | Application No. 2, Fig. 3.4. |
| 0.5 | 0.644 | |
| 0.4 | 0.631 | |
| 0.3 | 0.622 | |
| 0.2 | 0.616 | |
| 0.1 | 0.612 | |
| 0 | 0.611 | Small orifice flow: $\pi/(\pi+2)=0.611$. (Chapter I-7, section 3.2) |

Notes: Calculations by Von Mises in Rouse (1946).

## 2.4  Discussion

The flow net analysis is a very powerful tool. It is a graphical technique to solve the entire flow field in terms of the velocity potential and stream function. The results may also be expressed in terms of the two-dimensional velocity field. Each flow net element gives the magnitude and direction of the velocity vector at the centre of gravity of the element. Once the velocity field is known, the pressure field may be deduced from the Bernoulli principle.

A key issue is the simplicity of the flow net analysis. It may be applied to very complex flow situations, but the solution is a simple graphical exercise that does not require a computer and may be conducted on site and in the field. It is further applicable to nearly any two-dimensional flow, including complicated geometries. And its solution is unique.

In practice, the flow net analysis is particularly useful to predict the two-dimensional flow field and pressure distributions at the surface of submerged bodies in absence of separation. For example, the flow around a Joukowski airfoil at low angle of attack; the flow around a bridge pier; a two-dimensional orifice flow. Key features of the technique are the uniqueness of the solution for a given set of boundary conditions, the solution of the entire flow field (velocity vector and pressure) and the visual outcome of the results.

> **Remember**
>
> For a given set of boundary conditions, the flow net has an **unique solution**. This is a basic property of the Laplace equation. The graphical method converges to the same unique solution as analytical and numerical solutions of the same ideal fluid flow problem.

## 3  ANALOGIES

The "potential flow" theory applies not only to the irrotational motion of ideal fluid, but also to several systems. Herein, several analogies are presented: the electrical analogy, the Hele-Shaw cell flow analogy and the groundwater flow analogy. A further analogy is the thin elastic membrane analogy discussed in Vallentine (1969, p. 93).

## 3.1  Electrical analogy

The flow of an electric current in a two-dimensional conductor is analogous to an irrotational flow. The electrical potential is the counterpart of the velocity potential $\phi$.

In laboratory, a sheet of conducting material (e.g. Teledeltos paper) may be shaped so that its boundary is similar to that of the flow pattern. By means of a probe and a voltmeter, lines of constant potential are mapped out and plotted: these are equipotential lines. Lines orthogonal to the equipotential lines are streamlines.

### Notes

– This method has been extended to axially symmetrical three-dimensional flow problems. It was commonly used during the 20th century, but the interest vanished with the development of modern computer.
– The Teledeltos paper is a sheet of electrically conducting paper obtained by adding carbon black to paper pulp with a layer of insulating recording lacquer on one side and a thin aluminium layer on the other. The Teledeltos paper can give fairly accurate results from relative crude maps (Allen 1968).

## 3.2 Hele-Shaw analogy (viscous flow analogy)

In a viscous flow between two closely spaced, horizontal and parallel plates, the flow is laminar and the mean velocity across the space in any region is proportional to the pressure gradient:

$$V_x = -k \times \frac{\partial P}{\partial x} \qquad\qquad (3.12a)$$

$$V_y = -k \times \frac{\partial P}{\partial y} \qquad\qquad (3.12b)$$

where k is a function of the fluid properties and flow geometry. Hence the pressure is analogous to the velocity potential of irrotational flow:

$$\phi = -k \times P \qquad\qquad \text{Hele-Shaw cell analogy} \qquad (3.13b)$$

In practice, a transparent fluid is caused to flow slowly between the plates and dye is continuously injected into the fluid at regular intervals across the inflow boundary. An object placed between the plates causes the fluid to deviate in flowing around it, and the dyed portions of the fluid trace out streamlines for two-dimensional potential flow (Fig. 3.5).

---

**Discussion**

The Hele-Shaw-cell apparatus was initially presented in:

Hele-Shaw, H.J.S. (1898). "Investigation of the Nature of the Surface Resistance of Water and of Stream-line Motion under certain Experimental Conditions." *Trans. Inst. Naval Architects*, Vol. 40.

Figure 3.5A presents a Hele-Shaw cell apparatus with the dye injection system on the left and the flow from left to right around a square body made of rubber fabric. Figure 3.5B shows the streamline flow pattern for the flow past a blunt body, with the uniform flow from right to left. Note the dye dispersion downstream of the object.

The movies Foil01.mov and Foil03.mov show the operation of an Hele-Shaw cell apparatus with two configurations (App. F). The movie Foil01.mov illustrated an uniform flow (from right to left) past a 15% cambered foil perpendicular to the main flow direction. The dye is injected upstream of the foil and the video illustrates the dye motion around the object. The movie Foil03.mov presents an uniform flow past a thick rounded plate perpendicular to flow direction. The flow pattern may be compared with the flow normal to a flat plate with separation (Chapter I-7). Both videos illustrate some dye dispersion downstream of body, while rounded edges were used to prevent singularities.

---

**Note**

In laminar pipe flows, the application of the momentum principle in its integral form yields the longitudinal pressure loss:

$$\frac{\partial P}{\partial x} = -f \times \frac{1}{D} \times \rho \times \frac{V^2}{2}$$

(A) Top view of the apparatus with the dye injection system on the left, flow from left to right – The dye coloured streamlines flow arouund a square body – Flow from left to right

(B) Flow past a blunt body – Flow from left to right

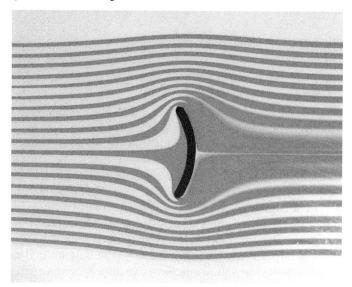

*Figure 3.5*  Hele-Shaw cell experiment.

where D is the pipe diameter, V is the cross-sectional averaged velocity and the Darcy-Weisbach friction factor f satisfies:

$$f = \frac{64 \times \mu}{\rho \times V \times D}$$

After transformation, it yields:

$$V = -\frac{1}{32} \times \frac{D^2}{\mu} \times \frac{\partial P}{\partial x}$$

The result for viscous pipe flow is consistent with Equation (3.12) using $k = -D^2/(32 \times \mu)$.

Note that the result may be extended to a non-circular cross-section using the equivalent pipe diameter ($D_H = 4 \times A/P_w$) instead of the pipe diameter.

## 3.3  Groundwater flow

For a one-dimensional flow, the velocity of seepage is given by the Darcy law for granular non-cohesive soils:

$$V = -K \times \frac{\partial H}{\partial s} \tag{3.14}$$

where K is the hydraulic conductivity (or coefficient of permeability, in m/s), H is the piezometric head (in m) and s is the streamwise coordinate. The Darcy law may be rewritten as a potential flow solution:

$$V_x = -\frac{\partial \phi}{\partial x} \tag{3.15a}$$

$$V_y = -\frac{\partial \phi}{\partial y} \tag{3.15b}$$

where the velocity potential $\phi$ is defined as:

$$\phi = K \times H \qquad\qquad \text{Groundwater flow analogy} \tag{3.16}$$

The hydraulic conductivity K depends not only on the permeability of the soil but also on the properties of the fluid. Dimensional analysis yields:

$$K = k \times \frac{\rho \times g}{\mu} \tag{3.17}$$

where k is the permeability (in $m^2$), $\rho$ the fluid density, g the gravity constant and $\mu$ the fluid dynamic viscosity.

In a groundwater, the continuity equation becomes:

$$\frac{dV_x}{dx} + \frac{dV_y}{dy} = -\frac{1}{K} \times \Delta\phi = 0$$

The velocity potential $\phi$ satisfies the Laplace equation. Applications include the study of percolation under a concrete dam, percolation in a simple earth/rockfill dam, seepage flow to a well, flow through permeable weirs, such as rubble mound weir, rockfill structures and some timber weirs (Michioku et al. 2005, Chanson 2006) (Fig. 3.6). A related application is the in-built spillway dam used in Australia since the 1950s (Olivier 1967, Chanson 2006).

*Figure 3.6* Seepage flow on the downstream face of a gabion stepped weir at Guaribraba, Capo Grande, Brazil (Courtesy of Officine Maccaferri) – In the absence of overflow on the stepped spillway, the seepage flow increases with downstream distance from the crest.

## Notes

– Henri Philibert Gaspard Darcy (1805–1858) was a French civil engineer. He performed numerous experiments of flow resistance in pipes (Darcy 1858) and in open channels (Darcy and Bazin 1865), and of seepage flow in porous media (Darcy 1856). He gave his name to the Darcy-Weisbach friction factor and to the Darcy law in porous media.

– Typical values of the hydraulic conductivity are:

| Soil type | K (m/s) | Remark |
| --- | --- | --- |
| Crushed rockfill | 1 E−1 to 1 E−2 | 2 to 600 mm range. Free draining. |
| Gravels | 1 E−3 to 1 E−5 | |
| Fine sand | 5 E−4 to 1 E−5 | |
| Silty sand | 2 E−5 to 1 E−6 | |
| Silt | 5 E−6 to 1 E−7 | |
| Clay | 1 E−7 to 1 E−10 | Impervious |

## Discussion

An example of massive seepage was the Quinson dam (France) built between 1866 and 1868. With a dam height of 20 m, the wall was built from stone blocks linked by means of metal clamps with lead joints. In 1888, the reservoir dried up because of large seepage under the dam (Colas des Francs 1975). In the 1970s, a new dam was constructed few metres downstream of the old structure (completion in 1974). Other examples of dams that were adversely affected by significant seepage in dam foundations include the 65 m high Saveh dam completed in Iran during the 14th century AD which was never filled. Figure 3.6 illustrates some seepage on the downstream face of a gabion weir.

In dam construction, a cutoff wall is an underground wall built to stop or reduce seepage in alluvial soils beneath the main dam wall (Fig. 3.1B). Examples of dam equipped with large cutoff walls include the Serre-Ponçon dam (France) where the cutoff wall extends 100 m beneath the natural river bed.

## Application

The water depth upstream of an earthfill embankment dam is 15 m (Fig. 3.7). The dam foundation is an impervious stratum. If the dam material is silty sand, calculate the seepage flow rate.

### Solution

First we must draw the flow net. Figure 3.7 presents the seepage flow net through the earthfill embankment. The upper surface of the flow (i.e. seepage line) is at atmospheric pressure, hence the flow is termed "unconfined". The free-surface, called seepage line, is a streamline. The upstream face of the dam is a line of constant potential because this is a line of constant piezometric head. But the downstream face of the dam is not an equipotential nor a streamline. In fact, the seepage face is a line of discontinuity.

The flow net analysis contained 5 streamlines and 4 stream tubes. Along the lowest stream tube, there are 23 equipotential tubes. The seepage flow rate per unit width is then:

$$q = \frac{4}{23} \times K \times H$$

where K is the hydraulic conductivity. For silty sand, K ∼ 1 E−5 m/s. It yields: q = 2.6 E−5 m²/s. That is, a seepage flow of 2.25 m³/day/m.

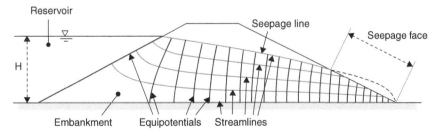

*Figure 3.7* Flow net in an earthfill dam.

**Notes**

– The seepage line may be located by trial using the empirical rule of Casagrande which approximates the free-surface line by a parabola issued from the dam toe. The start of the parabolic curve is drawn with a dashed line in Figure 3.7.
– The seepage is flow is calculated using the lowest stream tube because the seepage face is a discontinuity, and the upper stream tube is incomplete.
– Arthur Casagrande (1902–1981) was an Austrian-born American civil engineer who was a Professor in Geomechanics at Harvard University from 1932.

# Exercises

## 3.1  Cauchy-Riemann equations

- Determine the stream function for parallel flow with a velocity V inclined at an angle $\alpha$ to the x-axis.
- Find the velocity potential using the Cauchy-Riemann equations.

### Solution

By definition:

$$\frac{\partial \psi}{\partial y} = -V_x = -V \times \cos \alpha \psi = -V \times \cos \alpha \times y + f_1(x)$$

$$\frac{\partial \psi}{\partial x} = V_y = V \times \sin \theta \psi = V \times \sin \theta \times x + f_2(y)$$

Hence:

$$\psi = V \times \sin \theta \times x - V \times \cos \alpha \times y + \text{constant}$$

A streamline equation is:

$$y = \tan \theta \times x - \frac{\psi}{V \times \cos \alpha}$$

which is a straight line equation.

The Cauchy-Riemann equations are:

$$\frac{\partial \phi}{\partial x} = \frac{\partial \psi}{\partial y} \phi = -V \times \cos \alpha \times x + g_1(y)$$

$$\frac{\partial \phi}{\partial y} = -\frac{\partial \psi}{\partial x} \phi = -V \times \sin \theta \times y + g_2(x)$$

Hence:

$$\phi = -V \times \sin \theta \times y - V \times \cos \alpha \times x + \text{constant}$$

$$V_x = V_O \times \cos \alpha$$

$$V_y = V_O \times \sin \alpha$$

$$\psi = -V_O \times (y \times \cos \alpha - x \times \sin \alpha)$$

## 3.2  Flow net

For a two-dimensional seepage under a impervious structure with a cutoff wall (Fig. E3.1), the boundary conditions are:

$$H = 6\,\text{m}$$

$$K = 2.0\,\text{m/day}$$

(a)   What is the hydraulic conductivity in m/s? What type of soil is it?
(b)   Using the flow net, estimate the seepage flow par meter width of dam.

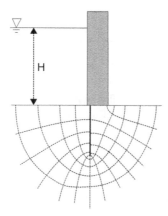

*Figure E3.1   Flow net beneath an impervious dam.*

### Solution

$$q = 4.6 \, \text{m}^2/\text{day}$$

## 3.3   Flow net

For a two-dimensional seepage under sheet piling with a permeable foundation (Fig. E3.2), the boundary conditions are:

$a = 9.4 \, \text{m}$   $b = 4.7 \, \text{m}$           $c = 4 \, \text{m}$
$H = 2.5 \, \text{m}$   $K = 2.0 \times 10^{-3} \, \text{cm/s}$

Using the flow net (with 5 stream tubes), estimate the seepage flow par meter width of dam.

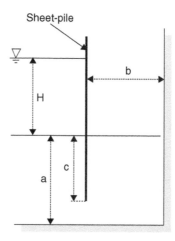

*Figure E3.2   Flow net beneath a sheet pile.*

## Solution

$q = 2.7\,\text{m}^2/\text{day}$

---

**Discussion**

The flow pattern may be analysed analytically using a finite line source (for the sheet pile) and the theory of images. The resulting streamlines are the equipotentials of the sheet-pile flow.

---

**Remember**

A velocity potential can be found for each stream function. If the stream function satisfies the Laplace equation the velocity potential also satisfies it. Hence the velocity potential may be considered as stream function for another flow case. The velocity potential $\phi$ and the stream function $\psi$ are called "*conjugate functions*" (Chapter I-2, paragraph 4.2).

---

### 3.4  Flow net

For a two-dimensional horizontal flow under a gate (Fig. 3.4), the boundary conditions are:

$a = 0.75\,\text{m} \quad b = 1\,\text{m} \quad P_O = 3.92 \times 10^4\,\text{Pa} \ (H_O = 4\,\text{m})$

Draw the flow net using 4 stream tubes (i.e. 5 streamlines). Determine the flow rate. Determine the pressure distribution on the gate.

## Solution

The problem is similar to Application No. 2, paragraph 2.3.

The pressure distribution on the gate is deduced from the Bernoulli principle. Note that the pressure is zero (i.e. atmospheric) at the orifice edge.

While the velocity nor flow rate are given, the flow velocity downstream of the orifice may be deduced from the Bernoulli equation. If the pressure upstream of the gate is much greater than the hydrostatic pressure downstream of the gate, the downstream flow velocity is about: $V \sim \sqrt{2 \times g \times H_o}$ where $H_o$ is the total head.

### 3.5  Flow net

For a two-dimensional seepage under an impervious dam with apron (Fig. 3.1), the boundary conditions are:

$H = 80\,\text{m} \quad K = 5 \times 10^{-5}\,\text{m/s}$

(a) Calculate the seepage flow rate. (b) In absence of the cutoff wall, sketch the flow net and determine the seepage flow. Determine the pressure distribution along the base of the dam and beneath the apron. Calculate the uplift forces on the dam foundation and on the apron.

## Solution

(a) The problem is similar to the flow net sketched in Figure 3.1B (paragraph 2.2). (b) In absence of cutoff wall, the streamlines are shorter and the seepage flow rate is greater. The uplift pressure on the dam foundation and apron become very significant. The apron structure would be subjected to high risks of uplift and damage.

## 3.6 Flow net

For a two-dimensional seepage through an earth dam (Fig. 3.7), the boundary conditions are:

$$H = 45\,m \qquad K = 2.0 \times 10^{-6}\,m/s$$

Using the flow net, estimate the seepage flow par meter width of dam. (Assume a homogeneous embankment.)

### Solution

See Application, paragraph 3.3.

## 3.7 Flow net

For a two-dimensional seepage under an impervious dam with a cutoff wall (Fig. E3.3), the boundary conditions are:

$$H_1 = 60\,m \quad H_2 = 5\,m \quad a = 60\,m \quad b = 100\,m$$
$$L = 70\,m \quad K = 1 \times 10^{-5}\,m/s$$

(A)   In absence of cutoff wall (i.e. a = 0 m), sketch the flow net; determine the seepage flow; determine the pressure distribution along the base of the dam; calculate the uplift force.
(B)   With the cutoff wall (i.e. a = 60 m): same questions: sketch the flow net; determine the seepage flow; determine the pressure distribution along the base of the dam; calculate the uplift force
(C)   Comparison and discuss the results.

*Use graph paper.*

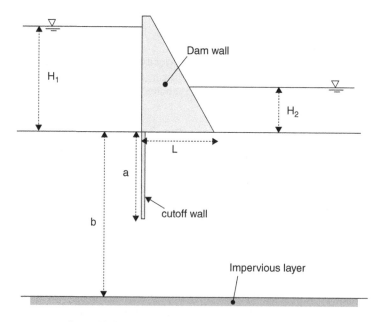

*Figure E3.3* Flow net beneath a dam with cutoff wall.

## 3.8   Flow net

A hydrodynamic study of the flow past a flat plate is conducted in water. The plate length is $L = 0.2$ m and the plate width equals the test section width (0.5 m). For a particular angle of incidence, dye injection in a water tunnel gives the flow pattern shown in Figure E3.4 on the next page. The mean flow is horizontal.

(a)   Complete the flow net by drawing the suitable equipotentials. (Draw the complete flow net on the examination paper.)

(b)   The upstream velocity is 6.5 m/s. Estimate the drag force and the lift force acting on the plate. *Indicate clearly the sign convention. You may have to draw more streamline and equipotentials to obtain a good accuracy.*

*Assume water at 20 Celsius ($\rho = 998.2 \, kg/m^3$, $\mu = 1.005 \, E-5 \, Pa.s$, $\sigma = 0.0736 \, N/m$).*

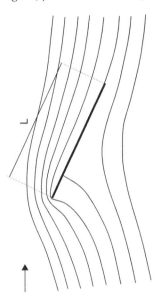

*Figure E3.4*   Flow visualisation and streamline patterns around an inclined flat plate (undistorted scale).

### Solution

The problem is solved by drawing the equipotentials and completing the flow net. Close to the foil, additional streamlines and equipotentials may be drawn to improve the estimates of the velocities next to the extrados and intrados of the foil. The pressure field is derived from the Bernoulli equation and the integration of the pressure distributions yields the lift and drag forces.

Note that the complete theory of lift and drag on airfoils, wings and hydrofoils is developed in Chapter I-6.

## 3.9   Flow net

Let us consider the water flow past a circular arc sketched in Figure E3.5. The chord is $L = 0.35$ m and the plate width equals 1 m. For a particular angle of incidence shown in Figure E3.5, conduct a graphical analysis of the flow field. The mean flow is horizontal.

(a)   Draw the flow net by drawing the suitable streamlines and equipotentials. (Draw the complete flow net with sufficient details in the vicinity of the circular arc.)

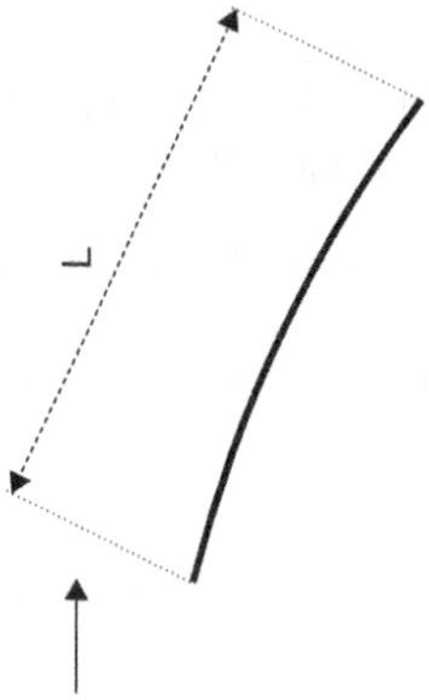

*Figure E3.5*  Flow past a circular arc (undistorted scale).

(b)  The upstream velocity is 15.2 m/s. Estimate the drag force and the lift force acting on the cambered plate. *Indicate clearly the sign convention.*

Assume water at 20 Celsius ($\rho = 998.2\,\text{kg/m}^3$, $\mu = 1.005\,\text{E}{-5}$ Pa.s, $\sigma = 0.0736$ N/m).

### Solution

The flow net solution can be compared with the theory of lift and drag. The problem may be solved analytically using the Joukowski transformation and theorem of Kutta-Joukowski. The complete theory of lift and drag on airfoils, wings and hydrofoils is developed in the Chapter I-6.

## 3.10  Flow analogy

A Hele-Shaw cell apparatus is built with a gap between plates $\delta = 1.5$ mm filled with Glycerol ($\rho = 1260\,\text{kg/m}^3$, $\mu = 1.4$ Pa.s). Estimate the hydraulic conductivity of the apparatus.

### Solution

In laminar flows between two parallel plates, the application of the momentum principle in its integral form yields an expression of the longitudinal head loss:

$$\frac{\partial H}{\partial x} = -\frac{f}{D_H} \times \frac{V^2}{2 \times g}$$

where $D_H$ is the equivalent pipe diameter (or hydraulic diameter), V is the cross-sectional averaged velocity and f is the Darcy-Weisbach friction factor:

$$f = \frac{64 \times \mu}{\rho \times V \times D_H}$$

Since $D_H = 2 \times \delta$, it yields:

$$V = -\frac{\rho \times \delta^2}{8 \times \mu} \times \frac{H}{\partial x}$$

By analogy with the Darcy law, this gives a hydraulic conductivity K:

$$K = \frac{\rho \times \delta^2}{8 \times \mu} = 2.5 \times 10^{-4}\, m/s$$

## 3.11  Flow analogy and flow net

Figure E3.6 shows the cross-section of an impervious dam above the pervious aquifer. A reservoir is located both upstream and downstream of the dam wall. The difference in reservoir elevation is 32 m and the dam length is 360 m. The aquifer consists of silty sand with a hydraulic conductivity of $1.7 \times 10^{-6}$ m/s.

(a)  On graph paper, sketch the flow net beneath the dam wall. Use at least 5 stream tubes. On graph paper, the sketch must be an undistorted copy of the sketch below (Fig. E3.6).
(b)  Calculate the seepage flow rate beneath the dam.
(c)  Calculate the uplift force acting on the dam foundation.
(d)  Calculate the sliding force exerted by the aquifer water pressure on the dam foundation.
(e)  Discuss and justify design techniques to reduce the seepage and the uplift pressure force.

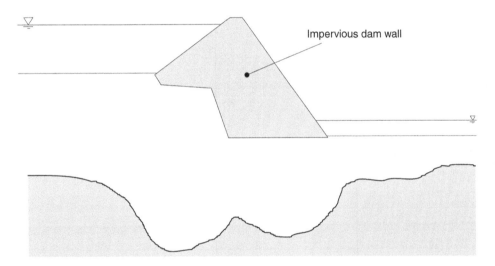

Impervious dam wall

Figure E3.6  Seepage beneath an impervious dam with an irregular impermeable aquifer base.

## Solution

The construction of the flow net with 5 stream tubes yields 10 equipotential tubes. Since the water level difference between the upstream and downstream reservoirs is 32 m, and denoting K the hydraulic conductivity, then: $\delta\phi = K \times 32/10$ (see paragraph 3.3). Since $\delta\phi = \delta\psi$, the seepage flow rate per unit width is: $q = 5 \times \delta\psi = 5 \times K \times 32/10$.

Numerical application: $K = 1.7 \times 10^{-6}$ m/s, $q = 2.7 \times 10^{-5}$ m²/s, $Q = q \times 360 = 0.0098$ m³/s. That is, 35 m³ per hour.

The application of the Bernoulli principle along the stream tube adjacent to the dam foundation yields the pressure distribution beneath the dam. The integration of the sum of pressure forces acting on the dam (foundation and slopes) in the vertical direction gives the uplift force, and the integration in the horizontal direction yields the sliding force.

The seepage flow and uplift force may be reduced using a cutoff wall to enlarge the streamline paths.

## 3.12 Flow net

Let us consider a two-dimensional flow in a Venturi-like contraction in a desalination plant (Fig. E3.7). The flow rate is 1.5 m²/s, the conduit internal size is 1.0 m upstream of the contraction and the upstream pressure is 135 kPa. *Assume that the flow is a two-dimensional irrotational flow of ideal fluid. The fluid is seawater at 20 Celsius, density: 1,024 kg/m³, dynamic viscosity: 0.00122 Pa.s.*

(a) On graph paper, sketch the flow net of the flow through the contraction at the same scale as the Figure E3.7. The flow direction is from left to right. Use at least 5 streamlines. On graph paper, the sketch must be a full-scale copy of the sketch below.

(b) The maintenance engineer is concerned about the velocity distributions and pressure field at the surface of the contraction.

    (b1) From your flow net, compute the velocity and pressure distribution along the surface of the contraction (line L-M-N shown in Fig. E3.7). *Plot your results on graph paper using dimensional axes: i.e., velocity in m/s, pressure in kPa, and distance in metres.*

    (b2) Where are located the points of maximum and minimum velocities, and maximum and minimum pressures on the line L-M-N?

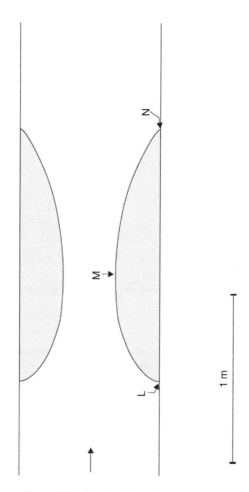

*Figure E3.7* Sketch of the flow contraction.

(b3)   Considering the entire flow motion through the contraction, what is the symbolic relationship between the pressure difference (upstream-minimum) across the contraction and the flow velocity?

(c)   An ideal fluid flow model is developed to predict the two-dimensional flow field in the contraction.

(c1)   Explain what standard flow patterns you would use to describe the flow in the Venturi-like contraction.

(c2)   Write the stream function and the velocity potential of the flow pattern as a function of the upstream velocity $V_o$ and the relevant contraction dimensions. *Do not use numbers. Express the results in a symbolic form.*

# Two-dimensional flows (2) basic flow patterns

## SUMMARY

The main equations for two-dimensional irrotational flows of ideal fluid are summarised. Their analytical solutions for simple flow patterns are developed. Then the flow fields associated with combinations of flow patterns are presented and complete analytical solutions are given.

## I  PRESENTATION

Although graphical solutions were presented in Chapter I-3, some analytical solutions of the entire flow field are presented herein for some basic flow patterns. The solution of the Laplace yields the algebraic expression of the velocity potential, and the stream function, from which the expression of the velocity vector may be derived. The pressure field is then deduced from the application of the Bernoulli principle. It must be noted that not all flow situations can be solved theoretically. The analysis of complicated ideal-fluid flows may require the use of graphical or numerical techniques.

For two-dimensional irrotational flows of ideal fluid, the main results may be summarised as follow.

### Irrotational flows

$$\overrightarrow{\text{curl}}\ \vec{V} = 0 \tag{4.1a}$$

For a two-dimensional flow, the condition to be irrotational is:

$$\frac{\partial V_y}{\partial x} - \frac{\partial V_x}{\partial y} = 0 \tag{4.1b}$$

### Velocity potential

$$\vec{V} = -\overrightarrow{\text{grad}}\ \phi \tag{4.2a}$$

$$V_x = -\frac{\partial \phi}{\partial x} \tag{4.2b}$$

$$V_y = -\frac{\partial \phi}{\partial y}$$

### Continuity equation

$$\text{div}\ \vec{V} = 0 \tag{4.3}$$

$$\vec{V} = -\overrightarrow{\text{curl}}\ \vec{\Psi} \tag{4.4a}$$

For a two-dimensional flow, the stream function vector becomes $\vec{\Psi} = (0, 0, \psi)$ where $\psi(x, y, t)$ is called the stream function:

$$V_x = -\frac{\partial \psi}{\partial y} \tag{4.4b1}$$

$$V_y = \frac{\partial \psi}{\partial x} \tag{4.4b2}$$

For irrotational flows, the condition of irrotationality becomes:

$$\Delta \psi = \frac{\partial^2 \psi}{\partial x^2} + \frac{\partial^2 \psi}{\partial y^2} = 0 \tag{4.5}$$

### Bernoulli equation

$$-\frac{\partial \phi}{\partial t} + \frac{V^2}{2} + \frac{P}{\rho} + g \times z = F(t) \tag{4.6}$$

where V is the velocity magnitude ($V = \sqrt{V_x^2 + V_y^2}$) and the z-axis is positive upward.

## 2  BASIC STEADY FLOW PATTERNS

### 2.1  Uniform flow

For a steady, uniform (i.e. parallel) and irrotational flow with a constant velocity $V_O$, inclined at an angle $\alpha$ to the x-axis (Fig. 4.1), the velocity field is:

$$V_x = V_O \times \cos \alpha \tag{4.7a}$$
$$V_y = V_O \times \sin \alpha \tag{4.7b}$$

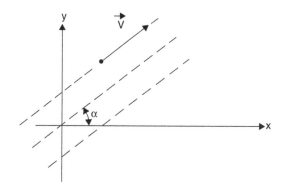

Figure 4.1  Sketch of an uniform flow.

Hence the velocity potential $\phi$ and stream function $\psi$ are:

$$\phi = -V_O \times (x \times \cos\alpha + y \times \sin\alpha) \tag{4.8a}$$

$$\psi = -V_O \times (y \times \cos\alpha - x \times \sin\alpha) \tag{4.9a}$$

In polar coordinates $(r, \theta)$, the velocity potential and stream function are respectively:

$$\phi = -V_O \times r \times \cos(\theta - \alpha) \tag{4.8b}$$

$$\psi = -V_O \times r \times \sin(\theta - \alpha) \tag{4.9b}$$

## 2.2  Sources – Sinks

A **source** is a point from which some fluid issues at an uniform rate in all directions (Fig. 4.2). The velocity field of a simple source at the origin is:

$$V_r = \frac{q}{2 \times \pi \times r} \tag{4.10a}$$

$$V_\theta = 0 \tag{4.10b}$$

$$\phi = -\frac{q}{2 \times \pi} \times Ln(r) \tag{4.11a}$$

$$\psi = -\frac{q}{2 \times \pi} \times \theta \tag{4.12a}$$

where q is the flow rate per unit width (Fig. 4.2A).

(A) Source at the origin

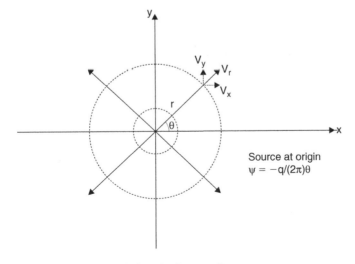

*Figure 4.2*  Sketch of source flow pattern.

(B) Source at a point $P(x_0, y_0)$

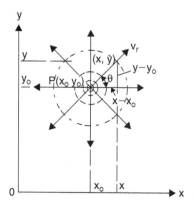

*Figure 4.2*  Continued.

In Cartesian coordinates:

$$V_x = \frac{q}{2 \times \pi} \times \frac{x}{x^2 + y^2} \tag{4.13a}$$

$$V_y = \frac{q}{2 \times \pi} \times \frac{y}{x^2 + y^2} \tag{4.13b}$$

$$\phi = -\frac{1}{2} \times \frac{q}{2 \times \pi} \times \text{Ln}(x^2 + y^2) \tag{4.11b}$$

$$\psi = -\frac{q}{2 \times \pi} \times \tan^{-1}\left(\frac{y}{x}\right) \tag{4.12b}$$

The strength of a source is the total flow q. The velocity at the origin is infinite and hence the origin is a singular point. A "negative source" is a **sink** $(q < 0)$.

### Source at a point P(x₀,y₀)

For a source at a point P of Cartesian coordinates $(x_O, y_O)$ (Fig. 4.2B), the velocity potential and stream function are respectively:

$$\phi = -\frac{1}{2} \times \frac{q}{2 \times \pi} \times \text{Ln}((x - x_O)^2 + (y - y_O)^2) \tag{4.14}$$

$$\psi = -\frac{q}{2 \times \pi} \times \tan^{-1}\left(\frac{y - y_O}{x - x_O}\right) \tag{4.15}$$

## 2.3  Vortex

A **vortex** is deduced from a source by selecting the stream function for the source as the velocity potential. An irrotational vortex of strength K located at the origin, is defined as (Fig. 4.3):

$$V_r = 0 \tag{4.16a}$$

$$V_\theta = \frac{K}{2 \times \pi \times r} \tag{4.16b}$$

$$\phi = -\frac{K}{2 \times \pi} \times \theta \tag{4.17}$$

$$\psi = \frac{K}{2 \times \pi} \times Ln(r) \tag{4.18}$$

The circulation $\Gamma$ around any closed path C that contains the vortex is:

$$\Gamma = \int_C \vec{V} \times \vec{\delta s} = 2 \times \pi \times r \times V_\theta = K \tag{4.19}$$

Basically the strength of the vortex equals the circulation around any closed path containing the vortex. The strength of a vortex is defined as: $K = 2 \times \pi \times V_\theta \times r$, and it is taken conventionally as positive for an anti-clockwise flow. In Figure 4.3A, the streamlines are shown for a positive strength $(K > 0)$. Note that the origin is a singular point.

### Notes

– The relationship between the strength of the vortex K and the speed of rotation $\omega$ is deduced from the expression of the tangential velocity:

$$V_\theta = \frac{K}{2 \times \pi \times r} = r \times \omega$$

and hence:

$$\omega = \frac{K}{2 \times \pi \times r^2}$$

– Combinations of vortices can be used to model numerically the turbulent flow field. Although friction and shear stress are not modelled, a combination of ideal-fluid vortices provides the same type of turbulent fluid motion as in real fluid flows.
– The vortex is the conjugate flow pattern of the source.

(A) Sketch of a vortex flow pattern

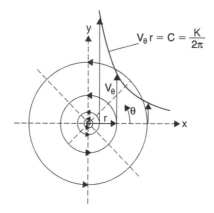

(B, Left) Air core of swirling vortex in a water column – The vortical motion was enhanced by swirling air injection at the bottom
(C, Right) Artificial tornado generated in an air column at Taipei Astronomical Museum

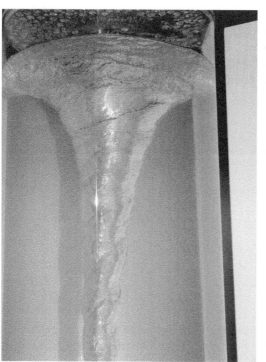
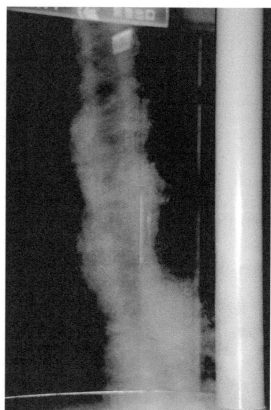

*Figure 4.3* Vortex flow.

**Discussion**

In Nature, vortical structures are associated with turbulent flows. Good examples include the bathtub, tornado and the whirlpool (Fig. 4.3B & 4.3C). Van Dyke (1982, p. 59) presented a superb illustration of the bathtub. Combinations of vortices are discussed in section 3.3 (in this chapter). Further illustrations include JSME (1988) and Homsy (2000, 2004).

In coastal zones, whirlpools are often observed at the edges of straits with large tidal currents. They are sometimes called Maelstrom, after the Norwegian current. The Maelstrøm is a strong tidal current of the Norwegian Sea in the Lofoten islands (Gjevik et al. 1997). The word maelstrom entered the English language via fiction novelists who exaggerated the current of the channel into a great whirlpool: e.g., Jules Vernes in "Twenty Thousand Leagues under the Seas". Notable oceanic whirlpools include those of Garofalo along the coast of Calabria in southern Italy, and of Messina in the strait between Sicily and peninsular Italy, the whirlpools near the Hebrides and Orkney islands, and those the Naruto strait between Awaji and Shikoku islands (Japan).

## 2.4  Doublet

A **doublet** is defined as the combination of a source and a sink of equal strength q which are allowed to approach each other, along a line called the doublet axis, in such a manner that the product of their strength $q/(2 \times \pi)$ and the distance between them $\delta s$, remains a constant in the limit (Streeter 1948, p. 44; Vallentine 1969, p. 110). The product: $(\mu = q/(2 \times \pi) \times \delta s$ is a constant called the strength of the doublet.

A doublet is defined by its axis direction and its strength $\mu$. For a doublet at the origin and with a horizontal axis, the velocity field, velocity potential and stream function are:

$$V_r = -\frac{\mu \times \cos\theta}{r^2} \tag{4.20a}$$

$$V_\theta = -\frac{\mu \times \sin\theta}{r^2} \tag{4.20b}$$

$$\phi = \frac{\mu \times \cos\theta}{r} \tag{4.21a}$$

$$\psi = -\frac{\mu \times \sin\theta}{r} \tag{4.22a}$$

In Cartesian coordinates:

$$\phi = \frac{\mu \times x}{x^2 + y^2} \tag{4.21b}$$

$$\psi = -\frac{\mu \times y}{x^2 + y^2} \tag{4.22b}$$

For a doublet at the origin and with horizontal axis, the equipotential lines (i.e. $\phi$ constant) are circles through the origin with centre on the x-axis, and streamlines (i.e. $\psi$ constant) are circles through the origin with centres on the y-axis. The origin is a singular point where velocity goes to infinity.

**Note**

A doublet is sometimes called a "dipole". This is the case in the software 2DFlow+ (App. E).

---

**Discussion: vortex doublet**

The vortex doublet, or vortex dipole, is deduced from a doublet by selecting the stream function for the doublet as a velocity potential. An irrotational vortex doublet of strength $\varphi$ located at the origin, is defined as:

$$\phi = -\frac{\varphi \times \sin\theta}{r}$$

$$\psi = \frac{\varphi \times \cos\theta}{r}$$

The vortex doublet is the combination of two vortices of equal strength K but opposite sign, which are allowed to approach each other along a line called the vortex doublet axis in such a manner that the product of their strength $K/(2 \times \pi)$ and the distance between them $\delta s$, remains a constant in the limit. The product $(\varphi = K/(2 \times \pi) \times \delta s)$ is constant and it is the strength of the vortex doublet.

For a vortex doublet at the origin and with horizontal axis, the streamlines (i.e. $\psi$ constant) are circles through the origin with centre on the x-axis, and equipotentials (i.e. $\phi$ constant) are circles through the origin with centres on the y-axis. The origin is a singular point where velocity goes to infinity. Basically, the flow net is that of a doublet with horizontal axis that is rotated by 90°.

Vortex doublets may be useful to study flow motion in the atmosphere and the oceans. For example, barotropic vortex doublets resemble jet streaks in the atmosphere.

---

**Notes**

- The vortex doublet is also called Lamb-Bachelor vortex dipole, after the fluid dynamicists Horace Lamb and George K. Bachelor.
- Sir Horace Lamb (1849–1934) was an English mathematician who made important contributions to acoustics and fluid dynamics. He was taught by George Gabriel Stokes (1819–1903) and James Clerk Maxwell (1831–1879). In 1875, Horace Lamb was appointed to the chair of mathematics at Adelaide SA, Australia where he stayed for 10 years before returning to England.
- George Keith Bachelor (1920–2000) was an Australian fluid dynamicist who worked on turbulence under Sir Geoffrey Ingram Taylor. He founded the Journal of Fluid Mechanics in 1956 and he was the journal editor for decades.
- The vortex doublet is the conjugate flow pattern of the doublet.

## 2.5 Flow around a corner

Considering a corner (angle $\alpha$) at the origin, the potential and stream functions are:

$$\phi = A \times r^{\pi/\alpha} \times \cos\left(\frac{\pi \times \theta}{\alpha}\right) \tag{4.23}$$

$$\psi = A \times r^{\pi/\alpha} \times \sin\left(\frac{\pi \times \theta}{\alpha}\right) \tag{4.24}$$

Figure 4.4 shows two examples of flow nets. Note that the corner itself is a flow singularity.

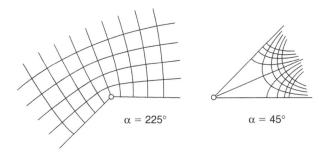

$\alpha = 225°$              $\alpha = 45°$

*Figure 4.4*  Sketch of a flow around a corner.

## 3   COMBINATIONS OF STEADY FLOW PATTERNS

### 3.1   Introduction

If two or more flow patterns are combined by superposition, the values of $\psi$ and $\phi$ at any point, other than a singular point, are simply the algebraic sum of the $\psi$-values and $\phi$-values of the component patterns at that point at any instant. This result derives from the principle of superposition and it is a basic property of the Laplace equation (Chapter I-2, paragraph 4.2).

### 3.2   Source and sink

If a source and sink of equal strength q are spaced apart by a distance $2 \times L$ on the x-axis, symmetrically about the origin, the velocity potential and stream function are for any point P(x, y):

$$\phi = -\frac{q}{2 \times \pi} \times \mathrm{Ln}\left(\frac{r_1}{r_2}\right) \tag{4.25}$$

$$\psi = -\frac{q}{2 \times \pi} \times (\theta_1 - \theta_2) \tag{4.26}$$

where the source of strength $+q$ is in $(+L, 0)$ and the sink of strength $-q$ is in $(-L, 0)$ (Fig. 4.5), and $r_1$, $r_2$, $\theta_1$ and $\theta_2$ are defined on Figure 4.5A. Basically $r_1$, $r_2$, $\theta_1$ and $\theta_2$ satisfy the following relationships:

$$r_1^2 = (x - L)^2 + y^2$$

$$r_2^2 = (x + L)^2 + y^2$$

$$\tan \theta_1 = \frac{y}{x - L}$$

$$\tan \theta_2 = \frac{y}{x + L}$$

The equipotential lines are circles of radius $L \times \mathrm{csch}\left(\frac{2 \times \pi \times \phi}{q}\right)$ with centres at $\left(L \times \coth\left(\frac{2 \times \pi \times \phi}{q}\right), 0\right)$. Streamlines are circles of radius $L \times \csc\left(\frac{2 \times \pi \times \psi}{q}\right)$ with centres at $\left(0, L \times \cot\left(\frac{2 \times \pi \times \psi}{q}\right)\right)$ (Fig. 4.5B).

(A) Definition sketch

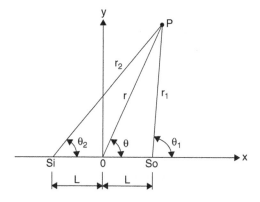

(B) Flow net

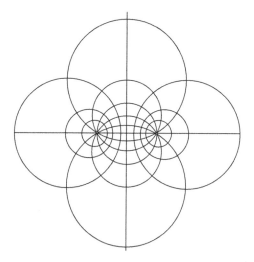

*Figure 4.5*   Flow pattern for a source and sink of equal strength.

## 3.3  Vortex pair

Two vortices of equal magnitude K, but opposite sign (i.e. opposite direction of rotation) constitute a vortex pair (Fig. 4.6). Considering the first vortex of strength $(+K)$ in $(+L, 0)$ and the second of strength $(-K)$ in $(-L, 0)$, we deduce by superposition:

$$\phi = -\frac{K}{2 \times \pi} \times (\theta_1 - \theta_2) \tag{4.27}$$

$$\psi = \frac{K}{2 \times \pi} \times \mathrm{Ln}\left(\frac{r_1}{r_2}\right) \tag{4.28}$$

where $r_1$, $r_2$, $\theta_1$ and $\theta_2$ are defined on Figure 4.6A.

(A) Definition sketch and streamlines

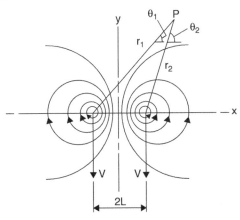

(B) Flow pattern seen by an observer travelling at the speed of the vortex centres

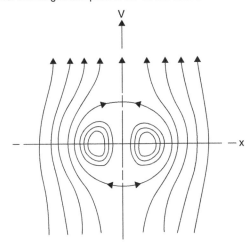

*Figure 4.6*  Flow pattern for a vortex pair.

Each vortex is affected by the movement of the fluid due to the other. Hence each vortex axis will move in a direction perpendicular to the line joining their centres: i.e., they will translate downward along the y-axis as shown in Figure 4.6. The magnitude of the velocity of each centre is:

$$V_{centre} = \frac{K}{2 \times \pi \times (2 \times L)} \tag{4.29}$$

Figure 4.6B illustrates the flow pattern seen by an observer moving with the vortex centres.

## Notes

– The induced motion of the vortex centres is a phenomenon observed in real-fluid motion.
– When the distance $2 \times L$ between vortices tends to zero, the resulting flow pattern is the vortex doublet (section 2.4, in this chapter).

### Application: "Meddies" in the Atlantic Ocean

The Mediterranean Sea looses more water by evaporation that it receives freshwater from rain and rivers. It is saltier than the Atlantic Ocean and the result is a deep, dense Mediterranean current flowing westward at the Straight of Gibraltar while fresher, lighter surface waters flow eastward into the closed sea. In the Antiquity, Phoenician sailors used heavy, ballasted sea-anchors sank in the bottom current to pull their boats into the Atlantic Ocean. Jacques Cousteau repeated the experiment with the Calypso in the 1980s (Cousteau and Paccalet 1987).

Once in the Atlantic Ocean, the denser current flows downward until it reach a layer of water with similar density. There the current becomes quasi-two-dimensional and develops as a vortex pair of about 400 to 800 m thickness and up to 200 km in diameter (McDowell and Rossby 1978, Richardson 1993). These organised structures may be felt as far as the Bahamas and some maintain their coherence for more than 2 years.

### Application

Considering a vortex pair in the Atlantic Ocean, separated by 30 km and with a circulation of 250,000 m²/s, estimate the time taken to cross the Atlantic Ocean if the induced motion of the vortex centres is a main phenomenon driving the meddy motion. *Assume the Atlantic Ocean to be about 3,200 km from East to West.*

### Solution

The velocity of each centre is:

$$V_{centre} = \frac{K}{2 \times \pi \times (2 \times L)} = \frac{250,000}{2 \times \pi \times 30,000} = 1.3 \, m/s$$

The vortex pair will take about 1.8 years to cross the Atlantic Ocean. For comparison, McDowell and Rossby (1978) suggested that some meddies may take up to 1 to 2 years to be cross the Ocean.

---

### Discussion: two vortices of equal magnitude and sign

A related application is the combination of two vortices of identical strength and sign. (Remember: in a vortex pair, the vortices have an opposite direction of rotation.) The superposition of two vortices of equal magnitude K and equal sign (i.e. same direction of rotation) is characterised by the following velocity potential and stream function:

$$\phi = -\frac{K}{2 \times \pi} \times (\theta_1 + \theta_2)$$

$$\psi = \frac{K}{2 \times \pi} \times Ln(r_1 \times r_2)$$

where one vortex is located at $(+L, 0)$ and the second at $(-L, 0)$, and $r_1$, $r_2$, $\theta_1$ and $\theta_2$ are defined on Figure 4.6A.

In this flow pattern, there is a stagnation point at the origin. Each vortex is affected by the movement of the fluid due to the other. Hence each vortex axis will move on a circle of radius L centred at the origin, and in the anti-clockwise direction for K > 0. The angular velocity of each centre (around the origin) is:

$$\omega_{centre} = \frac{K}{4 \times \pi \times L^2}$$

For a point far away from the vortices (i.e. $r \gg L$), $r_1 \approx r_2$ and $\theta_1 \approx \theta_2$. That is, the vortex pair is seen as one single vortex located at the origin and of strength $+2 \times K$.

The induced motion of the vortex centres is a phenomenon observed in real-fluid motion, although turbulence may induce a collapse of the vortices into one larger vortex: e.g., in a developing shear layer. A well-known geophysical example is the whirlpool formation. The movie P1040830.mov (App. F) shows some vortex pairing in the shear layer, leading to the development of large scale structures which are advected by the main flow.

## 3.4  A source and uniform flow (flow past a half-body)

Considering a source (flow rate $+q$) located at the origin in an uniform stream (velocity $V_O$ parallel to the x axis), the velocity potential and the stream function are respectively:

$$\phi = -V_O \times r \times \cos\theta - \frac{q}{2 \times \pi} \times Ln(r) \tag{4.30}$$

$$\psi = -V_O \times r \times \sin\theta - \frac{q}{2 \times \pi} \times \theta \tag{4.31}$$

The central streamline in the approaching uniform flow divides at the stagnation point S (Fig. 4.7). By definition, the velocity at the stagnation point S $(r_s, \pi)$ is zero:

$$V = V_O - \frac{q}{2 \times \pi \times r_s} = 0$$

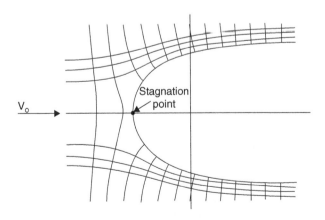

*Figure 4.7*  Flow past a half-Rankine body.

which yields:

$$r_s = \frac{q}{2 \times \pi \times V_O} \tag{4.32}$$

with $\theta_s = \pi$.

At the stagnation point, the stream function equals $\psi = -q/2$ (Eq. (4.31)). The equation of the streamline through the stagnation point is then given by:

$$\psi = -V_O \times r \times \sin \theta - \frac{q}{2 \times \pi} \times \theta = -\frac{q}{2} \tag{4.33a}$$

It yields:

$$r = \frac{q \times (\pi - \theta)}{2 \times \pi \times V_O \times \sin \theta} \qquad \text{Stream line through the stagnation point} \tag{4.33b}$$

Equation (4.33) is the equation of the **half-body** profile.

---

**Discussion**

A solid boundary is a streamline and there is no flow across a streamline. That is, the uniform flow "sees" the streamline through the stagnation point as a solid boundary (i.e. the half-body). There is no flow exchange between the uniform flow and the flow issued from the source, but on the streamline ($\psi = -q/2$) through the stagnation point.

---

At any point $P(x, y)$ the velocity magnitude is given by:

$$V^2 = V_x^2 + V_y^2 = V_O^2 + \frac{q^2}{4 \times \pi^2 \times r^2} + \frac{2 \times q \times V_O}{2 \times \pi \times r} \times \cos \theta \tag{4.34}$$

and the pressure may be deduced from the Bernoulli equation:

$$\frac{P - P_O}{\frac{1}{2} \times \rho \times V_O^2} = 1 - \frac{V^2}{V_O^2} \tag{4.35}$$

where $P_O$ is the reference pressure in the uniform flow.

Along the half-body (i.e. streamline $\psi = -q/2$), the pressure equals:

$$\frac{P - P_O}{\frac{1}{2} \times \rho \times V_O^2} = -\frac{\sin \theta}{\pi - \theta} \times \left( \frac{\sin \theta}{\pi - \theta} + 2 \times \cos \theta \right) \qquad \text{Pressure on the half-body} \tag{4.36}$$

Note that the pressure on the half-body equals the reference pressure $P_O$ for $\theta = +/-113°$.

**Notes**

– The "half-body" is an useful concept in studying the flow at the upstream end of symmetrical body long in comparison of their width: e.g., flow past bridge piers, flow at the bow of a ship, flow past a torpedo nose.

- The shape of the body may be altered by adjusting the strength of the source relative to the free-stream velocity $V_O$.
- Note that the value of the stream function along the streamline passing at the stagnation point is not zero. The definition of the stream function can be modified by adding a constant $(+q/2)$ such as $\psi = 0$ at the stagnation point, without affecting the flow pattern.
- The flow pattern is also called a "half-Rankine body".

## 3.5   A source and a vortex (spiral vortex)

The combination of a source (flow rate $+q$) and a vortex (strength $+K$) at the origin produces the pattern of an outward spiral flow (Fig. 4.8). The velocity potential and stream function equal:

$$\phi = -\frac{q}{2 \times \pi} \times Ln(r) - \frac{K}{2 \times \pi} \times \theta \tag{4.37}$$

$$\psi = -\frac{q}{2 \times \pi} \times \theta + \frac{K}{2 \times \pi} \times Ln(r) \tag{4.38}$$

The equation of a streamline (i.e. $\psi = $ constant) is:

$$Ln(r) = \frac{q \times \theta + 2 \times \pi \times \psi}{K} \tag{4.39a}$$

which yields:

$$r = \exp\left(\frac{2 \times \pi \times \psi}{K}\right) \times \exp\left(\frac{q}{K} \times \theta\right) \tag{4.39b}$$

Equation (4.39) is the equation of an equi-angular spiral.

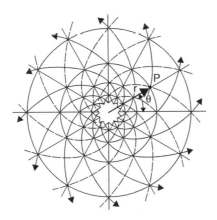

Figure 4.8   Flow pattern generated by a spiral vortex.

**Application**

The superposition of a source and vortex located both at the origin may be used to model the flow of the impeller of a radial pump. The impeller blades may be shaped like the streamlines.

*Application*

A centrifugal pump impeller has an outlet diameter of 0.45 m and an inlet diameter of 0.25 m. At the outlet, the radial and ortho-radial velocity components are respectively 1.9 m/s and 6.5 m/s.

(a) Calculate the strength of the source and vortex.
(b) Calculate the radial and ortho-radial velocity components at the impeller inlet.

*Solution*

(a) The radial and ortho-radial velocity components are:

$$V_r = -\frac{\partial \phi}{\partial r} = -\frac{1}{r} \times \frac{\partial \psi}{\partial \theta} = \frac{q}{2 \times \pi \times r}$$

$$V_\theta = \frac{\partial \psi}{\partial r} = -\frac{1}{r} \times \frac{\partial \phi}{\partial \theta} = \frac{K}{2 \times \pi \times r}$$

Hence $q = 5.37\,\text{m}^2/\text{s}$ and $K = 18.38\,\text{m}^2/\text{s}$.

(b) At the impeller inlet, the velocity components are: $V_r = 3.42\,\text{m/s}$ and $V_\theta = 11.7\,\text{m/s}$.

## 3.6 A doublet and uniform flow (flow past a cylinder)

The addition of the flow due to a doublet of strength $(-\mu)$ and an uniform flow (velocity $+V_O$ parallel to the x-axis) results in flow around a circular cylinder. The velocity potential and stream function of the flow superposition are respectively:

$$\phi = -V_O \times r \times \cos \theta + \frac{(-\mu) \times \cos \theta}{r} \tag{4.40a}$$

$$\psi = -V_O \times r \times \sin \theta - \frac{(-\mu) \times \sin \theta}{r} \tag{4.41a}$$

where the doublet strength $\mu$ is taken as positive.

The equation of the streamline $\psi = 0$ is given by:

$$\left( V_O \times r - \frac{-\mu}{r} \right) \times \sin \theta = 0$$

It may be rewritten as:

$$V_O \times r - \frac{-\mu}{r} = 0$$

which is the equation of a circular cylinder of radius R:

$$R = \sqrt{\frac{(-\mu)}{V_O}} \tag{4.42}$$

The streamline $\psi = 0$ is basically a circle. As any streamline in a steady flow is a possible solid boundary, the flow pattern is equivalent to an uniform flow past a circular cylinder (Fig. 4.9). The velocity potential and stream functions for an uniform flow around a circular cylinder of radius R may be expressed as:

$$\phi = -V_O \times \left(r + \frac{R^2}{r}\right) \times \cos\theta \tag{4.40b}$$

$$\psi = -V_O \times \left(r - \frac{R^2}{r}\right) \times \sin\theta \tag{4.41b}$$

The equipotential lines and streamlines are shown on Figure 4.9. On the surface of the cylinder (i.e. streamline $\psi = 0$), the velocity is tangential and equals to:

$$V_{r=R} = (V_\theta)_{r=R} = \left(\frac{\partial\psi}{\partial r}\right)_{r=R} = -2 \times V_O \times \sin\theta \tag{4.43}$$

Note the existence of two stagnation points at $\theta = 0$ and $\pi$ (Fig. 4.9).
Applying the Bernoulli principle, the pressure distribution on the cylinder is:

$$\frac{P - P_O}{\frac{1}{2} \times \rho \times V_O^2} = 1 - 4 \times \sin^2\theta \tag{4.44}$$

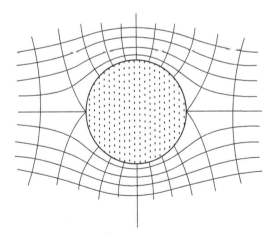

*Figure 4.9* Flow around a circular cylinder.

### Drag and lift forces on the cylinder

The drag force and the lift force on the circular cylinder are obtained by integrating respectively the x-component and the y-component of the pressure force on the cylinder:

$$\text{Drag} = -\int_0^{2\times\pi} P \times R \times \cos\theta \times d\theta \tag{4.45a}$$

$$\text{Lift} = -\int_0^{2\times\pi} P \times R \times \sin\theta \times d\theta \tag{4.46a}$$

The analytical integration demonstrates that the drag on the cylinder is zero:

$$\text{Drag} = +\frac{\rho \times R \times V_O^2}{2} \times \int_0^{2\times\pi} (1 - 4 \times \sin^2\theta) \times \cos\theta \times d\theta = 0 \tag{4.45b}$$

Similarly the lift force on the cylinder is zero:

$$\text{Lift} = 0 \tag{4.46b}$$

The result (i.e. no drag nor lift forces) conflicts with the practical experience. The contradiction is known as **d'Alembert's paradox** first published in 1752. In a real fluid flowing past a cylinder, viscous action causes the flow to separate from the downstream surface of the cylinder and forms a wake in which the pressure variation differs from the ideal-fluid calculations. In irrotational flow of ideal fluid, no separation occurs and the cylinder placed in an uniform flow is subjected to zero lift zero drag force.

### Notes

- Jean le Rond d'Alembert (1717–1783) was a French mathematician and philosopher. He was a friend of Leonhard Euler and Daniel Bernoulli. His paradox was published in 1752 in his "Essai d'une nouvelle théorie sur la résistance des fluides" (Alembert 1752).
- In Equations (4.45) and (4.46), the negative sign (−) is linked with the selection of the positive flow direction. A positive drag force corresponds to a larger pressure on the upstream perimeter of the cylinder than on its downstream surface. The drag force is positive in the direction of the flow motion (positive x-direction herein) and the lift force is positive along the positive y-direction.

---

### Discussion: real fluid flow past a circular cylinder

The flow of real fluid past a circular cylinder is one of the most complicated flow situations considering the simplicity of the geometry. For very low Reynolds numbers $Re = 2 \times V_O \times R/\nu$, the flow past the cylinder is very similar to that of an ideal fluid (Fig. 4.9), flow where $\nu = \mu/\rho$ is the kinematic viscosity of the fluid. For $Re > 1$ to 2, the boundary layers developing at the surface of the cylinder separate and a wake region develops (Fig. 4.10). For a range of Reynolds numbers from about 40 to 80, the separation points oscillate rearward and forward alternately. The oscillations cause vortices to be shed alternately from one side to the other of the cylinder, resulting in a pattern of vortices rotating in opposite directions behind the cylinder

(Fig. 4.10). This flow pattern is called a Karman vortex street, which was nicely illustrated by Van Dyke (1982, front cover & pp. 4–5, 100 & 130). The movement of the separation points causes an oscillating pressure at the cylinder surface which in turn may cause structural vibrations transverse to the free-stream. Figure 4.11 illustrate a street of vortices forming in the atmospheric boundary layer in the wake of a 1,300 m high volcanic island. The vortex street an area of 150 km in length by 70 km in width. A cloud-free effect is apparent at the vortex centres for the first few vortices.

For larger Reynolds numbers (Re > 90), the wake region expands. Separation occurs near 80° from the front stagnation point for 1 E+2 < Re < 1 E+5, with formation and shedding of vortices alternately from one side then the other side. The vortices behind the cylinder are well-organised and arranged in a zigzag pattern: i.e., the Karman vortex street (Fig. 4.10 & 4.11). The effects of vortex shedding can be devastating if the hydrodynamic frequency happens to coincide with the natural frequency of the structure: e.g., the Tacoma Narrows suspension bridge failure. Industrial chimneys are often protected from this phenomenon by spirals attached to the top, to induce flow separation at a fixed location and to prevent the oscillations (Fig. 4.12). The frequency $\omega_{\text{shedding}}$ of the hydrodynamic oscillations usually satisfies:

$$St = \frac{\omega_{\text{shedding}} \times 2 \times R}{V_O} \sim 0.2 \qquad\qquad 1\,E{+}2 < Re < 1\,E{+}5$$

where St is a dimensionless number called the Strouhal number. The Strouhal number is about 0.2 for a wider range of Reynolds number.

For very large Reynolds numbers (Re > 4 E+5), the boundary layers become turbulent and the separation points are moved further downstream to near 130°.

Figure 4.13 shows the changes in drag coefficient $C_D$ with Reynolds numbers where the drag coefficient is defined as:

$$C_D = \frac{\text{Drag force}}{\frac{1}{2} \times \rho \times V_O^2 \times L \times D}$$

where L and D are respectively the cylinder length and diameter (D = 2 × R). Figure 4.13 includes comparison with finite length cylinder results. For an infinitely long, smooth cylinder, note the decrease in drag coefficient around Re ∼ 4 E+5.

## Notes

- Theodore von Karman (or von Kármán) (1881–1963) was a Hungarian fluid dynamicist and aerodynamicist who worked in Germany between 1906 and 1929, and later in USA. He was a student of Ludwig Prandtl in Germany. He gave his name to the vortex shedding behind a cylinder (Karman vortex street).
- Vincenc Strouhal (1850–1922) was a Czech physicist. In 1878, he investigated first the 'singing' of wires (i.e. aeolian tones) caused by vortex shedding behind the wires.
- The Karman vortex street is also called Bénard-von Karman vortex street.
- In a boundary layer, a deceleration of fluid particles leading to a reversed flow within the boundary layer is called a separation. The decelerated fluid particles are forced outwards and the boundary layer is separated from the wall. At the point of separation, the velocity gradient normal to the wall is zero:

$$\left(\frac{\partial V_x}{\partial y}\right)_{y=0} = 0$$

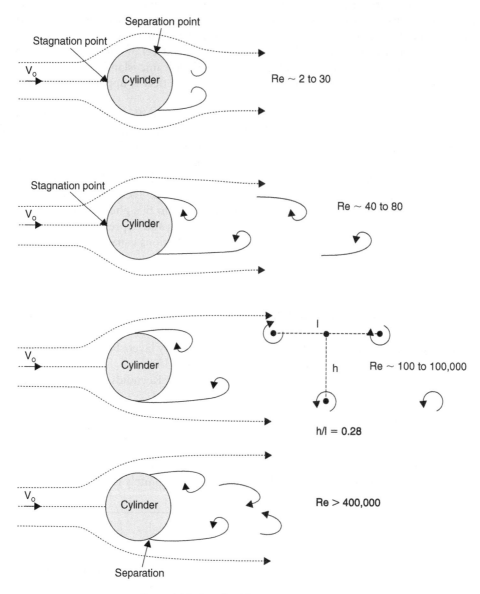

*Figure 4.10*   Real-fluid flow past a circular cylinder.

– In 1940, the first Tacoma Narrows Bridge opened over Puget Sound in Washington state (USA). Just four months after completion, the deck tore apart and collapsed under a moderate wind (i.e. 19 m/s) although the bridge was designed to withstand gales of 55 m/s. Motion pictures taken of the disaster show the deck rolling up and down and twisting wildly.
– Vortex shedding behind a cylinder may be prevented by installing a splitter plate in the wake downstream of the cylinder. The streamlining of the body (e.g. an airfoil) is another means to eliminate the formation of vortex shedding.

*Figure 4.11* Karman street of clouds downstream of the eastern Pacific island of Guadalupe off Mexico (Image courtesy NASA/GSFC/JPL, MISR Team and NASA Earth Observatory {http://earthobservatory. nasa.gov/IOTD/view.php?id=987}) – Guadalupe Island is a volcanic island rising 1,300 m above the sea surface, providing a significant disturbance to wind flow – Photograph taken on 7 December 2000.

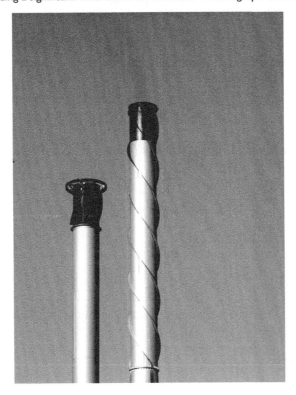

*Figure 4.12* Chimneys at Castlemaine XXXX beer factory in Brisbane in January 2003 – Note the different spirals around the chimneys to reduce wind load oscillations caused by Karman vortex street shedding.

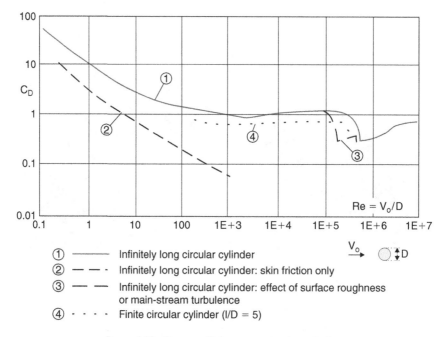

① ——— Infinitely long circular cylinder
② — — · Infinitely long circular cylinder: skin friction only
③ — — Infinitely long circular cylinder: effect of surface roughness or main-stream turbulence
④ · · · · Finite circular cylinder (l/D = 5)

*Figure 4.13* Drag coefficient on a circular cylinder.

– At large Reynolds number, the drag on a circular cylinder may be reduced by installing a small flat plate upstream. Minimum drag was achieved for a plate height $2/3 \times R$ placed $3 \times R$ upstream of the cylinder. The resulting drag of the system was 38% that of the bare cylinder for $1\,E+3 < Re < 1\,E+5$ (Prasad and Williamson 1997).

## 3.7 A doublet, a vortex and uniform flow (flow past a cylinder with circulation)

The addition of a vortex (strength $(+K)$), a doublet (strength $(-\mu)$) and an uniform flow (velocity $+V_O$ parallel to the x-axis) results in the flow around a circular cylinder with circulation (in rotation). The principle of superposition yields the velocity potential and stream function:

$$\phi = -V_O \times r \times \cos\theta + \frac{(-\mu) \times \cos\theta}{r} - \frac{K}{2 \times \pi} \times \theta \tag{4.47a}$$

$$\psi = -V_O \times r \times \sin\theta - \frac{(-\mu) \times \sin\theta}{r} + \frac{K}{2 \times \pi} \times Ln(r) \tag{4.48a}$$

where $V_O$, K and $\mu$ are positive. Both stream function and velocity potential may be rewritten in terms of the cylinder radius $R = \sqrt{+\mu/V_O}$ (paragraph 3.6):

$$\phi = -V_O \times \left(r + \frac{R^2}{r}\right) \times \cos\theta - \frac{K}{2 \times \pi} \times \theta \tag{4.47b}$$

$$\psi = -V_O \times \left(r - \frac{R^2}{r}\right) \times \sin\theta + \frac{K}{2 \times \pi} \times Ln(r) \tag{4.48b}$$

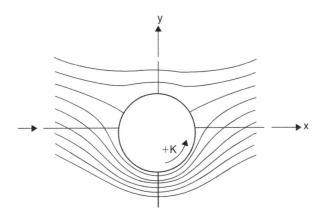

*Figure 4.14* Flow around a rotating cylinder.

The flow pattern corresponds to a rotating cylinder placed in an uniform flow, and it is sketched in Figures 4.14 and 4.15.

The effect of the circulatory flow (resulting from the vortex) is to increase the velocity of the flow on one side of the cylinder and to decrease it on the other (Fig. 4.14). The velocity at the surface of the cylinder is necessarily tangent to the cylinder because the cylinder is a boundary and hence a streamline. The velocity magnitude at the cylinder surface equals:

$$V_{r=R} = (V_\theta)_{r=R} = \left(\frac{\partial\psi}{\partial r}\right)_{r=R} = -2 \times V_O \times \sin\theta + \frac{K}{2 \times \pi \times R} \tag{4.49}$$

A stagnation point occurs when the velocity at the surface of the cylinder is zero: $V_{r=R} = 0$. It yields:

$$\sin\theta = \frac{K}{4 \times \pi \times R \times V_O} \qquad\qquad \text{Stagnation condition} \tag{4.50}$$

Equation (4.50) implies that there may be two, one or no stagnation points on the surface depending upon the value of the ratio $K/(4 \times \pi \times R \times V_O)$ (Fig. 4.15). When the ratio $K/(4 \times \pi \times R \times V_O)$ is less than unity, there are two stagnation points. When the circulation (i.e. vortex strength $\Gamma = K$) equals $4 \times \pi \times R \times V_O$, the two stagnation points coincide at $(r = R, \theta = +\pi/2)$. For larger values, there is one stagnation point away from the cylinder (Fig. 4.15).

When the pressure at infinity is $P_O$, the pressure $P$ at any point on the cylinder surface is deduced from the Bernoulli principle:

$$\frac{P - P_O}{\frac{1}{2} \times \rho \times V_O^2} = 1 - \left(-2 \times \sin\theta + \frac{K}{2 \times \pi \times R \times V_O}\right)^2 \tag{4.51}$$

Equation (4.51) is shown in Figure 4.16 for three dimensionless vortex strengths corresponding to the three flow patterns sketch in Figure 4.15.

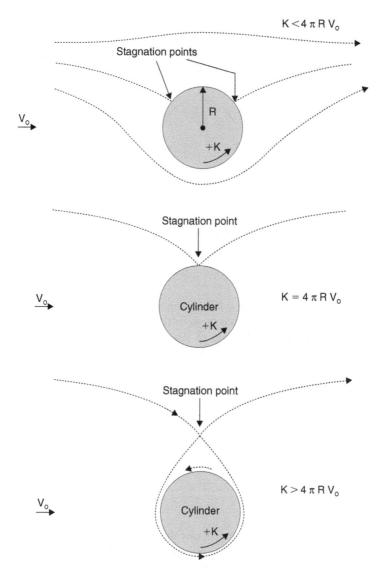

Figure 4.15  Flow patterns around a rotating cylinder.

As for the flow past a cylinder, the drag force (i.e. pressure force in the x-direction) is zero again:

$$\text{Drag} = 0 \tag{4.52}$$

However the lift force (i.e. pressure force in the y-direction) equals:

$$\text{Lift} = -\int_{0}^{2 \times \pi} P \times R \times \sin\theta \times d\theta = -\rho \times V_O \times K \tag{4.53}$$

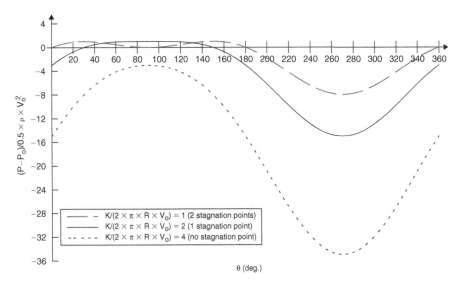

*Figure 4.16* Dimensionless pressure distribution on the surface of a rotating cylinder in an uniform flow.

If $V_O$ and K are positive, the lift force is negative. Further the magnitude of the lift force is directly proportional to the density of the fluid $\rho$, the approach velocity $V_O$ and the vortex circulation $\Gamma = K$. This thrust, which acts at right angles to the approach velocity, is referred to as the **Magnus effect** and it is independent of the cylinder size.

---

**Discussion**

Considering the flow past a rotating cylinder sketched in Figure 4.14, the rotation of the cylinder induces an acceleration of the fluid beneath the cylinder that is highlighted by the converging streamlines. Similarly some flow deceleration is seen next to the upper wall of the cylinder. In turn, the Bernoulli equation implies that the pressure next to the lower and upper walls must be respectively lower and greater than the static pressure. The resulting pressure force (i.e. lift force) is directed downward: i.e., a negative lift force in Figures 4.14 and 4.15.

The Magnus effect is the reason for top-spin tennis balls to drop very sharply toward the ground. The top-spin induces a lift force perpendicular to the ball direction acting downward. More generally, the trajectories of golf, baseball and ping-pong balls are affected by the ball rotation around its centre of gravity: e.g., slice, spin.

---

**Notes**

- H.G. Magnus (1802–1870) was a German physicist who investigated this effect in 1852:
  Magnus, H.G. (1853). "The Drift of Shells." *Poggendorf's Annals*, Germany.
- The strength of the vortex K is equal to the circulation $\Gamma$ around a closed path that contains the vortex.
- The lift force is not dependent upon the fluid viscosity.

*Figure 4.17* The Buckau Flettner rotor ship in November 1924 (Photograph German Federal Archive (Deutsches Bundesarchiv)).

– The rotation velocity $\omega$ of the cylinder is a function of the strength of the vortex (i.e. the circulation around the vortex):

$$\omega = \frac{K}{2 \times \pi \times R^2} = \frac{K \times V_O}{2 \times \pi \times \mu}$$

as the velocity at the surface of the cylinder resulting from a vortex is:

$$V_\theta = \frac{K}{2 \times \pi \times r} = r \times \omega$$

– The rotor ship, designed by Anton Flettner about 1924, used the Magnus effect as main propulsion technique. It was designed by mounting two large vertical circular cylinders on the ship and then mechanically rotating the cylinders to provide circulation. Figure 4.17 shows the Buckau ship after transformation. The ship, renamed Baden-Baden, successfully crossed the Atlantic Ocean. Each cylinder was 18.5 m long with a 2.8 m diameter. More recently a similar system was developed for the new ship "Alcyone" of J.Y. Cousteau (see Discussion).
– Anton Flettner (1885–1961) was a German engineer. He consulted Ludwig Prandtl and the Institute of Aerodynamics in Göttingen (Ergebnisse Aerodynamische Versuchanstalt, Göttingen) for the development of his rotorship (Flettner 1925).

---

**Discussion: Magnus effect in real fluid flows**

Historically, the Magnus effect was investigated because of the drift of spinning objects such as musket balls, bullets and ballistic missiles (Magnus 1853, Swanson 1961). Experimental studies of the Magnus effect demonstrated that Equation (4.53) predicts relatively well real-fluid lift forces (Fig. 4.18). In Figure 4.18, measured drag coefficients are also reported as functions of the ratio of the velocity at the surface of the cylinder ($R \times \omega$) to free-stream velocity $V_O$.

A related form of Magnus effect is the flow around a cylindrical body with circulation: e.g., the Turbosails™ of the Alcyone ship (Table 4.1). The propulsion system of the Alcyone was developed by the Frenchman L. Malavard for Jacques-Yves Cousteau (Fig. 4.19). Its consists of two identical fixed masts (i.e. Turbosail™). Boundary layer suction is generated on one side of the cylinder by a suction fan installed inside the mast, while a flap controls the flow separation downstream (Fig. 4.19B). The resulting effect is a large fluid circulation around the sail and a significant lift force ($C_L \sim 5$ to $6$). Optimum performances are obtained when the angle between wind and ship directions ranges between 50° and 140°.

| Reference | Cylinder length to radius ratio L/R | Re | Remarks |
|---|---|---|---|
| Reid (1924) | 26.6 | 3.9 to 11.6 E+4 | |
| Thom (1934) | 25.2 & 52 | 5.3 to 8.8 E+3 | End plates |
| Swanson (1956) | 4 | 5 E+4 | |

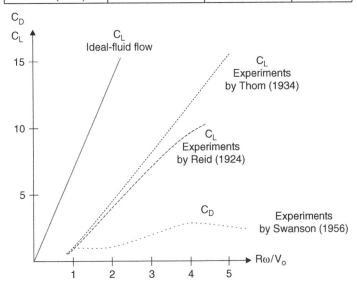

Figure 4.18  Drag and lift coefficients on a circular cylinder with rotation as a function of the ratio of the velocity at the wall to the free-stream velocity – Comparison between Equation (4.53) and experimental measurements (Reid 1924, Thom 1934, Swanson 1956).

Table 4.1  Characteristics of the Alcyone ship.

| Section (1) | Characteristics (2) | Remarks (3) |
|---|---|---|
| Hull | Built in 1984–85. Overall length: 31.1 m. Maximum draught: 2.34 m. Maximum width: 8.92 m. Half-load displacement: 76.8 t. Cruise speed: 10.5 knots | Designed by A. Mauric and J.C. Nahon (France) |
| Turbosails™ | 2 identical fixed masts equipped with a boundary layer suction system. Height: 10.2 m. Maximum chord: 2.05 m. Width: 1.35 m. Surface area: 21 m² (each). Suction fan: 25 hp. Performances: $C_L = 5$ to $6.5$ and $C_D = 1.2$ to $1.8$. | Developed by L. Malavard. Cousteau-Pechiney system built in aluminium. |

(A) Alcyone berthed at La Seyne sur Mer on 25 April 2010 (Courtesy of Stéphane Saissi)

(B) Real fluid flow around a rotating cylinder (Left) and around the Turbosail™ (Right)

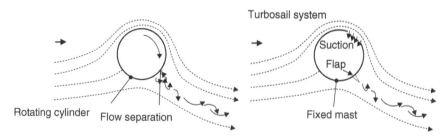

*Figure 4.19* The sail system of the Alcyone ship.

## Notes

– For a real fluid flow, the lift and drag forces per unit width are expressed as:

$$\text{Lift} = C_L \times 2 \times R \times \rho \times \frac{V_O^2}{2}$$

$$\text{Drag} = C_D \times 2 \times R \times \rho \times \frac{V_O^2}{2}$$

where the lift and drag coefficients, $C_L$ and $C_D$ respectively, are deduced from Figure 4.18.
- Lucien Malavard (1910–1990) was a Professor in fluid mechanics at the University Paris VI Pierre et Marie Curie, and a member of the French Académie des Sciences.
- A major difference between the Alcyone ship and Flettner rotorship is the fixed masts in the Alcyone ship (Fig. 4.17 and 4.19). The efficiency of both systems is basically identical.
- Boundary layer suction is a technique commonly used in aeronautics to re-attach the developing boundary layer on the extrados of a foil. The re-attachment enables a greater lift force.

## 3.8  A source, a sink and uniform flow (flow past a Rankine body)

The flow past a **Rankine body** is the pattern resulting from the combinations of a source and sink of equal strength (q) in an uniform flow (velocity $+V_O$ parallel to the x-axis). The velocity potential and stream function are:

$$\phi = -V_O \times r \times \cos\theta - \left( \frac{q}{2 \times \pi} \times \text{Ln}\left( \frac{r_1}{r_2} \right) \right) \tag{4.54}$$

$$\psi = -V_O \times r \times \sin\theta - \left( \frac{q}{2 \times \pi} \times (\theta_1 - \theta_2) \right) \tag{4.55}$$

where the subscript 1 refers to the source of strength $(+q)$ located at $(-L, 0)$, and the subscript 2 refers to the sink of strength $(-q)$ located at $(+L, 0)$ (Fig. 4.20).

**Note**

This configuration of the source and sink is the *opposite* of the flow pattern presented in the later paragraph "Source and sink". In this section on the Rankine body, the strength q is positive, and the source is located at $(-L, 0)$ and the sink at $(+L, 0)$.

The profile of the Rankine body is given by the equation of the streamline $\psi = 0$:

$$\psi = -V_O \times r \times \sin\theta + \frac{q}{2 \times \pi} \times (\theta_2 - \theta_1) = 0 \tag{4.56a}$$

which becomes:

$$r = \frac{q \times (\theta_2 - \theta_1)}{2 \times \pi \times V_O \times \sin\theta} \tag{4.56b}$$

The length of the body equals the distance between the stagnation points (Fig. 4.20B). The stagnation points are located on the x-axis and the velocity is zero:

$$V = V_O + \frac{q}{2 \times \pi \times r_1} - \frac{q}{2 \times \pi \times r_2} = V_O + \frac{q}{2 \times \pi} \times \left( \frac{1}{r_s - L} - \frac{1}{r_s + L} \right) = 0$$

Hence the total length of the Rankine body is:

$$L_{body} = 2 \times r_s = 2 \times L \times \sqrt{1 + \frac{q}{\pi \times L \times V_O}} \tag{4.57}$$

where $2 \times L$ is the distance between the source and the sink.

The half-width of the body h is deduced from the profile equation at the point (h, π/2):

$$h = \frac{q \times (\theta_2 - \theta_1)}{2 \times \pi \times V_O} \tag{4.58a}$$

(A) Definition sketch

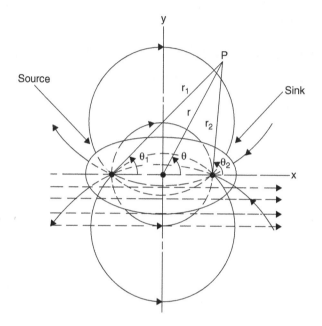

(B) Flow pattern

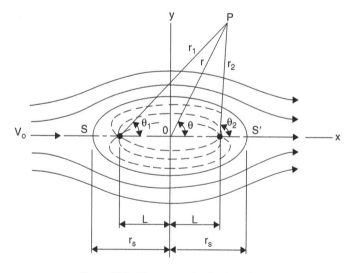

*Figure 4.20* Flow around a Rankine body.

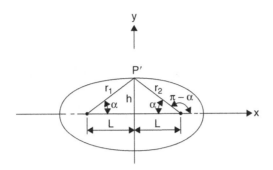

*Figure 4.21* Definition of the Rankine body.

where: $\theta_1 = \alpha$ and $\theta_2 = \pi - \alpha$ (Fig. 4.21). The angle $\alpha$ satisfies:

$$\alpha = \frac{\pi}{2} - \frac{\pi \times h \times V_O}{q}$$

as well as:

$$\tan \alpha = \frac{h}{L}$$

The half-width of the body is the solution of the equation:

$$h = L \times \cot\left(\frac{\pi \times V_O}{q} \times h\right) \tag{4.58b}$$

Note that h appears in the right and left hand-sides of Equation (4.58b) which must be solved by iteration.

## Notes

- The flow past a cylinder (uniform flow and a doublet) can be regarded as a special case of the flow past a Rankine body.
- William J.M. Rankine (1820–1872) was a Scottish engineer and physicist. He developed the theory of sources and sinks.
- The shape of the Rankine body may be altered by varying the distance between source and sink (i.e. $2 \times L$) or by varying the strength of the source and sink. Other shapes may be obtained by using a source and a sink of different strength, and by the introduction of additional sources and sinks. William Rankine developed ship contours in this way. Figure 4.22 illustrates some example of older ship hulls for which the flow around the hull could be studied using Rankine's method.

(A) Half ship hull scale model of a Hai sailing boat designed by the Finish naval architect Gunnar L. Stenback in 1930 – Named 'Requin' in French

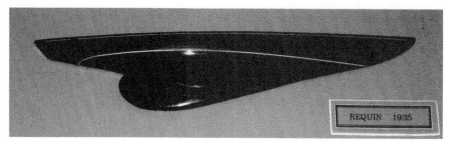

REQUIN 1935

(B) Pilot boat "La Pauline" (1899) sailing in front of Pléneuf-Val-André (France) on 11 October 2008

*Figure 4.22* Old ship hulls.

## 4  APPLICATIONS OF STEADY FLOW PATTERN COMBINATIONS

### 4.1  Introduction

When two or more flow patterns are combined by superposition, the values of stream function and velocity potential at any point, other than a singular point, are simply the algebraic sum of the $\psi$-values and $\phi$-values of the component patterns at that point, at any instant. Flow patterns may be based upon uniform flow, sink, source, vortex, doublet, ...

### 4.2  Sink and source in uniform flow

In the previous section, we considered a source and a sink in an uniform stream and the resulting pattern was a Rankine body. We now consider a source located at $(+L, 0)$, a sink located at $(-L, 0)$ and an uniform flow (velocity $+V_O$ parallel to the x-axis). The source and the sink are of equal strength q. The velocity potential and the stream function are respectively:

$$\phi = -V_O \times r \times \cos\theta + \left( -\frac{q}{2 \times \pi} \times Ln\left(\frac{r_1}{r_2}\right) \right) \tag{4.59a}$$

$$\psi = -V_O \times r \times \sin\theta + \left( -\frac{q}{2 \times \pi} \times (\theta_1 - \theta_2) \right) \tag{4.60a}$$

where the subscript 1 refers to the source, the subscript 2 to the sink, and q is positive.
   In Cartesian coordinates, it becomes:

$$\phi = -V_O \times x - \frac{q}{4 \times \pi} \times Ln\left(\frac{(x-L)^2 + y^2}{(x+L)^2 + y^2}\right) \tag{4.59b}$$

$$\psi = -V_O \times y - \frac{q}{2 \times \pi} \times \left( \tan^{-1}\left(\frac{y}{x-L}\right) - \tan^{-1}\left(\frac{y}{x+L}\right) \right) \tag{4.60b}$$

The streamline pattern will belong to one of three possible types, depending on the magnitude of q, L and $V_O$ (Fig. 4.23). These three possible flow patterns are:

(a)   *Two stagnation points on the x-axis*
    For a relatively weak source/sink strength (i.e. $q/(\pi \times L \times V_O) < 1$), two stagnation points are located on the horizontal axis at $(-x_s, 0)$ and $(+x_s, 0)$ respectively. Their positions are given by:

$$V_x = -\frac{d\phi}{dx} = +V_O - \frac{q}{2 \times \pi} \times \left( \frac{x-L}{(x-L)^2 + y^2} - \frac{x+L}{(x+L)^2 + y^2} \right) = 0$$

and substituting the appropriate values of $x = +x_s$ and $y = 0$:

$$x_s = L \times \sqrt{1 - \frac{q}{\pi \times L \times V_O}} \tag{4.61}$$

(b)   *One stagnation point*
    There is an unique stagnation point at the origin $(0, 0)$ for:

$$\frac{q}{\pi \times L \times V_O} = 1 \tag{4.62}$$

(c)   *Two stagnation points on the y-axis*
    For $q/(\pi \times L \times V_O) > 1$, there are two stagnation points located on the vertical axis at $(0, +y_s)$ and $(0, -y_s)$ with:

$$y_s = L \times \sqrt{\frac{q}{\pi \times L \times V_O} - 1} \tag{4.63}$$

In this case, some flow occurs from the source to the sink. The total flow rate recirculating between the source and the sink is given by:

$$q_r = 2 \times (\psi_s - \psi_{(0,0)})$$   (4.64a)

where $\psi_s$ and $\psi_{(0,0)}$ are respectively the values of the stream function at the stagnation point $(0, y_s)$ and at the origin $(0, 0)$. Their values are:

$$\psi_s = -\frac{q}{2} - V_O \times L \times \sqrt{\frac{q}{\pi \times L \times V_O} - 1} + \frac{q}{\pi} \times \tan^{-1}\left(\sqrt{\frac{q}{\pi \times L \times V_O} - 1}\right)$$

$$\psi_{(0,0)} = -\frac{q}{2}$$

Hence, the recirculation flow from the source to sink equals:

$$\frac{q_r}{q} = \frac{2}{\pi} \times \left(-\frac{\sqrt{\frac{q}{\pi \times L \times V_O} - 1}}{\frac{q}{\pi \times L \times V_O}} + \tan^{-1}\left(\sqrt{\frac{q}{\pi \times L \times V_O} - 1}\right)\right)$$   (4.64b)

**Note**

In the first case $(q/(\pi \times L \times V_O) < 1)$, the source does not interfere with the sink. In the third case $(q/(\pi \times L \times V_O) > 1)$, some fluid injected at the source is trapped into the sink: that is, the recirculating discharge $q_r$.

---

**Application**

Considering the flow through a porous medium confined between parallel impervious surfaces, the analogy with irrotational flow is used in the x-y plane. The above flow pattern would represent a well (sink) and a recharge well (source) in an aquifer carrying a natural flow of groundwater (Fig. 4.23). The well is upstream and the recharge well is downstream. If the recharge well injects polluted waters, will the well placed upstream be affected?

The answer is: NO if $\frac{q}{\pi \times L \times V_O} < 1$.

---

## 4.3   Finite line source

A line source is defined as a line over which infinitesimal point sources are continuously distributed (Streeter 1948, p. 61). Considering a line source located between $(-L, 0)$ and $(+L, 0)$, the strength of the line source is q and the flow out from the line per unit length is $q/(2 \times L)$.

At any point $P(x, y)$, the contribution to the stream function of an elemental length $\delta l$, which may be regarded as a point source of strength $q \times \delta l/(2 \times L)$, is:

$$\delta \psi = -\frac{q}{2 \times \pi} \times \frac{\delta l}{2 \times L} \times \alpha$$

where $\alpha$ is the angle between the x-axis and the line joining P to the elemental source.

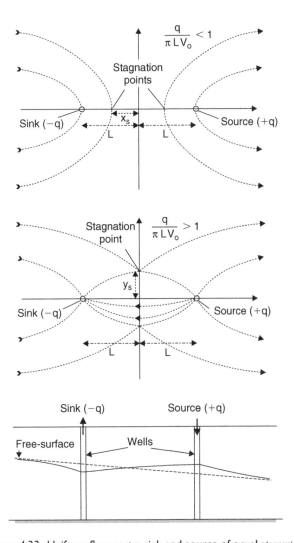

*Figure 4.23* Uniform flow past a sink and source of equal strength.

The stream function of the line source results from the superposition of the infinitesimal sources. It is:

$$\psi = -\frac{q}{4 \times \pi \times L} \times \int_{-L}^{+L} \alpha \times \delta l \qquad (4.65a)$$

where:

$$l = x - y \times \cot \alpha$$

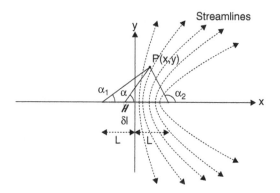

*Figure 4.24*  Finite line source.

and hence:

$$\delta l = \frac{y}{\sin^2 \alpha} \delta \alpha$$

This gives the stream function of a finite line source:

$$\psi = -\frac{q}{2 \times \pi} \times \frac{y}{2 \times L} \times [-\alpha \times \cot \alpha + Ln(\sin \alpha)]_{\alpha=\alpha_1}^{\alpha=\alpha_2} \qquad (4.65b)$$

where $\alpha_1$ is the angle between the x-axis and the line joining P to the point $(-L, 0)$, and $\alpha_2$ the angle between the x-axis and the line joining P to the point $(+L, 0)$ (Fig. 4.24). Streamlines are shown in Figure 4.24. Note the symmetry around the y-axis. The streamlines are hyperbolae with foci at the ends of the lined source.

**Notes**

– The potential function of a finite line source is:

$$\phi = -\frac{q}{4 \times \pi \times L} \times \int_{-L}^{+L} Ln(\xi) \times \delta l$$

where:

$$\xi^2 = (x - 1)^2 + y^2$$

– Note the following relationship:

$$\int \frac{X}{\sin^2 X} \times dX = -X \times \cot(X) + Ln(\sin(X))$$

– Values of the function $(-\alpha \times \cot \alpha + Ln(\sin \alpha))$ are summarised in Table 4.2.

*Table 4.2* Values of the function $(-\alpha \times \cot\alpha + \text{Ln}(\sin\alpha))$.

| $\alpha$ (deg.) (1) | $\alpha$ (radians) (2) | $-\alpha \times \cot\alpha + \text{Ln}(\sin\alpha)$ (3) | $\alpha$ (deg.) (1) | $\alpha$ (radians) (2) | $-\alpha \times \cot\alpha + \text{Ln}(\sin\alpha)$ (3) |
|---|---|---|---|---|---|
| 0 | 0 | $-\infty$ | 90 | $\pi/2$ | 0 |
| 7.5 | $\pi/24$ | $-3.0305$ | 97.5 | | 0.2154 |
| 15 | $\pi/12$ | $-2.3287$ | 105 | | 0.4564 |
| 22.5 | $\pi/8$ | $-1.9086$ | 112.5 | | 0.7341 |
| 30 | $\pi/6$ | $-1.6$ | 120 | | 1.0654 |
| 37.5 | $5 \times \pi/24$ | $-1.3493$ | 127.5 | | 1.476 |
| 45 | $\pi/4$ | $-1.132$ | 135 | $3 \times \pi/4$ | 2.0096 |
| 52.5 | $7 \times \pi/24$ | $-0.9346$ | 142.5 | | 2.7449 |
| 60 | $\pi/3$ | $-0.7484$ | 150 | | 3.8413 |
| 67.5 | $3 \times \pi/8$ | $-0.5672$ | 157.5 | | 5.6759 |
| 75 | $5 \times \pi/12$ | $-0.3854$ | 165 | | 9.3959 |
| 82.5 | $11 \times \pi/12$ | $-0.1982$ | 180 | $\pi$ | $+\infty$ |

## Applications of the finite line source

1   Uniform flow below an impervious dam
    Considering the seepage flow under an impervious dam (e.g. concrete dam), the base of
    the dam is a streamline. To the left of the base, the top surface of the porous medium
    is the equipotential corresponding to the upstream head $H_{U/S}$, and to the right of the
    base, there is another equipotential which corresponds to the downstream head $H_{D/S}$. This
    is analogous to a finite line source: i.e., the base of the dam. With the differences that
    the streamlines of the line source are the equipotential lines of the seepage flow, and the
    equipotential lines of the finite line source represent the streamlines of the flow under the
    impervious dam.
       The flow pattern is deduced by superposing a finite line source and the appropriate
    boundary conditions.

> **Discussion**
>
> Remember: a velocity potential can be found for each stream function. If the stream
> function satisfies the Laplace equation the velocity potential also satisfies it. Hence the
> velocity potential may be considered as stream function for another flow case. The veloc-
> ity potential $\phi$ and the stream function $\psi$ are called *"conjugate functions"* (Chapter I-2,
> paragraph 4.2).

2   Flow past a streamlined body
    Considering the superposition of (a) a point source at the origin $(0, 0)$ with a strength $+q$, (b)
    a finite line sink from $(0, 0)$ to $(+1, 0)$ of strength $-q$, and (c) a uniform horizontal flow of
    velocity $+V_O$, the resulting flow pattern provides a closed surface with a streamlined airship
    profile (Vallentine 1969, pp. 250–251). Applications include gun shells, rockets, wings and
    airships. The variations of the source strength and the characteristics of the line sink (strength,
    length), can be used to produce a variety of aerofoils.

**Note**

The flow past a streamlined body is an extension of the flow past a Rankine body (paragraph 3.8, in this chapter).

## 4.4   Finite vortex line

A vortex line is defined as a line over which infinitesimal point vortices are continuously distributed. It is the conjugate flow pattern of a line source. Considering a line vortex from $(-L, 0)$ to $(+L, 0)$, the strength of the vortex line is K and the strength per unit length is $K/(2 \times L)$.

At any point P $(x, y)$, the value of the velocity potential function due to an elemental length $\delta l$, which may be regarded as a vortex point of strength $K \times \delta l/(2 \times L)$, is:

$$\delta \phi = -\frac{K}{2 \times \pi} \times \frac{\delta l}{2 \times L} \times \alpha$$

where $\alpha$ is the angle between the x-axis and the line joining P to the elemental source.

The velocity potential resulting from the superposition of the infinitesimal vortices is:

$$\phi = -\frac{K}{4 \times \pi \times L} \times \int_{-L}^{+L} \alpha \times \delta l \tag{4.66a}$$

where:

$$\delta l = \frac{y}{\sin^2 \alpha} \delta \alpha$$

This yields the velocity potential of a finite vortex line:

$$\phi = -\frac{K}{2 \times \pi} \times \frac{y}{2 \times L} \times [-\alpha \times \cot \alpha + \text{Ln}(\sin \alpha)]_{\alpha=\alpha_1}^{\alpha=\alpha_2} \tag{4.66b}$$

where $\alpha_1$ is the angle between the x-axis and the line joining P to the point $(-L, 0)$, and $\alpha_2$ the angle between the x-axis and the line joining P to the point $(+L, 0)$. The equipotential lines are the streamlines of a finite line source from $(-L, 0)$ to $(+L, 0)$ (Fig. 4.24).

## 4.5  Vortex street

Considering the real-fluid flow past a cylinder or a blunt body, some organised vortex shedding is observed within a range of Reynolds number (Fig. 4.10). Eddies are detached alternately from each side of the cylinder at a rate that is practically constant for a given free-stream velocity $V_O$. The resulting flow pattern consisting of a double row of vortices is known as a vortex street or Karman street of vortices.

The real-fluid flow past a circular cylinder may be modelled by an uniform flow past a doublet with two rows of vortices of equals strength and opposite sign (Fig. 4.25A). Each vortex is affected by the movement of the fluid due to the others while it is advected downstream (i.e. x-axis on Fig. 4.25).

(A) Flow past a circular cylinder with two rows of vortices

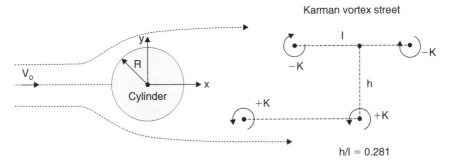

(B) Inverted Karman vortex street behind a fish

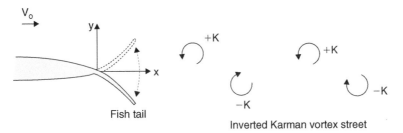

*Figure 4.25*  Definition sketch of Karman vortex street and inverted Karman vortex street.

## Notes

– Theodore von Karman demonstrated that the only stable arrangement is that for which the vortices are symmetrically staggered and the ratio of row spacing h to vortex spacing l is $h/l = 0.281$ (Fig. 4.25A).
– For the real-fluid flow past a circular cylinder with Reynolds numbers $2 \times V_O \times R/\nu$ between 1 E+3 and 1 E+5, experimental results showed that:

$$h = 2.4 \times R$$

while the shedding frequency $\omega$ is:

$$\frac{\omega_{shedding} \times 2 \times R}{V_O} \sim 0.2$$

where R is the cylinder radius and $V_O$ is the free-stream velocity (Fig. 4.25A) (Vallentine 1969, p. 266).

## Discussion

Karman vortex streets are encountered in many real-fluid flows. Examples include the wake behind blunt bodies and non-streamlined objects. A related flow pattern is the wake behind some swimming fish: e.g., mullet, tuna, great blue shark. That is, the inverted Karman vortex street illustrated in Figure 4.25B.

During active propulsion, both fish body and tail contribute to the thrust production. The undulating fish body generates some circulation shed in the wake while the remainder is generated by the tail. The movies Shark01.mov, Ray05.mov and Penguin03.mov show some example of swimming motion (App. F). The mullets typically shed one vortex per half tailbeat when the tail reaches its most lateral position. The resulting inverted Karman vortex street is sketched in Figure 4.25B. Compare the vortex rotational direction with the Karman vortex street (Fig. 4.25A). Another intriguing pattern is the stingray swimming motion (Movie Ray05.mpov, App. F). The thrust is generated by undulatory waves passing down the enlarged pectoral fins. The swim speed increases with increasing fin-beat frequency and wave speed.

### Notes

− The inverted Karman vortex street is also called reversed Karman vortex street.
− Relevant references on swimming fish hydrodynamics include Barrett et al. (1999) and Zhu et al. (2002).
− Another type of vortex street is the series of vortical structures in a developing shear layer. Whirlpools are a good example. The movie P1040830 illustrates the development of large vortical structures, or eddies, in a shear layer (App. F). During the ebb flow, separation at the end of the jetty creates a shear layer development. The video highlights the formation of large vortices by vortex pairing as the eddies are advected in the shear layer (App. F).

## 5   METHOD OF IMAGES

### 5.1   Introduction

The pattern resulting from a combination of two sources of equal strength includes one straight streamline midway between them which is perpendicular to the line joining them (Fig. 4.5B). Two equal sinks, two vortices of opposite sign and equal strength, two cylinders aligned across the flow, also produce a similar line of symmetry. Since the line of symmetry is a streamline, it can be considered as a boundary wall, without affecting the other half of the pattern.

Problems of potential flows involving straight or circular boundaries can be solved by the method of images. In this paragraph, a few examples are presented to illustrate the application of this powerful method.

### 5.2   Flow due to a source near an infinite plane

Considering a source of strength q located at $(−L, 0)$, in the presence of a wall $(x = 0)$, the wall must be a streamline. In order to ensure that this condition is satisfied, a mirror image of the source is superposed on that due to the source itself (Fig. 4.26). The image of the source is a source of strength q located at $(+L, 0)$.

Using the principle of superposition, the resulting potential and stream functions are respectively:

$$\phi = -\frac{q}{2 \times \pi} \times \mathrm{Ln}(r_1 \times r_2) \tag{4.67}$$

$$\psi = -\frac{q}{2 \times \pi} \times (\theta_1 + \theta_2) \tag{4.68}$$

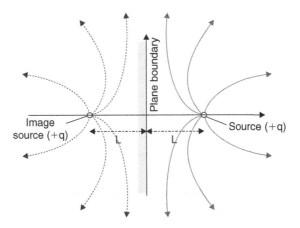

*Figure 4.26*  Source near a plane boundary.

where:

$$r_1^2 = (x - L)^2 + y^2$$

$$r_2^2 = (x + L)^2 + y^2$$

$$\tan \theta_1 = \frac{y}{x - L}$$

$$\tan \theta_2 = \frac{y}{x + L}$$

That is, the subscript 1 refers to the source, the subscript 2 to the image source, and q is positive (Fig. 4.26).

## 5.3  Sink near a boundary

### 5.3.1  Sink near an impervious wall

A practical example of a sink adjacent to a wall is the case of percolation (groundwater flow) to a well near a vertical and impervious boundary (e.g. building foundation). Using the theory of images, the flow pattern is that of two equal-strength sink with the boundary midway between them and perpendicular to the line joining the sinks (Fig. 4.26).

The effect of the boundary is to reduce the potential and hence the height of the water table from that due to the well alone, by an amount due to the effect of the image sink.

### 5.3.2  Sink near a reservoir boundary

A different case is a well located near a reservoir. A river or a reservoir boundary is a source of flow. The effect on the flow pattern can be determined by the addition of an image source to the normal sink pattern. Figure 4.27 illustrates two cases: a sink next to a reservoir boundary without and with an uniform flow.

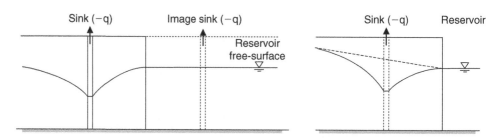

*Figure 4.27* Well in an aquifer next to a reservoir (Left) Without uniform flow (Right) with uniform flow.

## 5.4    Flow into a sink midway between two parallel infinite planes

A more complicated pattern is the flow between to parallel infinite planes because it is necessary to use an infinite number of images.

Considering two planes located at $(y = L)$ and $(y = -L)$, with a sink of strength q (with q > 0) located at $(0, 0)$, the stream function is:

$$\psi = -\sum_i \frac{-q}{2 \times \pi} \times \theta_i$$

$$= -\frac{-q}{2 \times \pi} \times \left( \tan^{-1}\left(\frac{y}{x}\right) + \tan^{-1}\left(\frac{y - L}{x}\right) + \tan^{-1}\left(\frac{y + L}{x}\right) + \cdots \right) \qquad (4.69)$$

## 5.5    Cylinder near a wall

The flow pattern of an uniform flow past a cylinder results from the combination of a doublet and uniform flow. Introducing a solid boundary parallel to the uniform flow direction, the uniform flow is in the x-direction $(+V_O)$ and the cylinder of radius R located at $(0, +L)$ with the x-axis being a solid boundary (Fig. 4.28). The flow pattern is deduced by superposing an image cylinder located at $(0, -L)$, provided that $L > 2 \times R$. This condition ensures that the doublet itself (representing the cylinder) is not distorted by the proximity of the wall.

If the x-axis is the streamline $\psi = 0$, the stream function is:

$$\psi = -V_O \times \left( r \times \cos\theta + \frac{R^2}{r_1} \times \cos\theta_1 + \frac{R^2}{r_2} \times \cos\theta_2 \right) \qquad (4.70a)$$

where:

$$r_1^2 = (y - L)^2 + x^2$$

$$r_2^2 = (y + L)^2 + x^2$$

$$\tan\theta_1 = \frac{y - L}{x}$$

$$\tan\theta_2 = \frac{y + L}{x}$$

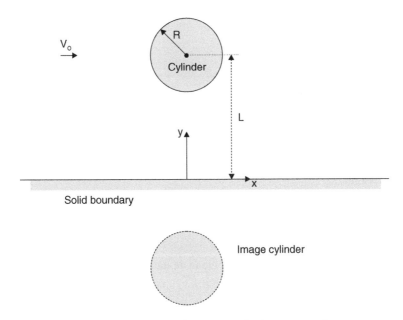

*Figure 4.28*  Uniform flow past a cylinder near a wall.

In Cartesian coordinates, the stream function becomes:

$$\psi = -V_O \times \left( x + \frac{R^2 \times y}{x^2 + (y - L)^2} + \frac{R^2 \times y}{x^2 + (y + L)^2} \right) \tag{4.70b}$$

## Note

If the distance between the cylinder axis and the wall is less than $2 \times R$, the doublet which represents the "cylinder" is distorted by the proximity of the image doublet.

# Exercises

## 4.1 Flow pattern (1)

In two-dimensional flows, what is the nature of the flow given by:

$$\phi = 7 \times x + 2 \times \text{Ln}(r)$$

Draw the flow net and deduce the main characteristics of the flow field.

### Solution

The velocity potential of a source located at the origin in an uniform stream is:

$$\phi = -V_O \times x - \frac{Q}{2 \times \pi} \times \text{Ln}(r)$$

The flow pattern described above is the flow past a half-body. The strength of the source and the flow velocity are:

$$Q = -12.566 \, \text{m}^2/\text{s} \quad V_O = -7 \, \text{m/s}$$

## 4.2 Doublet in uniform flow (1)

Select the strength of doublet needed to portray an uniform flow of ideal fluid with a 20 m/s velocity around a cylinder of radius 2 m.

### Solution

A doublet and uniform flow is analogous to the flow past a cylinder of radius:

$$R = \sqrt{\frac{-\mu}{V_O}}$$

where $\mu$ is the strength of the doublet. Hence:

$$\mu = -V_O \times R^2 = -80 \, \text{m}^3/\text{s}$$

## 4.3 Flow pattern (2)

Considering a vortex (strength $+K$) and a doublet (strength $+\mu$) at the origin:

1 What are the potential and stream functions?
2 What are the velocity components in Cartesian and polar coordinates?
3 Sketch the streamlines and equipotential lines.

## Solution

The velocity potential and stream function are:

$$\phi = -\frac{K}{2 \times \pi} \times \theta + \frac{\mu \times \cos \theta}{r} \qquad \psi = \frac{K}{2 \times \pi} \times Ln(r) - \frac{\mu \times \sin \theta}{r}$$

In Cartesian coordinates:

$$\phi = -\frac{K}{2 \times \pi} \times \tan^{-1}\left(\frac{y}{x}\right) + \frac{\mu \times x}{x^2 + y^2} \qquad \psi = \frac{K}{4 \times \pi} \times Ln(x^2 + y^2) - \frac{\mu \times y}{x^2 + y^2}$$

The velocity components are:

$$V_r = -\frac{\mu \times \cos \theta}{r^2} \qquad V_\theta = \frac{K}{2 \times \pi \times r} - \frac{\mu \times \sin \theta}{r^2}$$

In Cartesian coordinates:

$$V_x = \frac{\frac{K}{2 \times \pi} \times y + \mu \times (1 - 2 \times x^2)}{x^2 + y^2} \qquad V_y = -\frac{\frac{K}{2 \times \pi} \times x + 2 \times \mu \times x \times y}{x^2 + y^2}$$

A streamline is such as:

$$\phi = -\frac{K}{2 \times \pi} \times \theta + \frac{\mu \times \cos \theta}{r} \qquad \psi = \frac{K}{2 \times \pi} \times Ln(r) - \frac{\mu \times \sin \theta}{r}$$

## 4.4  Source and sink

A source discharging $0.72 \, m^2/s$ is located at $(-1, 0)$ and a sink of twice the strength is located at $(+2, 0)$. For a remote pressure (far away) of $7.2 \, kPa$, $\rho = 1{,}240 \, kg/m^3$, find the velocity and pressure at $(0, 1)$ and $(1, 1)$

## 4.5  Doublet in uniform flow (2)

We consider the air flow ($V_O = 9 \, m/s$, standard conditions) past a suspension bridge cable ($\varnothing = 20 \, mm$).

(a)  Select the strength of doublet needed to portray the uniform flow of ideal fluid around the cylindrical cable.
(b)  In real fluid flow, calculate the hydrodynamic frequency of the vortex shedding.

## Solution

(a)  A doublet and uniform flow is analogous to the flow past a cylinder of radius:

$$R = \sqrt{\frac{-\mu}{V_O}}$$

where $\mu$ is the strength of the doublet. Hence:

$$\mu = -V_O \times R^2 = 9 \, E{-}4 \, m^3/s$$

(b)    The Reynolds number of the flow is 1.1 E−4. For that range of Reynolds number, the vortex shedding behind the cable is characterised by a well-defined Karman street of vortex. The hydrodynamic frequency satisfies:

$$St = \frac{\omega_{shedding} \times 2 \times R}{V_O} \sim 0.2$$

It yields: $\omega_{shedding} = 90\,Hz$. If the hydrodynamic frequency happens to coincide with the natural frequency of the structure, the effects may be devastating: e.g., Tacoma Narrows bridge failure on 7 November 1940.

## 4.6   Flow pattern (2)

In two-dimensional flow we now consider a source, a sink and an uniform stream.

For the pattern resulting from the combinations of a source (located at $(-L, 0)$) and sink (located at $(+L, 0)$) of equal strength Q in uniform flow (velocity $+V_O$ parallel to the x-axis):

(a)    Sketch streamlines and equipotential lines;
(b)    Give the velocity potential and the stream function.
        This flow pattern is called the flow past a Rankine body. W.J.M. Rankine (1820–1872) was a Scottish engineer and physicist who developed the theory of sources and sinks. The shape of the body may be altered by varying the distance between source and sink (i.e. $2 \times L$) or by varying the strength of the source and sink. Other shapes may be obtained by the introduction of additional sources and sinks and Rankine developed ship contours in this way.
(c)    What is the profile of the Rankine body (i.e. find the streamline that defines the shape of the body)?
(d)    What is the length and height of the body?
(e)    Explain how the flow past a cylinder can be regarded as a Rankine body. Give the radius of the cylinder as a function of the Rankine body parameter.

### Solution

The flow past a Rankine body is the pattern resulting from the combinations of a source and sink of equal strength in uniform flow (velocity $+V_O$ parallel to the x-axis):

$$\phi = -V_O \times r \times \cos\theta - \left( +\frac{q}{2 \times \pi} \times Ln\left(\frac{r_1}{r_2}\right) \right)$$

$$\psi = -V_O \times r \times \sin\theta - \left( +\frac{q}{2 \times \pi} \times (\theta_1 - \theta_2) \right)$$

where the subscript 1 refers to the source, the subscript 2 to the sink and q is positive for the source located at $(-L, 0)$ and the sink located at $(+L, 0)$.

The profile of the Rankine body is the streamline $\psi = 0$:

$$\psi = -V_O \times r \times \sin\theta + \frac{q}{2 \times \pi} \times (\theta_1 - \theta_2) = 0$$

$$r = \frac{q \times (\theta_1 - \theta_2)}{2 \times \pi \times V_O \times \sin\theta}$$

The length of the body equals the distance between the stagnation points where:

$$V = V_O + \frac{q}{2 \times \pi \times r_1} - \frac{q}{2 \times \pi \times r_2} = V_O + \frac{q}{2 \times \pi} \times \left( \frac{1}{r_s - L} - \frac{1}{r_s + L} \right) = 0$$

and hence:

$$L_{body} = 2 \times r_s = 2 \times L \times \sqrt{1 + \frac{q}{\pi \times L \times V_O}}$$

The half-width of the body h is deduced from the profile equation at the point (h, $\pi/2$):

$$h = \frac{q \times (\theta_1 - \theta_2)}{2 \times \pi \times V_O}$$

where: $\theta_1 = \alpha$ and $\theta_2 = \pi - \alpha$ and hence:

$$\alpha = \frac{\pi}{2} - \frac{\pi \times h \times V_O}{q}$$

But also:

$$\tan \alpha = \frac{h}{L}$$

So the half-width of the body is the solution of the equation:

$$h = L \times \cot \left( \frac{\pi \times V_O}{q} \times h \right)$$

## 4.7  Flow pattern (3)

In two-dimensional flow we consider again a source, a sink and an uniform stream. But, the source is located at $(+L, 0)$ and the sink is located at $(-L, 0)$ (i.e. opposite to a Rankine body flow pattern). They are of equal strength q in an uniform flow (velocity $+V_O$ parallel to the x-axis).

Derive the relationship between the discharge q, the length L and the flow velocity such that no flow injected at the source becomes trapped into the sink.

## 4.8  Rankine body (1)

(a)  We consider a Rankine body (Length = 3 m, Breadth = 1.2 m) in an uniform velocity field (horizontal direction, V = 3 m/s) using a source and sink of equal strength.

Calculate the source/sink strength and the distance between source and sink. Draw the flow net on a graph paper.

(b)  Using 2DFlowPlus, investigate the flow pattern resulting by addition of one source (strength +5) at $(X = -3, Y = 0)$ and one sink (strength −5) at $(X = +3, Y = 0)$ in an uniform flow ($V_\infty = 1$). Plot the pressure and velocity field.

Note: the software 2DFlowPlus is presented in Appendix E.

### *Solution*

(b) See 2DFlowPlus File Rankin1.xyb.

## 4.9   Source and sink in uniform flow

Using the software 2DFlowPlus, investigate the flow pattern resulting by addition of one source (strength +5) at $(X=+3, Y=0)$ and one sink (strength −5) at $(X=−3, Y=0)$ in an uniform flow $(V_\infty = 1)$. Plot the pressure and velocity field. Where are located the stagnation points?

### Solution

The stagnation points are located on the x-axis. There is no recirculation (contamination) flow from the source to the sink because $q/(\pi \times L \times V_O)=0.53 < 1$.

　　See 2DFlowPlus File So_Si1.xyb.

## 4.10   Rankine body (2)

(a)   Plot the flow net for a Rankine body (Length = 3, Breadth = 1.35) in an uniform velocity field (horizontal direction, $V = 3.5$) with a source and sink of equal strength.

(b)   Using 2DFlowPlus, investigate the flow pattern resulting by addition of one vortex $(K=−0.1)$ at $(X=+3.5, Y=0)$ and another $(K=+0.1)$ at $(X=+4.5, Y=0)$. Describe in words the physical nature of the flow.

### Solution

(a)   The length of the Rankine body satisfies:

$$L_{body} = 2 \times r_s = 2 \times L \times \sqrt{1 + \frac{q}{\pi \times L \times V_O = 3}} = 3 \qquad \text{Rankine body length}$$

The half-width of the body h is:

$$h = L \times \cot\left(\frac{\pi \times V_O}{q} \times h\right) = 0.675 \qquad \text{Rankine body half-width}$$

This system of two equations with two unknowns (q, L) gives:

$$q = 1.0 \, m^2/s$$

$$L = 1.45 \, m$$

where L is the distance between the origin and the sink.

(b)   The addition of vortex is similar to a wake region behind Rankine body.

See 2DFlowPlus File Rankin2.xyb.

## 4.11   Magnus effect (1)

Two 15-m high rotors 3 m in diameter are used to propel a ship. Estimate the total longitudinal force exerted upon the rotors when the relative wind velocity is 25 knots, the angular velocity of the rotors is 220 revolutions per minute and the wind direction is at 60° from the bow of the ship.

　　Perform the calculations for (a) an ideal fluid with irrotational motion and (b) a real fluid.

　　(c) What orientation of the vector of relative wind velocity would yield the greatest propulsion force upon the rotorship? Calculate the magnitude of this force.

*Assume a real fluid flow. The result is trivial for ideal fluid with irrotational flow motion.*

(d) Determine how nearly into the wind the rotorship could sail. That is, at wind angle would the resultant propulsion force be zero ignoring the wind effect on the ship itself.

## Solution

The tangential velocity of the rotors is:

$$R \times \omega = 2 \times \pi \times 220/60 \times 1.5 = 34.56 \, m/s$$

The relative velocity of the wind is:

$$V_O = 25 \times 1852/3600 = 12.86 \, m/s$$

Note: if the wind comes from starboard, the rotation of the rotor masts must be in the trigonometric positive direction to propel the ship forward.

(a) For an ideal fluid with irrotational motion, the lift force per unit length on each rotor is:

$$Lift = -\rho \times V_O \times K$$

where the vortex strength K satisfies:

$$\frac{K}{2 \times \pi \times R} = R \times \omega \Rightarrow K = 325.7 \, m^2/s$$

It yields:

$$Total \, Lift = 2 \times 15 \times 1.2 \times 12.86 \times 325.7 = 150.8 \, kN$$

In the direction of flow motion, the total force is:

$$Total \, force = Total \, Lift \times \cos 30° = 131 \, kN$$

(b) For a real fluid flow, the lift and drag forces (per unit width on each rotor) are estimated as:

$$Lift = C_L \times 2 \times R \times \rho \times \frac{V_O^2}{2}$$

$$Drag = C_D \times 2 \times R \times \rho \times \frac{V_O^2}{2}$$

where the lift and drag coefficients, $C_L$ and $C_D$ respectively, are deduced from Figure 4.18.
The total lift and drag forces are hence:

$$Total \, Lift \sim 2 \times 15 \times 7 \times 2 \times 1.5 \times 1.2 \times 12.86^2/2 = 62.5 \, kN$$

$$Total \, Drag \sim 2 \times 15 \times 2 \times 2 \times 1.5 \times 1.2 \times 12.86^2/2 = 17.9 \, kN$$

In the direction of flow motion, the total force is:

$$Total \, force = Total \, Lift \times \cos 30° - Total \, drag \times \sin 30° = 45.2 \, kN$$

## 4.12   Magnus effect (2)

A infinitely long rotating cylinder (R = 1.1 m) is placed in a free-stream flow ($V_O$ = 15 knots) of water. The cylinder is rotating at 45 rpm and the ambient pressure far away is 1.1 E+5 Pa.

(a)   Calculate and plot the pressure distribution on the cylinder surface.
(b)   Find the maximum and minimum pressures on the cylinder surface.
(c)   Find the location $\theta$ where the pressure on the cylinder surface is the fluid pressure far away from the cylinder.

### Solution

(a)   The flow pattern has two stagnation points.
(b)   The minimum and maximum pressures at the cylinder surface are respectively −7.24 E+4 and +1.40 E+5 Pa for $\theta$ = 270° and 20° (and 160°) respectively.

Notes: (1) The minimum pressure is sub-atmospheric and may lead to some cavitation. (2) There are two locations the pressure is maximum which correspond both the location of a stagnation point.

(c)   The pressure on the cylinder surface equals the ambient pressure for $\theta$ = 55°, 115°, 190° and 350°.

## 4.13   Whirlpools

Whirlpools may be approximated by a series of vortices of same signs advected into an uniform flow.

(a)   Consider two vortices of equal strength K = +1 located at (−2,) and (+2, 0). Estimate how far away the effect of the vortices is perceived to be that of an unique vortex. What would be the strength of that vortex?
(b)   Consider two vortices of equal strength K = +1 located at (−2,) and (+2, 0) in a horizontal uniform flow V = +0.03. What are the stream function and velocity potential of the resulting flow motion.

### Solution

(a)   See 2DFlowPlus File Whirlpl1.xyb.
(b)   See 2DFlowPlus File Whirlpl2.xyb.

## 4.14   Wind force on a Nissen hut

A 45 km/h wind flows over a Nissen hut which has a 3.5 m radius and is 54.9 m long (Fig. E4.1). The upstream pressure and temperature are 1.013 E+5 Pa and 288.2K respectively, and they are equal to the pressure and temperature inside the Nissen hut.

(a)   Calculate the lift and drag forces on the building.
(b)   Find the location $\theta$ on the building roof where the pressure is 1.013 E+5 Pa.

*Assume an irrotational flow motion of ideal fluid in parts (a) and (b).*

(c)   Calculate the drag force for a real fluid flow.

Notes: The Nissen hut is a building made from a semi-circle of corrugated steel. A variant was the Quonset hut used extensively during World War 2 by the Commonwealth and US military for army camps and air bases. The design was named after Major Peter Norman Nissen, 29th Company Royal Engineers who experimented with hut design and constructed three prototype semi-circular huts in April 1916. The building was not only economical but also portable.

(A) Wind flow past a Nissen hut

(B) Nissen hut in Port Lincoln, South Australia – Converted into a church in the early 1950s, the hut was demolished in the late 1960s

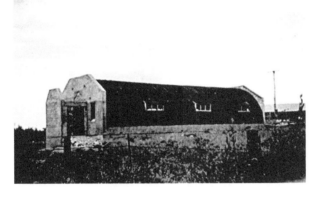

*Figure E4.1   Nissen hut.*

## Solution

(a)   Using the method of images, the flow field is that a doublet in an uniform flow. The pressure on the roof is

$$P = P_O + \frac{1}{2} \times \rho \times V_O^2 \times (1 - 4 \times \sin^2\theta)$$

where $P_O = 1.013\ E{+}5\ Pa$, $V_O = 12.5\ m/s$ and $\rho = 1.225\ kg/m^3$.

The flow direction is positive from left to right, and the doublet is centred at the origin. The lift and drag forces on the building are respectively:

$$\text{Lift} = -\int_0^\pi (P - P_O) \times R \times \sin\theta \times d\theta$$

$$\text{Drag} = -\int_0^\pi (P - P_O) \times R \times \cos\theta \times d\theta$$

The integration yields:

$$\text{Lift} = +2 \times \rho \times V_O^2 \times R = 1.3\ kN/m$$

$$\text{Drag} = 0$$

The total lift force on the 54.9 m long Nissen hut is 72 kN.

**Note**

$$\int (1 - 4 \times \sin^2\theta) \times d\theta = \sin(2 \times \theta) - \theta$$

$$\int (1 - 4 \times \sin^2\theta) \times \sin\theta \times d\theta = 2 \times \cos(\theta) - \frac{1}{3} \times \cos(3 \times \theta)$$

$$\int (1 - 4 \times \sin^2\theta) \times \cos\theta \times d\theta = \frac{1}{3} \times \sin(3 \times \theta)$$

(b) The pressure on the building roof is 1.013 E+5 Pa for $\theta = +30°$ ($\pi/6$) and $+150°$ ($5 \times \pi/6$).

(c) For a real fluid flow, separation takes place on the downstream of the building. The Reynolds number $\rho \times V_O \times D/\mu$ equals 5.6 E+6, and the corresponding drag coefficient is about 0.8 (Fig. 4.13). The total drag force equals hence:

$$\text{Drag force} = \frac{1}{2} \times C_D \times \frac{1}{2} \times \rho \times V_O^2 \times D \times L = 14\,\text{kN}$$

where $D = 2 \times R$.

## 4.15    Magnus effect airplane

Two rotating cylinders are used instead of conventional wings to provide the lift to an airplane. Calculate the length of each cylinder wing for the following design conditions:

Cruise speed: 320 km/h
Cruise altitude: 2,000 m
Airplane mass: 8 E+6 kg
Cylinder radius: 1.8 m
Cylinder rotation speed: 500 rpm

**Solution**

The air density at 2,000 m altitude is about 1.1 kg/m³.

The ideal fluid flow pattern is the superposition of an uniform flow ($V_O = 88.9$ m/s), a doublet (strength $\mu$) and a vortex strength K.

The cylinder radius and doublet strength are linked as:

$$R = \sqrt{\frac{(-\mu)}{V_O}}$$

while the vortex strength and rotation speed ($\omega = 52.36$ rad/s) satisfy:

$$\omega = \frac{K}{2 \times \pi \times R^2} = \frac{K \times V_O}{2 \times \pi \times \mu}$$

The lift force per unit length of wing is

$$\text{Lift} = -\rho \times V_O \times K$$

The right wing must rotate in the anti-clockwise direction as seen by the pilot and $\omega$ must be negative to provide a positive lift.

The calculations for cruise conditions imply that each wing is 376 m long (!).

Note: The Magnus effect lift force decreases with decreasing speed and it becomes fairly small at take-off and landing conditions.

## 4.16  Wind flow past columns

The facade of a 150 m tall building is hindered by two cylindrical columns (a service column and a structural column) (Fig. E4.2). There are concerns about the wind flow around the columns during storm conditions ($V_O = 35$ m/s).

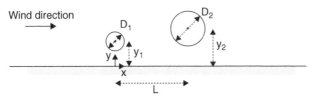

*Figure E4.2*  View in elevation of the building facade and cylindrical columns.

The column dimensions are:

|  | Column 1 | Column 2 | Units |
|---|---|---|---|
| D | 0.8 | 2.0 | m |
| x (centre) | 0 | 2.8 | m |
| y (centre) | 1.0 | 2.0 | m |

You will assume that the wind flow around the building is a two-dimensional irrotational flow of ideal fluid. The atmospheric conditions are: $P = P_{atm} = 105$ Pa; $T = 20$ Celsius.

(a)  On graph paper, sketch the flow net for a 35 m/s wind flow. Indicate clearly on the graph the discharge between two streamlines, the x-axis and y-axis, and their direction.

(b)  The developer is concerned about the maximum wind speeds next to the building facade that may damage the cladding and service column (column 1).

(b1)  From your flow net, compute the wind velocity and the pressure at:

$$x = 0\,\text{m}, \quad y = 0.4\,\text{m}$$
$$x = 2.8\,\text{m}, \quad y = 1.0\,\text{m}$$
$$x = 1.4\,\text{m}, \quad y = 0.8\,\text{m}$$

(b2)  Where is located the point of maximum wind velocity and minimum ambient pressure? Show that location on your flow net.

(b3)  Discuss your results. Do you think that these results are realistic? Why?

(c)  Explain what standard flow patterns you would use to describe the flow around these two columns.

(d)  Write the stream function and the velocity potential as a function of the wind speed $V_O$ (35 m/s), the two cylinder diameters $D_1$ and $D_2$, and their locations $(x_1, y_1)$ and $(x_2, y_2)$. Do not use numbers. Express the results as functions of the above symbols.

(e)  For a real fluid flow, what is (are) the drag force(s) on each cylinder (Height: 150 m) neglecting interactions between the columns?

### Solution

The flow pattern is solved using the method of images. Each column is represented by a doublet in an uniform flow ($V_O = 35$ m/s) and the building facadce is a line of symmetry.

# Complex potential, velocity potential and Joukowski transformation

## SUMMARY

A brief introduction of the conformal transformation is presented. Complex potential and complex velocity are defined. Later the transformation of the circle is developed.

## 1 INTRODUCTION

### 1.1 Presentation

For a two-dimensional flow, the position of any point P(x, y) may be represented by the complex number z:

$$z = x + i \times y \tag{5.1}$$

where $i = \sqrt{-1}$. The complex number z is shown in the Argand-Cauchy diagram in Figure 5.1.

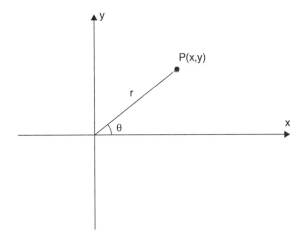

*Figure 5.1* Argand-Cauchy diagram.

**Discussion**

In the complex number z, the first part x is called the **real part** and the second part y is called the **imaginary part** of the complex number. The **modulus** (i.e. absolute value) of the complex number z, denoted r, is defined as:

$$r = \sqrt{x^2 + y^2}$$

The **argument** $\theta$ of the complex number is the position vector measured from the positive x-axis in an anti-clockwise direction (Fig. 5.1):

$$\theta = \tan^{-1}\left(\frac{y}{x}\right)$$

An alternative mode of expressing a complex number is:

$$z = x + i \times y = r \times (\cos\theta + i \times \sin\theta) = r \times e^{i \times \theta}$$

**Notes**

– The Argand diagram, or Argand-Cauchy diagram, was devised by the Swiss mathematician Jean Robert Argand in about 1806. It is named also after the French mathematician Augustin Louis de Cauchy (1789–1857).
– The complex number i may be regarded as the complex number with a modulus of unity and argument $\pi/2$:

$$i = e^{i \times (\pi/2)}$$

– The multiplication of a complex number by i does not affect its modulus but increases the argument by $\pi/2$:

$$i \times z = i \times x - y = r \times e^{i \times (\theta + \pi/2)}$$

The division of a complex number by i decreases the argument by $\pi/2$:

$$\frac{1}{i} \times z = y - i \times x = r \times e^{i \times (\theta - \pi/2)}$$

## 1.2   Complex potential

The **complex potential** W is defined as:

$$W(z) = \phi + i \times \psi \tag{5.2}$$

where $\phi$ and $\psi$ satisfy the Cauchy-Riemann equations (Streeter 1948, p. 86). $\phi$ and $\psi$ are the velocity potential and stream function respectively.

The complex potential W(z) defines a point in an Argand-Cauchy diagram which has $\phi$ as abscissa and $\psi$ as ordinate. The diagram is referred to as W-diagram, while the Argand-Cauchy diagram in terms of $z = x + i \times y$ is called the z-diagram.

---

**Discussion**

For each point $(x, y)$ in the z-diagram, there is a corresponding point $(\phi, \psi)$ in the W-diagram, since the velocity potential $\phi$ and stream function $\psi$ are functions of x and y. Similarly, for each line in the z-diagram, there is a corresponding line in the W-diagram. In the W-diagram, the streamlines are horizontal lines while the equipotential lines are vertical lines (e.g. Fig. 5.2A).

If the complex potential has one and only one value for each value of z and if the derivative of the complex potential has one value, the function $W = f(z)$ is said to be **regular**, holomorphic or analytic (Streeter 1948, p. 90).

---

*Properties of the complex potential*

Basic properties of the complex potential include:

$$\frac{\partial W}{\partial x} = -\frac{1}{i} \times \frac{\partial W}{\partial y} \tag{5.3a}$$

$$\frac{\partial W}{\partial y} = \frac{1}{i} \times \frac{\partial W}{\partial x} \tag{5.3b}$$

$$\frac{\partial \phi}{\partial x} = \frac{\partial \psi}{\partial y} \qquad \text{Cauchy-Riemann equation} \tag{5.4a}$$

$$\frac{\partial \phi}{\partial y} = -\frac{\partial \psi}{\partial x} \qquad \text{Cauchy-Riemann equation} \tag{5.4b}$$

$$\Delta W = \Delta(\phi + i \times \psi) = \Delta \phi + i \times \Delta \psi \tag{5.5}$$

---

**Application No. 1**

The complex potential equals: $W = z^3$.

(a)  Estimate the velocity potential and stream function.
(b)  Determine the point in the W-diagram corresponding to $(x = 1.5, y = 3.2)$ in the z-diagram.
(c)  Draw the streamlines and equipotentials in the z-diagram and W-diagram.

*Solution*

(a)  If $W = z^3$, it yields:

$$W = (x + i \times y)^3 = x^3 + 3 \times x^2 \times i \times y + 3 \times x \times (i)^2 \times y^2 + (i \times y)^3$$
$$= (x^3 - 3 \times x \times y^2) + i \times (3 \times x^2 \times y - y^3)$$

Equating the real and imaginary parts, it gives:

$$\phi = x \times (x^2 - 3 \times y^2)$$
$$\psi = y \times (3 \times x^2 - y^2)$$

(b)   For $z = 1.5 + i \times 3.2$, the complex potential is:

$$W = -42.7 - i \times 11.2$$

(c)   In the z-diagram, the equation of the streamlines is:

$$\frac{1}{3} \times \left( \frac{\psi}{y} + y^2 \right) = x^2$$

The equation of the equipotentials is:

$$\frac{1}{3} \times \left( x^2 - \frac{\phi}{x} \right) = y^2$$

The results are plotted in Figure 5.2.

---

**Application No. 2**

The complex potential is $W = z^2$. (a) Find the velocity potential and stream function. (b) Estimate the inverse transformation $z = f^{-1}(W)$.

*Solution*

The complex potential equals:

$$W(z) = z^2 = r^2 \times e^{i \times 2 \times \theta} = (x^2 - y^2) + i \times x \times y$$

Equating the real and imaginary parts, the velocity potential and stream function are:

$$\phi = x^2 - y^2$$
$$\psi = 2 \times x \times y$$

The inverse transformation is:

$$z = f^{-1}(W) = \sqrt{W}$$

That is:

$$r = \sqrt{R} \quad \text{and} \quad \theta = \frac{\Theta}{2}$$

where R and $\Theta$ are respectively the modulus and argument of the complex potential.

(A) W-diagram

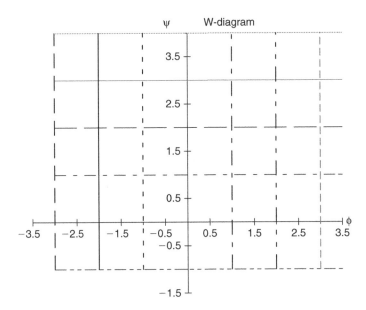

(B) z-diagram

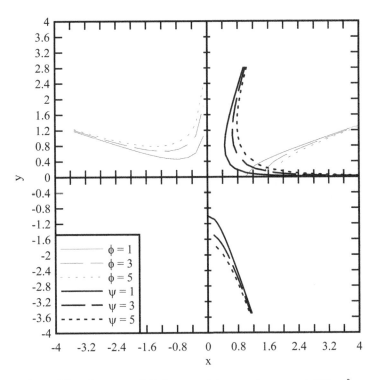

*Figure 5.2*   z-diagram and W-diagram of the complex potential $W = z^3$.

## 1.3   Complex velocity

The **complex velocity** is defined as:

$$w(z) = \frac{dW}{dz} \tag{5.6a}$$

Note the limits of the differentiation. If $dz \approx \delta z = \delta x$, the complex velocity tends to:

$$w(z) \approx \frac{\delta\phi}{\delta x} + i \times \frac{\delta\psi}{\delta x} = -V_x + i \times V_y \quad \text{for } \delta z = \delta x$$

If $dz \approx \delta z = i \times \delta y$, the complex velocity tends to:

$$w(z) \approx \frac{\delta(\phi + i \times \psi)}{i \times \delta y} = \frac{1}{i} \times \frac{\partial\phi}{\partial y} + \frac{\partial\psi}{\partial y} = -V_x + i \times V_y \quad \text{for } \delta z = i \times \delta y$$

That is, the complex velocity equals:

$$w(z) = \frac{dW}{dz} = -V_x + i \times V_y \tag{5.6b}$$

**Notes**

- If $i = \sqrt{-1}$, it satisfies;

$$i = -\frac{1}{i}$$

The complex number $1/i$ is the conjugate of the complex number $i$.
- Stagnation points are those points where the velocity is zero. That is:

$$w = \frac{dW}{dz} = 0$$

The modulus of the complex velocity equals:

$$|w| = \sqrt{V_x^2 + V_y^2} = |V| \tag{5.7}$$

where $|V|$ is the magnitude of the velocity. The argument of the complex velocity is:

$$\text{Arg}(w) = \pi - \tan^{-1}\left(\frac{V_y}{V_x}\right) \tag{5.8}$$

An alternative mode of expressing the complex velocity is:

$$w = |V| \times \exp\left(i \times \left(\pi - \tan^{-1}\left(\frac{V_y}{V_x}\right)\right)\right) \tag{5.6c}$$

## Note

Note the sign convention. If the local flow direction in the z-diagram is $\theta_V$ (i.e. $\theta_V = \tan^{-1}(V_y/V_x)$), the complex velocity may be rewritten:

$$w = |V| \times \exp(i \times (\pi - \theta_V))$$

$\theta_V$ is also called the argument of the velocity vector in the z-diagram.

## 1.4  Conformal transformation

The W-diagram has always a flow net that consists of a grid parallel to the $\phi$-lines and $\psi$-lines. The vertical lines are the equipotentials and the horizontal lines are the streamlines. As a result the flow represented by the W-diagram is that of an infinite fluid flowing uniformly in the horizontal direction (Streeter 1948). The particular configuration in the z-plane depends entirely upon the transformation $W = f(z)$, as the flow net in the W-plane is always the same. The solution of two-dimensional flow problems is attacked by the indirect method of investigating various functions to determine the shape of boundaries to which they might apply.

The relationship $W = f(z)$ is called the **conformal transformation**. The transformation from the z-diagram to the W-diagram is called **conformal mapping**. The inverse function relating W to z (i.e. $z = f^{-1}(W)$) is said to be the **inverse transformation**. Many useful flow nets are obtained by the inverse transformation. It must be emphasised that the z-flow net is entirely different from the W-flow net. But, in some cases, the inverse function cannot be solved explicitly for W.

---

**Discussion**

The conformal transformation is a technique to solve the entire flow field for ideal fluid with irrotational flow motion. It is well-suited to two-dimensional flows for which the z-diagram corresponds to the two-dimensional plane, and it may be extended to axisymmetrical flow situations.

---

## 2  SIMPLE TRANSFORMATIONS

The complex potential may be simply derived for the two-dimensional steady flow patterns developed in Chapter I-4 as $W = \phi + i \times \psi$. Basic results are presented first. Later more advanced flow patterns are detailed.

Application of the principle of superposition shows that, if two or more flow patterns are combined by superposition, the values of $\psi$ and $\phi$ at any point other than a singular point, are simply the algebraic sum of the $\psi$-values and $\phi$-values of the component patterns at that point, at any instant. This result which is a basic property of the Laplace equation (Chapter I-2, paragraph 4.2) does apply to the complex potential.

## 2.1   Basic transformations

### Uniform flow

For an uniform flow, the application of the basic flow pattern (Chapter I-4) yields:

$$W = V_O \times z \tag{5.7}$$

If $V_O$ is a real number, the flow is parallel to the x-axis. That is, the angle between the flow direction and the x-axis is: $\alpha = 0$. If $V_O$ is a complex number, the angle $\alpha$ satisfies:

$$\tan \alpha = \frac{V_y}{V_x}$$

The modulus of the complex potential equals:

$$R^2 = \phi^2 + \psi^2 = V_{O_x}^2 + V_{O_y}^2 = |V_O|^2 \tag{5.8a}$$

The argument of the complex potential is:

$$\tan \Theta = \frac{\psi}{\phi} = \tan(\theta - \alpha) \tag{5.8b}$$

### Source/Sink

For a source at the origin, the application of the basic flow patterns (Chapter I-4) shows that:

$$W = -\frac{q}{2 \times \pi} \times Ln(z) \tag{5.9}$$

The complex velocity is:

$$w(z) = \frac{dW}{dz} = -V_x + i \times V_y = -\frac{q}{2 \times \pi} \times \frac{1}{r} \times e^{-i \times \theta} \tag{5.10}$$

For a source located at $z_O = x_O + i \times y_O$, the complex potential is:

$$W = -\frac{q}{2 \times \pi} \times Ln(z - z_O) \tag{5.11}$$

---

**Remark**

$$Ln(z) = Ln(r) + i \times \theta = \frac{1}{2} \times Ln(x^2 + y^2) + i \times \tan^{-1}\left(\frac{y}{x}\right) \quad \text{for } 0 \leq \theta < 2 \times \pi$$

### Vortex

For a vortex at the origin, the application of the basic flow patterns (Chapter I-4) yields:

$$W = +i \times \frac{K}{2 \times \pi} \times Ln(z) \tag{5.12}$$

where K is the vortex strength which is equal to the circulation around the vortex centre. For a vortex located at $z_O = x_O + i \times y_O$, the complex potential is:

$$W = +i \times \frac{k}{2 \times \pi} \times Ln(z - z_O) \tag{5.13}$$

### Doublet

For a doublet at the origin, the complex potential is:

$$W = \frac{\mu}{z} \tag{5.14}$$

where $\mu$ is the doublet strength. In turn the inverse transformation is:

$$z = \frac{\mu}{W} \tag{5.15}$$

The modulus of the complex potential is:

$$R = \frac{\mu}{r} \tag{5.15a}$$

The argument of the complex potential equals:

$$\Theta = -\theta \tag{5.15b}$$

### Note

A complex number at the power n equals:

$$z^n = r^n \times e^{i \times n \times \theta}$$

Hence the inverse of a complex number is:

$$\frac{1}{z} = \frac{1}{r} \times e^{-i \times \theta}$$

For a doublet located at $z = z_O$, the application of the basic flow patterns (Chapter I-4) shows that:

$$W = \frac{\mu}{z - z_O} \tag{5.16}$$

where $\mu$ is the doublet strength.

### Spiral vortex

A spiral vortex is the combination of a source of strength q and a vortex of strength K both located at $z = z_O$. The complex potential equals:

$$W = -\frac{1}{2 \times \pi} \times (q - i \times K) \times \text{Ln}(z - z_O) \qquad (5.17)$$

Further basic transformations are listed in Table 5.1.

Table 5.1  Summary of basic transformations.

| Flow pattern | Complex potential $W = f(z)$ | Remarks |
|---|---|---|
| Uniform flow | $V_O \times z$ | $z = \dfrac{R}{|V_O|} \times e^{i \times (\Theta - \alpha)}$ |
| Source (strength q) at origin | $-\dfrac{q}{2 \times \pi} \times \text{Ln}(z)$ | $w = -\dfrac{q}{2 \times \pi} \times \dfrac{1}{r} \times e^{-I \times \theta}$ |
| Source (strength q) at $z = z_O$ | $-\dfrac{q}{2 \times \pi} \times \text{Ln}(z - z_O)$ | |
| Vortex (strength K) at origin | $+i \times \dfrac{K}{2 \times \pi} \times \text{Ln}(z)$ | |
| Vortex (strength K) at $z = z_O$ | $+i \times \dfrac{K}{2 \times \pi} \times \text{Ln}(z - z_O)$ | |
| Spiral vortex (source: q; vortex: K) at $z = z_O$ | $\dfrac{-I}{z \times \pi} \times (q - i \times K) \times \text{Ln}(z - z_O)$ | |
| Flow at a wall angle $\alpha$ | $A \times z^{\pi/\alpha}$ | for $\alpha < \pi/2$ |
| Doublet (strength $+\mu$) at origin | $\dfrac{\mu}{z}$ | $z = \dfrac{\mu}{W}$ |
| Doublet (strength $+\mu$) at $z = z_O$ | $\dfrac{\mu}{z - z_O}$ | $z = z_O + \dfrac{\mu}{W}$ |
| Source (strength $+q$) at $z = -z_O$ and sink (strength $-q$) at $z = +z_O$ | $-\dfrac{q}{2 \times \pi} \times \text{Ln}\left(\dfrac{z + z_O}{z - z_O}\right)$ | |
| Flow past a cylinder of radius R at the origin | $V_O \times \left(z + \dfrac{R^2}{z}\right)$ | $w(z) = V_O \times \left(1 - \dfrac{R^2}{z^2}\right)$ |
| Flow past a cylinder of radius R with circulation K the origin | $V_O \times \left(z + \dfrac{R^2}{z}\right) + i \times \dfrac{K}{2 \times \pi} \times \text{Ln}(z)$ | $z = \dfrac{W \pm \sqrt{w^2 - 4 \times V_O^2 \times R^2}}{2 \times V_O}$ |
| Flow past a half-body (source of strength q at $z = z_O$) | $V_O \times z - \dfrac{q}{2 \times \pi} \times \text{Ln}(z - z_O)$ | |
| Flow past a Rankine body (source at $z = -z_O$; sink at $z = +z_O$) | $V_O \times z - \dfrac{q}{2 \times \pi} \times \text{Ln}\left(\dfrac{z + z_O}{z - z_O}\right)$ | |

Notes: $W = R \times e^{i \times \Theta}$; w: complex velocity.

## 2.2   Other transformations

### Flow past a circular cylinder

The complex potential of an uniform flow (velocity $V_O$) past a cylinder of radius R at the origin is deduced by superposition of a doublet and an uniform flow:

$$W = V_O \times z + \frac{\mu}{z} \tag{5.18a}$$

where the strength of the doublet is: $\mu = V_O \times R^2$. After transformation, it yields:

$$W = V_O \times \left(z + \frac{R^2}{z}\right) \tag{5.18b}$$

The complex velocity is:

$$w(z) = \frac{dW}{dz} = V_O \times \left(1 - \frac{R^2}{z^2}\right) \tag{5.19}$$

For $z = +R$ and $-R$, the complex velocity is zero: i.e., the stagnation points.

### Note

The inverse transformation is:

$$z = \frac{W \pm \sqrt{W^2 - 4 \times V_O^2 \times R^2}}{2 \times V_O}$$

### Flow past a rotating cylinder

The complex potential of an uniform flow (velocity $V_O$) past a rotating cylinder of radius R at the origin is:

$$W = V_O \times \left(z + \frac{R^2}{z}\right) + i \times \frac{K}{2 \times \pi} \times Ln(z) \tag{5.20}$$

### Flow through an opening

For the flow through an opening (Fig. 5.3), the inverse transformation is:

$$z = L \times \cosh(W) \tag{5.21}$$

where L is the half-width of the opening, cosh is the hyperbolic cosine function and W is the complex potential (Vallentine 1969, pp. 161–162). The substitution of x, y, $\psi$ and $\phi$ in the inverse transformation gives:

$$x + i \times y = L \times \cosh(\phi + i \times \psi) = L \times (\cosh(\phi) \times \cos(\psi) + i \times \sinh(\phi) \times \sin(\psi))$$

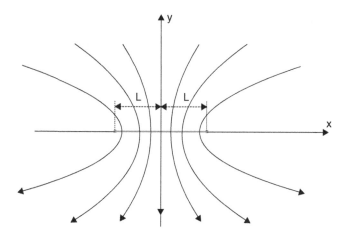

*Figure 5.3* Flow through an opening.

That is:

$$x = L \times \cosh(\phi) \times \cos(\psi)$$

$$y = L \times \sinh(\phi) \times \sin(\psi)$$

Since $\sin^2(\psi) + \cos^2(\psi) = 1$, the equations of the equipotentials are:

$$\frac{x^2}{L^2 \times \cosh^2(\phi)} + \frac{y^2}{L^2 \times \sinh^2(\phi)} = 1 \qquad\qquad \text{Equipotentials} \qquad (5.22)$$

The equipotentials are a family of ellipses with foci at $(+L, 0)$ and $(-L, 0)$.

Noting that $\cosh^2(\phi) - \sinh^2(\phi) = 1$, the streamline equations are:

$$\frac{x^2}{L^2 \times \cos^2(\psi)} + \frac{y^2}{L^2 \times \sin^2(\psi)} = 1 \qquad\qquad \text{Streamlines} \qquad (5.23)$$

The streamlines are confocal hyperbolae with foci at $(+L, 0)$ and $(-L, 0)$.

**Remarks**

1.  The flow pattern is similar to that of a finite line source/sink (Chapter I-4, paragraph 4.3). In Figure 5.3, the upper flow region is similar to a finite line sink (of length $2 \times L$) while lower flow region is that of a finite line source of equal length and strength as the sink.

2. Properties of the hyperbolic functions are given in Appendix D.
3. Important properties of the sinusoidal and hyperbolic functions are:

$$\sin^2(x) + \cos^2(x) = 1$$

$$\cosh^2(x) - \sinh^2(x) = 1$$

### Discussion

The flow through an opening (width $2 \times L$) may be regarded as the flow through a slot. In practice, separation would occur at a sharp-edge orifice. But as any streamline may be regarded as a solid boundary, the flow pattern may be used for smooth converging nozzles. Care is required however in the region of diverging flow as separation may occur.

An application of the flow pattern is the ebb flow through a narrow inlet mouth. During retreating tides, the ebb flow comes all around the inlet mouth in the form of a two-dimensional potential flow (Fig. 5.4). Figure 5.4 illustrates the flow pattern at the Imagire inlet of Hamanako Lake on the Enshu coast along the Pacific Ocean (Japan). There is some tidal dissymmetry of the flow through the narrow inlet mouth, because the flood (rising tide) enters into the estuary like an orifice flow, forming a confined jet. The resulting flow dissymmetry between ebb and flood flows may induce residual circulation in the estuary, sometimes called "tidal pumping" (Fischer et al. 1979).

### *Flow into a rectangular channel*

Considering the inverse transformation:

$$z = \exp(-W) - W \tag{5.24}$$

the resulting flow pattern is that of the flow into a rectangular channel (Fig. 5.5) (Vallentine 1969, pp. 162–163). Substituting of $x, y, \psi$ and $\phi$ into the inverse transformation gives:

$$x = \exp(-\phi) \times \cos\psi - \phi$$

$$y = -\exp(-\phi) \times \sin\psi - \psi$$

The streamline $\psi = 0$ gives $y = 0$: i.e., the x-axis. The streamline $\psi = -\pi$ is defined by $y = +\pi$ and $x = \exp(-\phi) - \phi$. The latter condition yields $-\infty < x \leq -1$ for $-\infty < \phi < +\infty$. Basically the streamline $\psi = -\pi$ is the upper wall of the open-ended channel sketched in Figure 5.5. The streamline $\psi = +\pi$ is the lower wall of the opening.

### Remark

The flow into the two-dimensional channel is assumed to be ideal: i.e., without separation (Fig. 5.5). The flow into a two-dimensional channel with separation is called the flow through a Borda's mouthpiece (Chapter I-7, paragraph 3.3).

(A) Sketch of ebb flow pattern at Hamanako Lake inlet

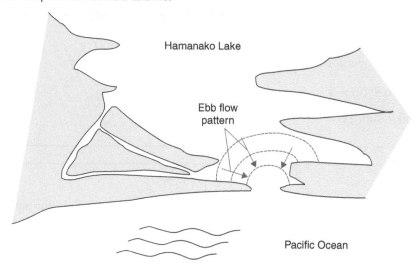

(B) Aerial photograph Hamanako Lake inlet in 1997 – The jetties and three detached breakwaters were constructed to protect the ocean beach against the erosion

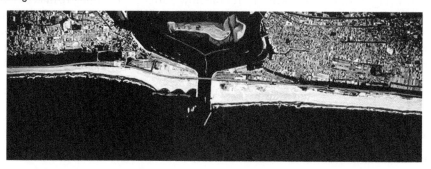

*Figure 5.4* Ebb flow pattern at the inlet mouth of Hamanako Lake (Japan).

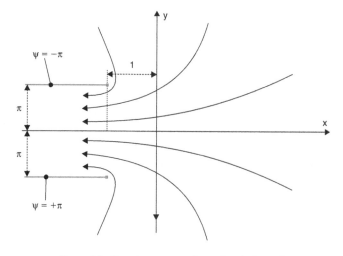

*Figure 5.5* Flow into a two-dimensional channel.

## 3  TRANSFORMATION OF THE CIRCLE

### 3.1  Presentation

The flow pattern around a circular cylinder may be itself transformed into other flow patterns. The technique, called transformation of the circle, may be used to study flows past plates, streamlined struts, arc and airfoils (Streeter 1948, pp. 144–155; Vallentine 1969, pp. 164–177). In simple terms, the transformation of the circle is an intermediate step of the conformal transformation. The latter is replaced by a succession or series of simple transformations, one of which being the transformation of the circle (e.g. Fig. 5.6).

For such relevant flow patterns, the transformation from the z-diagram into the W-diagram is done in two steps, by introducing an intermediary diagram, denoted the z′-diagram:

z-diagram ⇒ z′-diagram ⇒ W-diagram

where the second transformation is defined as:

$$W = V_O \times \left( z' + \frac{R^2}{z'} \right)$$
   Transformation of the circle   (5.25)

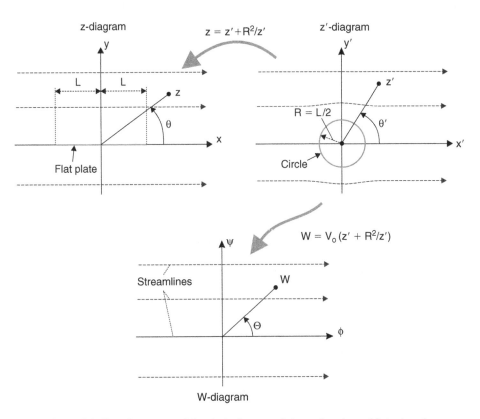

Figure 5.6  Transformation of the circle: flow parallel to a flat plate of finite length.

and $z' = x' + i \times y'$. R is the radius of an imaginary circle that is selected to satisfy the relevant boundary conditions ([1]). Considering flows past simple objects, some transformations from the z-diagram into the z'-diagram may be solved analytically, yielding the conformation transformation $W = f(z)$. The profile of these objects is the profile of the R-circle (Eq. (5.25)) in the z-diagram.

**Note**

- The transformation of the circle is based upon the intermediate transformation $W = V_O \times (z' + R^2/z')$, that is a complex potential for flow past a cylinder. The transformation $W = V_O \times (z' + R^2/z') + i \times K/(2 \times \pi) \times Ln(z')$ is called transformation of Joukowski or Kutta-Joukowski transformation. It is the transformation of the circle with some circulation (Chapter I-6).
- The complex potential of an uniform flow past a circular cylinder is:

$$W = V_O \times \left( z + \frac{R^2}{z} \right)$$ (5.18b)

where R is the cylinder radius.

## 3.2    Application: flow parallel to a flat plate of finite length

Figure 5.6 illustrates one example: i.e., a flow parallel to a flat plate of finite length $2 \times L$. The z-diagram, the z'-diagram and the W-diagram are sketched.

The transformation from the z'-diagram to the W-diagram is given by Equation (5.25). Note that the inverse transformation is:

$$z' = \frac{W \pm \sqrt{W^2 - 4 \times V_O^2 \times R^2}}{2 \times V_O}$$

Let us consider the transformation from the z-diagram to the z'-diagram:

$$z = z' + \frac{R^2}{z'}$$ (5.26)

where $R = L/2$ and L is the half-plate width. The stagnation points in the z'-diagram occur at $z' = +R$ and $z' = -R$ for which $z = +2 \times R$ and $z = -2 \times R$ respectively. More, for $z' = R \times e^{i \times \theta'}$ (i.e. circle perimeter), Equation (5.26) gives: $z = 2 \times R \times \cos \theta'$ for $0 \leq \theta' \leq 2 \times \pi$. Thus the circle of radius R in the z'-diagram becomes a line between $(-2 \times R, 0)$ and $(+2 \times R, 0)$ in the z-diagram.

Equations (5.25) and (5.26) are shown in Figure 5.6.

## 3.3    Flow normal to a flat plate

The flow field normal to a flat plate may be derived from the above transformation by adding one more intermediate transformation: i.e., a rotation by $-\pi/2$ (Fig. 5.7).

---

[1]In fact the transformation of the R-circle into the z-diagram will give the profile of an object in the flow.

(A) Transformation

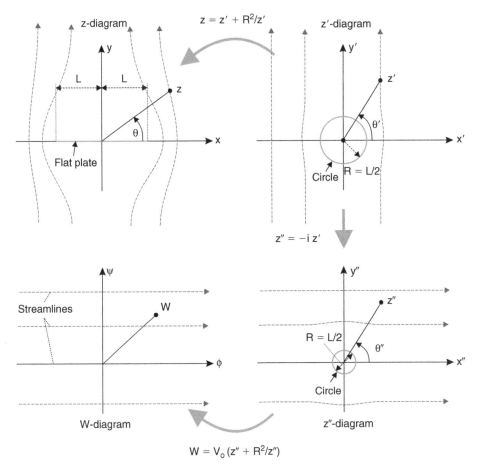

*Figure 5.7*   Flow normal to a flat plate of finite length.

Considering the basic transformations:

$$W = V_O \times \left( z'' + \frac{R^2}{z''} \right)$$

(5.27a)

$$z'' = \frac{1}{i} \times z' = -i \times z'$$

(5.27b)

$$z = z' + \frac{R^2}{z'}$$

(5.27c)

Equation (5.27b) is a rotation by $-\pi/2$. Equation (5.27c) satisfies $z = 2 \times R \times \cos \theta'$ for $z' = R \times e^{i \times \theta'}$ (i.e. circle perimeter) and $0 \le \theta' \le 2 \times \pi$. In other terms, the circle of radius R in the z'-diagram is the flat plate normal to the flow in the z-diagram.

(B) Streamline pattern in a Hele-Shaw cell apparatus: flow from top to bottom – Note some dye dispersion behind the plate

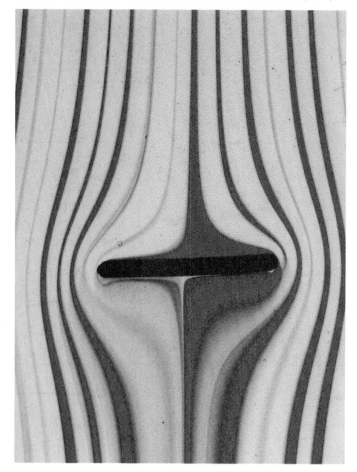

*Figure 5.7* Continued.

By elimination of $z'$ and $z''$ in Equation (5.27a), the complex potential becomes:

$$W = V_O \times \left( -i \times z' + i \times \frac{R^2}{z''} \right) = -i \times V_O \times \left( -z' + \frac{R^2}{z'} \right)$$

Combining with Equation (5.27c):

$$W^2 + V_O \times z^2 = -4 \times V_O^2 \times R^2$$

it yields:

$$W = i \times V_O \times \sqrt{z^2 + 4 \times R^2} = i \times V_O \times \sqrt{z^2 + L^2} \qquad (5.28)$$

where L is the half-plate length.

---

**Discussion**

The flow normal to a flat plate is a good illustration of the use of the conformal transformation. The conformal transformation $W = f(z)$ is decomposed into simple intermediary transformations (Fig. 5.7A). The result may be analytically solved in terms of the complex potential and hence the entire flow field. Figure 5.7B shows the streamline pattern. The movie Foil03.mov present a movie of the Hele-Shaw cell analogy (App. F).

For a horizontal flow normal to a flat, vertical plate (Fig. 5.8), the transformation is obtained by rotating the z-diagram (in the above application) by $-\pi/2$ for a flow from left to right. Such a rotation is achieved by dividing z by i in Equation (5.28). The complex potential is then:

$$W = i \times V_O \times \sqrt{-z^2 + L^2} \qquad \text{Horizontal flow normal to a flat plate} \qquad (5.29)$$

The above development assumed no flow separation behind the plate. The latter case in discussed in Chapter I-7 (section 3.4).

---

## 3.4　Flow past an ellipse

The flow past an ellipse is based upon the following basic transformations:

$$W = V_O \times \left( z' + \frac{R^2}{z'} \right) \qquad (5.30a)$$

$$z = z' + \frac{R_1^2}{z'} \qquad (5.30b)$$

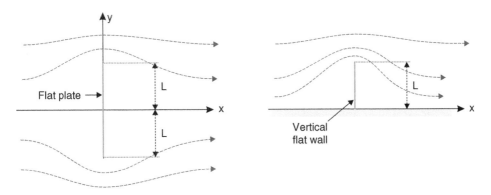

*Figure 5.8*　Horizontal flow normal to a flat plate of finite length.

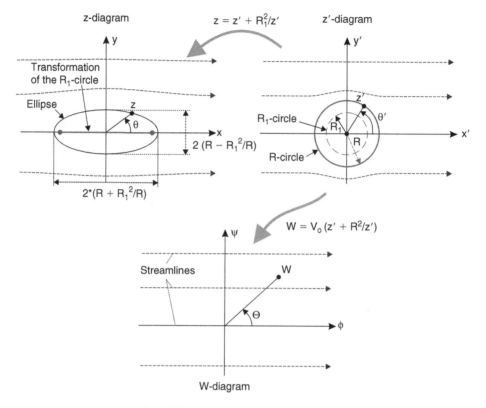

*Figure 5.9* Horizontal flow past an ellipse.

where R and $R_1$ are the radii of two concentric circles in the z'-diagram (Fig. 5.9). Equation (5.30a) is the transformation of the circle. In Equation (5.30b), the ratio $R_1/R$ defines the shape of the ellipse. For $R_1/R = 1$, Equation (5.30) yields the flow parallel to a plate.

The profile of the R-circle in the z-diagram is an ellipse. Combining Equations (5.30a) and (5.30b), the profile of the ellipse is defined as:

$$z = r \times e^{i \times \theta} = R \times e^{i \times \theta'} + \frac{R_1^2}{R} \times e^{-i \times \theta'} \qquad \text{Surface of ellipse for } 0 \le \theta' \le 2 \times \pi \qquad (5.31)$$

In Cartesian coordinates, the equations of the ellipse are:

$$x = \left(R + \frac{R_1^2}{R}\right) \times \cos \theta' \quad \text{for } 0 \le \theta' \le 2 \times \pi$$

$$y = \left(R - \frac{R_1^2}{R}\right) \times \sin \theta' \quad \text{for } 0 \le \theta' \le 2 \times \pi$$

For $R_1/R < 1$, the transformations are sketched in Figure 5.9. The transformation of the $R_1$-circle from the z'-diagram into the z-diagram is a flat plate of length $4 \times R_1$. The R-circle, from the

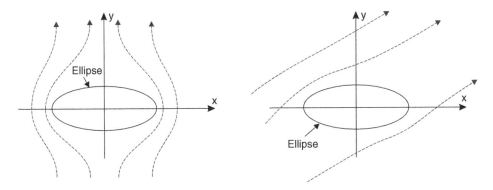

*Figure 5.10* Flow past an ellipse (a) for $V_O = i \times |V_O|$ and (b) for $V_O = |V_O| \times e^{i \times \theta}$.

$z'$-diagram, is transformed into an ellipse in the z-diagram. The foci of the ellipse are the ends of the flat plate of length $4 \times R_1$: i.e., $(-2 \times R_1, 0)$ and $(+2 \times R_1, 0)$.

**Notes**

- The length of the ellipse is $2 \times (R + R_1^2/R)$.
- The total height of the ellipse is: $2 \times (R - R_1^2/R)$.
- The flow past an ellipse is said to be a Joukowski profile (Vallentine 1969, p. 171).
- In the $z'$-diagram, the profile of the R-circle is $z' = R \times e^{i \times \theta'}$ with $0 \le \theta' \le 2 \times \pi$.

*Discussion*

Figure 5.9 was drawn for a real velocity $V_O$. If the velocity $V_O$ is an imaginary number (i.e. $V_O = i \times |V_O|$), the horizontal velocity component is zero and the approach flow velocity is vertical (Fig. 5.10). A more general case is shown in Figure 5.10 for a complex velocity.

### 3.5  Flow past a streamlined strut

Considering the basic transformations:

$$W = V_O \times \left( z'' + \frac{R^2}{z''} \right) \tag{5.32a}$$

$$z'' = z' + \Delta x \tag{5.32b}$$

$$z = z' + \frac{R_1^2}{z'} \tag{5.32c}$$

Equation (5.32b) is a translation by $\Delta x$ in the x-direction where $\Delta x = R - R_1$ and the R- and $R_1$-circles are tangent in one point. The conformal transformation is sketched in Figure 5.11.

In the z-diagram, the R-circle becomes a streamlined strut and its profile is defined by:

$$z = r \times e^{i \times \theta} = R \times e^{i \times \theta''} - \Delta x + \frac{R_1^2}{R} \times \frac{1}{e^{i \times \theta''} - \frac{\Delta x}{R}} \quad \text{Surface or strut } 0 \le \theta'' \le 2 \times \pi \tag{5.33}$$

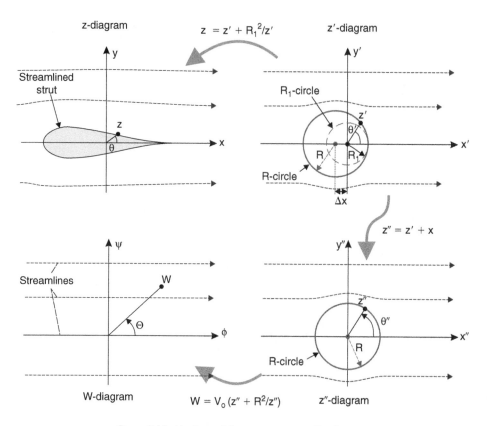

*Figure 5.11* Horizontal flow past a streamlined strut.

Dominant features of the streamlined strut are an upstream round nose corresponding to $\theta'' = \pi$ and a cusp of zero angle at the rear end for $\theta'' = 0$.

**Discussion**

The flow past a streamlined strut is another Joukowski profile. The transformation is very similar to that of the flow past an ellipse but the $R_1$-circle is not centred upon the centre of the R-circle.

The entire surface of the strut may be derived from Equation (5.33). For $\theta'' = 0$, the corresponding point in the z-diagram is the downstream sharp edge. For $\theta'' = \pi$, the corresponding point is the downstream round nose. The length or chord of the strut is:

$$\text{Length} = 2 \times \left( R + \frac{R_1^2}{R} \times \frac{1}{\left(1 - \frac{\Delta x}{R}\right) \times \left(1 + \frac{\Delta x}{R}\right)} \right).$$

**Remarks**

1. The flow past a streamlined strut is also called flow past a symmetrical Joukowski streamlined strut (Vallentine 1969, p. 172).
2. For $\Delta x = 0$, the strut becomes an ellipse (see above).
3. Note that the R-circle and $R_1$-circle are tangent for $\theta'' = 0$ (i.e. sharp downstream edge of strut).
4. In the $z''$-diagram, the profile of the R-circle is $z'' = R \times e^{i \times \theta''}$.

## 3.6 Flow past a circular arc

A related flow pattern is one for which the R-circle is translated along the imaginary y-axis:

$$W = V_O \times \left( z'' + \frac{R^2}{z''} \right) \tag{5.34a}$$

$$z'' = z' - i \times \Delta y \tag{5.34b}$$

$$z = z' + \frac{R_1^2}{z'} \tag{5.34c}$$

where $\Delta y = \sqrt{R^2 - R_1^2}$. That is, $\Delta y$ is selected such as the R-circle intersects the $R_1$-circle on the $x'$-axis in the $z'$-diagram (Fig. 5.12).

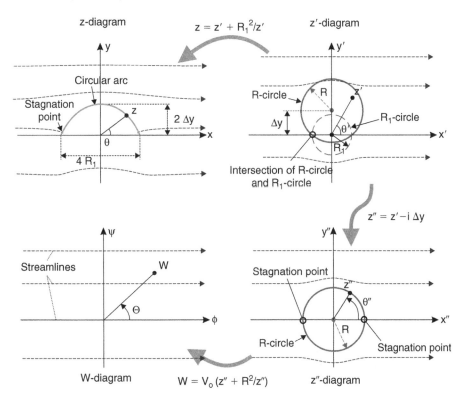

*Figure 5.12* Horizontal flow past a circular arc.

The intermediate transformations are sketched in Figure 5.12. The resulting flow pattern is the flow past a circular arc and the transformation of the R-circle into the z-diagram is the profile of the arc. The surface of the arc is given by the transformation of the R-circle into the z-diagram (Fig. 5.12):

$$z = r \times e^{i \times \theta} = R \times e^{i \times \theta''} + i \times \Delta y + \frac{\frac{R_1^2}{R}}{e^{i \times \theta''} + i \times \frac{\Delta y}{R}} \quad \text{Surface of arc } 0 \leq \theta'' \leq 2 \times \pi \quad (5.35)$$

The entire surface of the arc may be drawn by varying $\theta''$ between 0 and $2 \times \pi$. Two stagnation points occur at locations corresponding to the stagnation points of the flow past a cylinder in the $z''$-diagram ($\theta'' = 0$ and $\pi$) (Fig. 5.12).

---

**Remarks**

1.  Note that the R-circle and $R_1$-circles are not tangent anywhere but they must intersect on the $x'$-axis in the $z'$-diagram (Fig. 5.12):

$$\Delta y = \sqrt{R^2 - R_1^2}$$

2.  The two stagnation points correspond to the intersection of the R-circle with the $x''$-axis in the $z''$-diagram: i.e., $\theta'' = 0$ and $\pi$ (Fig. 5.12).
3.  The arc length in the flow direction (i.e. chord) equals:

Chord $= 4 \times R_1$

and the camber of the arc is:

Camber $= 2 \times \Delta y$

Both terms are defined in Figure 5.13.
4.  The relationship between chord and radius of curvature is:

Chord $= 4 \times R_1 = 4 \times R \times \cos(\varepsilon)$
Arc radius $= R \times \sec(\varepsilon)$

where sec is the secant function: $\sec(\varepsilon) = 1/\cos(\varepsilon)$ (see Appendix D). It yields:

Arc radius $= \dfrac{R^2}{R_1}$

5.  The ends of the arc are singularities where the flow velocity is theoretically infinite.

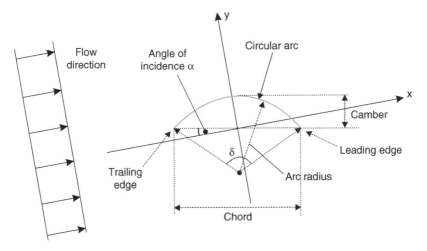

*Figure 5.13*  Horizontal flow past an inclined circular arc: definition sketch.

---

**Discussion**

Considering the circular arc sketched in Figure 5.13, the angle of arc opening is $\delta$. Basic trigonometry gives the following relationships between the chord, arc length, camber and opening angle:

$$\text{Chord} = 2 \times (\text{Arc radius}) \times \sin\left(\frac{\delta}{2}\right)$$

$$\text{Arc length} = \delta \times (\text{Arc radius})$$

$$\cos\left(\frac{\delta}{2}\right) = \frac{\text{Arc radius} - \text{Camber}}{\text{Arc radius}}$$

After transformation, it yields an unique relationship between the chord, radius of curvature and angle of opening:

$$\text{Chord} = 2 \times \text{Camber} \times \sqrt{\frac{1 + \cos\left(\frac{\delta}{2}\right)}{1 - \cos\left(\frac{\delta}{2}\right)}}$$

If the arc chord and camber are known, the above relationship yields and angle of opening and hence the radius of curvature (Table 5.2).

---

### 3.7   Flow past a Joukowski airfoil

A Joukowski airfoil profile is defined as:

$$W = V_O \times \left(z'' + \frac{R^2}{z''}\right) \tag{5.36a}$$

*Table 5.2*  Geometric properties of a circular arc.

| Angle of opening δ deg.<br><br>(1) | Camber<br>Chord<br>(2) | Camber<br>Curvature radius<br>(3) |
| --- | --- | --- |
| 0 | 0 | 0 |
| 1 | 0.002182 | 3.81E−05 |
| 3 | 0.006545 | 0.000343 |
| 5 | 0.01091 | 0.000952 |
| 10 | 0.02183 | 0.003805 |
| 15 | 0.032772 | 0.008555 |
| 20 | 0.043744 | 0.015192 |
| 30 | 0.065826 | 0.034074 |
| 40 | 0.088163 | 0.060307 |
| 50 | 0.110847 | 0.093692 |
| 60 | 0.133975 | 0.133975 |
| 75 | 0.169727 | 0.206647 |

$$z' = z'' + \Delta R \times e^{i \times \Delta} \tag{5.36b}$$

$$z = z' + \frac{R_1^2}{z'} \tag{5.36c}$$

Equation (5.36b) is a translation of the R-circle centre by the complex number $\Delta R \times e^{i \times \Delta}$ while the R- and $R_1$-circles intersect at one point on the x'-axis. The conformal transformation is sketched in Figure 5.14.

In the z-diagram, the R-circle is a streamlined airfoil whose shape is defined by:

$$z = R \times e^{i \times \theta''} - \Delta R \times e^{i \times \Delta} + \frac{R_1^2}{R} \times \frac{1}{e^{i \times \theta''} - \frac{\Delta R}{R} \times e^{i \times \Delta}} \quad \text{Surface of airfoil } 0 \leq \theta'' \leq 2 \times \pi \tag{5.37}$$

The dominant features of the Joukowski airfoil are an upstream round nose corresponding to $\theta'' = \pi$, a cusped tailing edge for $\theta'' = 0$ and some camber.

---

**Discussion**

Figure 5.14 and Equation (5.36) define the profile of a Joukowski airfoil. The shape is named after the Russian mathematician Nikolai Egorovich Joukowski (1847–1921) [2]. Simple examples of the Joukowski profile include the flow past an ellipse, past a streamlined strut and past an circular arc (see above).

Some basic properties of the Joukowski airfoil include:

(a)  At the tailing edge, the angle of intersection between the upper and lower surfaces of the airfoil is zero. (This is unfeasible in practice for construction reasons.)
(b)  The chord length is greater than $4 \times R_1$. In fact, the Joukowski airfoil profile is based upon a circular arc of chord $4 \times R_1$ for which $\Delta y = \Delta R \times \sin \Delta$.
(c)  The thickness of the foil increases with increasing distance $\Delta R$.
(d)  The camber of the foil increases with increasing distance $\Delta R \times \sin \Delta$.

---

[1]His name is also spelled Zhukovsky or Zhukoskii.

## Notes

- The Joukowski airfoil (Fig. 5.14) is also called a cambered Joukowski airfoil. The symmetrical Joukowski airfoil is the streamlined strut (see above, paragraph 3.5).
- The Joukowski profile is an extension of the circular arc. It is based upon the circular arc of chord $4 \times R_1$ and defined by $\Delta y = \Delta R \times \sin\Delta$.

### Extension of the Joukowski airfoil

A modified profile may be obtained by replacing Equation (5.36c) by:

$$\frac{z + n \times R_1}{z - n \times R_1} = \left(\frac{z' + n \times R_1}{z' - n \times R_1}\right)^n \tag{5.38}$$

The resulting profile has a finite angle $\lambda$ at the trailing edge where:

$$\lambda = \pi \times (2 - n)$$

For $n = 2$, Equation (5.38) yields:

$$z = z' + \frac{R_1^2}{z'} \tag{5.36c}$$

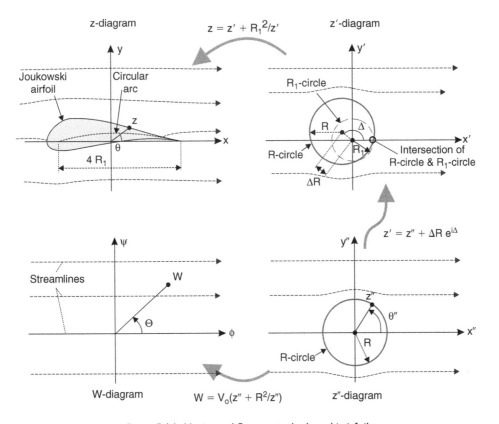

Figure 5.14   Horizontal flow past a Joukowski airfoil.

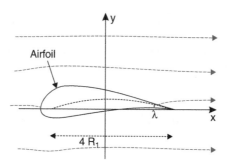

*Figure 5.15*  Horizontal flow past a modified Joukowski airfoil.

**Remarks**

1. The extended Joukowski airfoil is attributed to von Karman and Trefftz (Glauert 1924, Milne-Thomson 1958).
2. A generalised form of the Joukowski airfoil can be derived for n slightly less than 2 (Glauert 1924).

# Exercises

## 5.1 Complex numbers (1)

Find the modulus and argument of the following complex numbers:

$$z = 3 + 2 \times i$$
$$z = (1 + i) \times (4 - 2 \times i)$$
$$z = Ln(i)$$
$$z = Ln(1 + i)$$

### Solution

$$|z| = 3.61, \theta = 33.7°$$
$$|z| = 8.72, \theta = 18.43°$$
$$|z| = 1, \theta = \pi/2$$
$$|z| = 1.61, \theta = 77.6°$$

## 5.2 Complex potential

For the following complex potential, draw the equipotentials and streamlines in the {x, y} plane:

$$W = z^2$$
$$W = 2 \times Ln(z + i)$$
$$W = 5/z$$
$$W = 3 \times z - 2 \times Ln(z - 1)$$

### Solution

Flow past a 90° angle at the origin
  Source located at $z = -i$
  Doublet at origin
  Flow past a half-body at $z = 1$

## 5.3 Complex numbers (2)

Demonstrate the following relationships:

$$\sinh(i \times x) = i \times \sin(x)$$
$$\cosh(x) = \cos(i \times x)$$
$$d(\operatorname{sech}^{-1}(x)) = \frac{-dx}{x \times \sqrt{1 - x^2}}$$

## 5.4   Complex velocity (1)

Considering the flow past a horizontal cylinder placed at the origin, demonstrates that:

$$z = \frac{W \pm \sqrt{W^2 - 4 \times V_O^2 \times R^2}}{2 \times V_O}$$

For the same flow pattern, calculate the complex velocity.

## 5.5   Complex velocity (2)

Considering the ebb flow from a bay into a narrow inlet $(z = 16 \times \cosh(W))$, calculate the flow velocity at the inlet throat.

## 5.6   Application (1)

A kite has an effective surface area of $1.2\,m^2$ and its weight is $0.40\,kg$. The kite is rigged so that its surface is held at an angle $(\alpha + \delta) = 62°$ (Fig. 5.E1). During a wind of $42\,km/h$, the pull on the cord is $3\,kg$ and the angle of incidence of the kite $\alpha$ is $15°$. Calculate the corresponding coefficients of lift and drag.

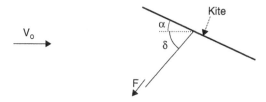

*Figure 5.E1*   Sketch of a flying kite.

## 5.7   Application (2)

A glider flies with a a line of descent of $4°$ to the horizontal and it reaches a maximum speed of $150\,km/h$. If the glider's gross weight is $132\,kg$, calculate the magnitude of the corresponding lift and drag forces.

### Solution

This is a problem of simple geometry. At maximum speed, the forces acting on the glider are at equilibrium. These are the weight force, the lift and drag. It yields:

$$\text{Lift} \times \sin 5° = \text{Drag} \times \cos 5°$$
$$9.80 \times 132 = \text{Lift} \times \cos 5° + \text{Drag} \times \sin 5°$$

Hence:

$$\text{Lift} = 1289\,N$$
$$\text{Drag} = 112.7\,N$$

# Joukowski transformation, theorem of Kutta-Joukowski and lift force on airfoil

## SUMMARY

The transformation of Joukowski is presented. The calculations of lift force on air foil are developed. The theorem of Kutta-Joukowski is introduced.

## I PRESENTATION

In the previous chapter, the horizontal flow past a circular arc and a Joukowski airfoil was presented (Chap. I-5, sections 3.6 & 3.7). In each case, there was no circulation in the z-diagram and the lift force on the object was zero. This approximation is incorrect in practice. Some lift is experienced on the airfoil and circular arc, more generally on any profile that is not symmetrically formed or situated with respect to the flow direction. Physically this is related to the flow next to the trailing edge of the airfoil or arc (Fig. 6.1). In absence of circulation, a stagnation point is observed on the nose and on the upper surface next to the trailing edge (Fig. 6.1A). At the rear end, the velocity is non-zero and the cusp of the trailing edge induces a discontinuity in terms of velocity. That is, the trailing edge is a singularity.

### Joukowski hypothesis

At the trailing edge of an airfoil, there is a discontinuity in velocity in the general cases. The Joukowski hypothesis states that **the circulation around an airfoil always adjusts itself so that velocity is finite at the trailing edge** (Streeter 1948). This hypothesis is well substantiated experimentally.

Figure 6.1 illustrates the addition of circulation to a horizontal flow past a circular arc, shifting the rear stagnation to the trailing edge and hence removing the singularity (Fig. 6.1B).

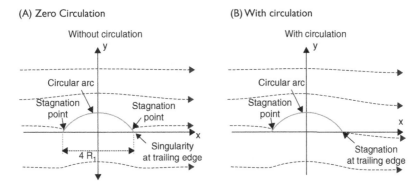

*Figure 6.1* Flow past a circular arc.

### Discussion

The Joukowski hypothesis eliminates the singularity at the trailing edge. More importantly it provides a means of calculating analytically the circulation and the lift force. All the developments are based upon the key assumptions of (1) an infinitely wide object (i.e. two-dimensional flow) and (2) the absence of flow separation. The latter is true for small angles of incidences.

Practically, the Joukowski hypothesis allows the addition of some circulation around an object to remove flow singularity at the rear end and to calculate the lift force. The simplest example is the flow past a circular cylinder (Chapter I-4). By adding some circulation, a lift force develops: i.e., the Magnus effect (Chapter I-4, section 3.7).

In Chapter I-2 (section 2.2), it was shown that rotation or lack of rotation of the fluid particles is a property of the fluid itself and not its position in space. Under the same assumptions underlying irrotational flow motion of ideal fluid, Kelvin's circulation theorem implies that the circulation around a vortex does not change with time regardless of the fluid motion. In a real fluid, separation is observed at the trailing edge of the flow past a foil (e.g. Fig. 6.1B). Viscous action produces an anticlockwise vortex developing in the wake of the foil with a circulation of its own. As Kelvin's circulation theorem suggests zero circulation, an artificial circulation must be introduced clockwise in Fig. 6.1: i.e., of equal magnitude but opposite direction to the vortex at the trailing edge. This artificial circulation ensures zero total circulation around the foil to satisfy Kelvin's circulation theorem (see section 3.1, in this chapter).

---

**Remark**

William Thomson (1824–1907), Baron Kelvin of Largs, was a British physicist. He contributed to the development of the second law of thermodynamics, the absolute temperature scale (measured in Kelvin), the dynamical theory of heat, fundamental work in hydrodynamics ...

---

## 2  TRANSFORMATION OF JOUKOWSKI

### 2.1  Presentation

The transformation of Joukowski is an extension of the transformation of the circle (Chapter I-5, section 3). In the Joukowski transformation, the velocity potential equals:

$$W = V_O \times \left( z' + \frac{R^2}{z'} \right) + \frac{i \times K}{2 \times \pi} \times Ln(z') \tag{6.1}$$

The main difference with the transformation of the circle is the last term: i.e., a circulation term. Vallentine (1969, pp. 177–183) discussed some aspects of the transformation.

### Notes

– The transformation of Joukowski is called after the Russian mathematician Nikolai Egorovich Joukowski (1847–1921). It is also called the Kutta-Joukowski transformation. Joukowski applied the transformation for the study of flow past airfoils.
– Martin Wilhelm Kutta (1867–1944) was a German engineer, known also for his contribution to the numerical solution of differential equations: i.e., the Runge-Kutta method.

– For the flow pattern sketched in Figure 6.1, the artificial circulation must be clockwise to bring the stagnation point at the trailing edge. And K must be negative in Equation (6.1).

## 2.2  Flow around a circular arc (with circulation)

Considering the basic transformations:

$$W = V_O \times \left( z''' + \frac{R^2}{z'''} \right) + \frac{i \times K}{2 \times \pi} \times \text{Ln}(z''') \qquad (6.2a)$$

$$z'' = z''' \times e^{i \times \alpha} \qquad (6.2b)$$

$$z'' = z' - i \times \Delta y \qquad (6.2c)$$

$$z = z' + \frac{R_1^2}{z'} \qquad (6.2d)$$

the first transformation (Eq. (6.2a)) is the transformation of Joukowski: i.e., that of a circular cylinder (radius R) with circulation (strength K) (Fig. 6.2). The second transformation is a rotation by an angle $\alpha$ that will be the angle of incidence (Eq. (6.2b)). The third transformation (Eq. (6.2c)) is a translation of the R-circle along the imaginary axis (see also Chapter I-5, section 3.6). In the z'-diagram, the $R_1$-circle is defined with its origin at $x' = 0$ and $y' = 0$, and its radius equals: $R_1 = R - \Delta y$. The $R_1$-circle and R-circle intersect on the x'-axis. The fourth transformation (Eq. (6.2d)) transforms the R-circle into a circular arc of chord $4 \times R_1$ and camber $2 \times \Delta y$.

With this conformal transformation, the R-circle in the $z'''$-diagram is transformed into a circular arc with an angle of incidence or angle of attack $\alpha$. There are two stagnation points corresponding to (See Chapter I-4, section 3.7):

$$\sin \theta''' = \frac{K}{4 \times \pi \times R \times V_O} \qquad \text{Stagnation points} \qquad (6.3)$$

The principal problem is that the flow pattern does have a singularity at the trailing edge (Fig. 6.2), unless an appropriate value of the circular K is selected to make the trailing edge a stagnation point.

---

**Remarks**

1.  For the flow sketched in Figure 6.2, the circulation K is negative and the angle of incidence $\alpha$ is positive.
2.  The intersections of the R-circle and $R_1$-circle in the z'-diagram correspond to the leading and trailing edges of the arc in the z-diagram (Fig. 6.3).
3.  The angle of incidence or angle of attack is the angle between the approaching flow velocity vector and the chordline.

---

If the stagnation point is brought extremely close to the trailing edge, it remains a stagnation point and this point is valid for the evaluation of the circulation K. When the rear stagnation point coincides with the trailing edge on an airfoil with a cusped trailing edge, the velocity at the

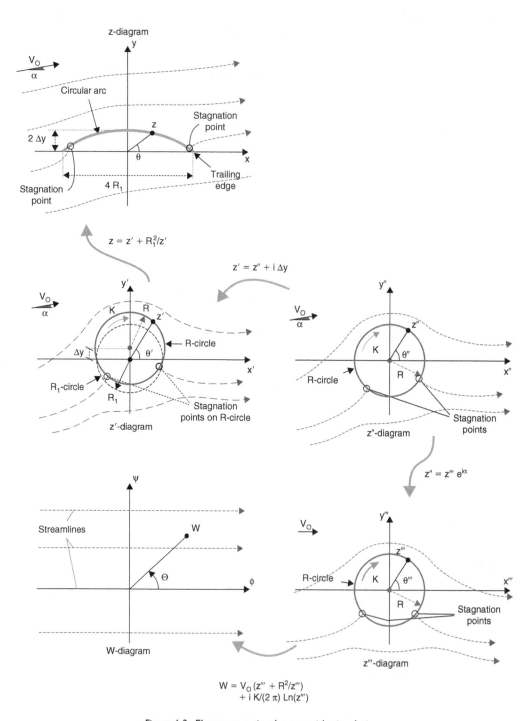

*Figure 6.2* Flow past a circular arc with circulation.

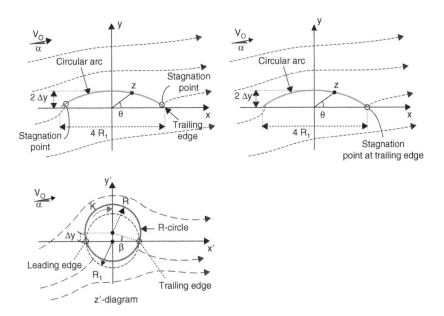

*Figure 6.3* Flow past a circular arc with circulation – On the right sketch, the stagnation is located at the trailing edge of the circular arc.

stagnation point is no longer zero (Vallentine 1969) (Fig. 6.3). In other words, the vortex strength K is selected to achieve a finite velocity at the trailing edge of the circular arc.

The leading and trailing edges of the circular arc are the points $(-R_1, 0)$ and $(+R_1, 0)$ in the $z'$-diagram (Fig. 6.3). That is, the intersection of the R-circle with the $R_1$-circle. The trailing edge of the arc is a stagnation point if the value of K satisfies that the complex velocity for the flow past the profile is zero:

$$w = \frac{dW}{dz} = 0 \tag{6.4}$$

where

$$\frac{dW}{dz} = \frac{dW}{dz'''} \times \frac{dz'''}{dz''} \times \frac{dz''}{dz'} \times \frac{dz'}{dz}$$

Using Equation (6.2), it yields:

$$\frac{dW}{dz} = \left( V_O \times \left( 1 - \frac{R^2}{z'''^2} \right) + \frac{i \times K}{2 \times \pi \times z'''} \right) \times e^{-i \times \alpha} \times \frac{1}{1 - \dfrac{R_1^2}{z'^2}} = 0$$

If the trailing edge $(z' = R_1)$ is a stagnation point, the circulation must satisfy:

$$V_O \times \left( 1 - \frac{R^2}{z'''^2} \right) + \frac{i \times K}{2 \times \pi \times z'''} = 0$$

where:

$$z''' = (R_1 - i \times \Delta y) \times e^{-i \times \alpha} = R \times e^{-i \times (\beta + \alpha)}$$

and $\tan \beta = \Delta y / R_1$. After substitution, it yields the value of the circulation K which makes the trailing edge a stagnation point:

$$K = -4 \times \pi \times R \times V_O \times \sin(\alpha + \beta) \qquad (6.5)$$

---

**Remarks**

1.  By geometry, the angle $\beta$ satisfies:

$$\sin \beta = \frac{\Delta y}{R} \quad \text{and} \quad \cos \beta = \frac{R_1}{R}$$

since the trailing edge is at the intersection of the $R_1$- and R-circles, and $R^2 = R_1^2 + \Delta y^2$ (Fig. 6.3).

2.  It can be shown in a similar fashion that the leading edge of the circular arc is a stagnation point if:

$$K = -4 \times \pi \times R \times V_O \times \sin(\alpha + \beta)$$

where $\tan \beta = \Delta y / R_1$.

3.  The chord of the arc equals $4 \times R_1$ while the camber is $2 \times \Delta y$ (Chapter I-5, section 3.6).

---

### 2.3   Flow around a Joukowski airfoil (with circulation)

Considering the basic transformations:

$$W = V_O \times \left( z''' + \frac{R^2}{z'''} \right) + \frac{i \times K}{2 \times \pi} \times \text{Ln}(z''') \qquad (6.6a)$$

$$z'' = z''' \times e^{i \times \alpha} \qquad (6.6b)$$

$$z' = z'' + \Delta R \times e^{i \times \Delta} \qquad (6.6c)$$

$$z = z' + \frac{R_1^2}{z'} \qquad (6.6d)$$

the resulting conformal transformation is the flow past a Joukowski airfoil (Fig. 6.4). Note that the third transformation (Eq. (6.6c)) is translation of the R-circle by the complex number $\Delta R \times e^{i \times \Delta}$ with the R- and $R_1$-circles intersecting at one point on the $x'$-axis (see also Chapter I-5, section 3.7).

In the $z'''$-diagram, there are two stagnation points on the rotating R-circle for:

$$\sin \theta''' = \frac{K}{4 \times \pi \times R \times V_O} \qquad \text{Stagnation points} \qquad (6.3)$$

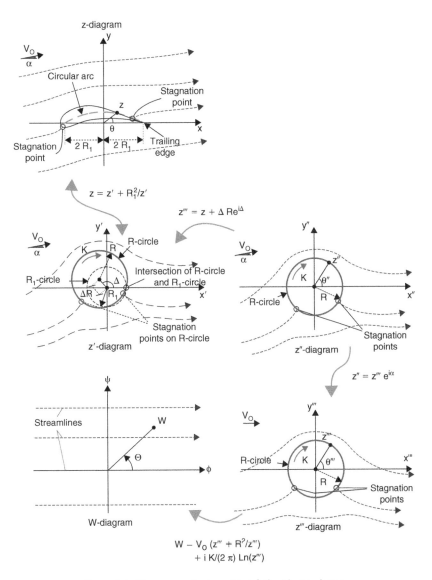

*Figure 6.4* Flow past a Joukowski airfoil with circulation.

The vortex strength K is selected to achieve a finite velocity at the trailing edge of the circular arc: i.e., the rear stagnation point coincides with the trailing edge on the airfoil. The trailing edge of the arc is a stagnation point if the value of K satisfies:

$$\frac{dW}{dz} = 0 \qquad\qquad (6.4)$$

Using Equation (6.6), the circulation must satisfy:

$$V_O \times \left(1 - \frac{R^2}{z'''^2}\right) + \frac{i \times K}{2 \times \pi \times z'''} = 0$$

where:

$$z''' = (R_1 - i \times \Delta y) \times e^{-i \times \alpha} = R \times e^{-i \times (\beta + \alpha)}$$

and the angle $\beta$ satisfies (Fig. 6.5):

$$R \times \sin \beta = \Delta R \times \sin \Delta$$

After substitution, the circulation K equals.

$$K = -4 \times \pi \times R \times V_O \times \sin(\alpha + \beta) \tag{6.7}$$

---

**Remarks**

1.  The above development is very close to that for the flow past a circular arc, but for the definition of the angle $\beta$ (Fig. 6.5):

$$\sin \beta = \frac{\Delta R}{R} \times \sin \Delta$$

2.  The Joukowski airfoil profile derives from that of a circular arc of chord $4 \times R_1$ and camber $2 \times \Delta y$, where: $\Delta y = R_1 \times \sin \Delta = R \times \cos \beta$.

---

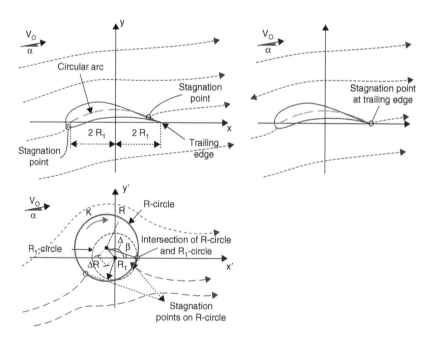

*Figure 6.5* Flow past a Joukowski airfoil with circulation – On the right, the stagnation is at the trailing edge.

# 3   KUTTA-JOUKOWSKI LAW

## 3.1   Kutta-Joukowski law

The Kutta-Joukowski law states that the magnitude of the lift force per unit width is $\rho \times V_O \times K$ where $\rho$ is the fluid density, $V_O$ is the free-stream velocity magnitude and K is the circulation around the solid. The law applies to circular cylinders and to airfoils, and also to any shape in two-dimensional irrotational flow (Vallentine 1969, pp. 179–181 & 281–283). The proof is based upon the application of the Blasius theorem and Cauchy integral theorem to two-dimensional irrotational flow.

A key issue in the application of Kutta-Joukowski law is the appropriate value for the circulation K. In the case of an airfoil, it is the value of K which makes the trailing edge a stagnation point. In real-fluid flows, skin friction and flow separation may be significant and the observed lift may differ from that predicted by Kutta-Joukowski law. For streamlined profiles (e.g. airfoil), the law is however in fair agreement with experimental observations for angles of attack up to 10°.

---

**Remarks**

1.   In 1906, the Russian mathematician Nikolai Egorovich Joukowski (1847–1921) published two papers in which he gave a mathematical expression for the lift on an airfoil. Today the theorem is known as the Kutta-Joukowski theorem, since W. Kutta pointed out that the equation appears also in his 1902 dissertation.
2.   The Kutta-Joukowski law forms the basis of the circulation theory of lift on aircraft wings, thrust of propeller blades, and transverse forces on unsymmetrical solid bodies and rotating balls and cylinders moving through a fluid. Figure 6.6 illustrates some modern aircrafts whose performances at low speed may be deduced from the Kutta-Joukowski law.

---

### *Blasius theorem*

Considering a two-dimensional object, the fluid pressure force acting on the object may be expressed as a complex number:

$$F_p = (F_p)_x + i \times (F_p)_y$$

where the subscript x and y refer to the x- and y-components of the pressure force.

The moment about the origin due to fluid pressure on the object is:

$$M_p = (M_p)_x + i \times (M_p)_y = \iint_{CS} P \times e^{-i \times 2 \times \theta} \times z \times dz$$

where P is the pressure and CS is the control surface or surface of the object. Basically the integration of the pressure force is carried out completely around the periphery of the object.

The Blasius theorem states that:

$$(F_p)_x - i \times (F_p)_y = \frac{i}{2} \times \rho \times \iint_{CS} \left( \frac{dW}{dz} \right)^2 \times dz \tag{6.8}$$

(A) L-1011 Tristar lifting off the Meadows Field Runway at Bakersfield, California, on its first flight on 21 May 1997 (Courtesy of Dryden Flight Research Center, Photo Number EC97-44077-3) – The experiment seeked to reduce fuel consumption of large jetliners by improving the aerodynamic efficency of their wings at cruise conditions

(B) Concorde F-BVFF "Fox Fox" at Roissy airport (France)

Figure 6.6  Photographs of modern aircrafts.

$$(M_p)_x + i \times (M_p)_y = -\frac{\rho}{2} \times \iint_{CS} z \times \left(\frac{dW}{dz}\right)^2 \times dz \tag{6.9}$$

where $\rho$ is the fluid density and $W$ is the complex potential.

---

**Remarks**

1.  H. Blasius (1883–1970) was German scientist, student and collaborator of Ludwig Prandtl.
2.  The Blasius theorem is derived from the Bernoulli equation. It is valid for steady flows (Streeter 1948, pp. 137–139).

### Cauchy integral theorem

The Cauchy theorem states that the integral of a regular complex function f(z) around any closed object is zero:

$$\oint\oint_{CS} f(z) \times dz = 0 \tag{6.10}$$

> **Remark**
>
> Augustin Louis de Cauchy (1789–1857) was a French engineer from the 'Corps des Ponts-et-Chaussées'. He devoted himself later to mathematics and he taught at Ecole Polytechnique, Paris, and at the Collège de France. He worked with Pierre-Simon Laplace and Joseph Louis Lagrange. In fluid mechanics, he contributed greatly to the analysis of wave motion.

## 3.2  Lift force on a circular arc

Considering the flow past a circular arc (section 2.2, in this chapter), the magnitude of the lift force per width of airfoil is $\rho \times V_O \times K$. It yields:

$$Lift = -\rho \times V_O \times K = 4 \times \pi \times R \times \rho \times V_O^2 \times \sin(\alpha + \beta) \tag{6.11}$$

where $\alpha$ is the angle of incidence and the angle $\beta$ is defined in Figure 6.3. Note that the lift force is positive for the flow sketched in Figures 6.2 and 6.3.

More generally the lift force is expressed as:

$$Lift = C_L \times Chord \times \frac{1}{2} \times \rho \times V_O^2 \tag{6.12}$$

where the lift coefficient $C_L$ is dimensionless. For a circular arc, the lift coefficient equals:

$$C_L = 2 \times \pi \times \sin(\alpha + \beta) \tag{6.13}$$

for an arc of chord:

$$Chord = 4 \times R$$

Since the angle $\beta$ is positive, the lift force is positive for $\alpha > -\beta$. That is, even for zero angle of attack.

### Discussion

The lift force is directly proportional to the density of the fluid $\rho$, the square of the approach velocity $V_O$ and the vortex circulation K (see also Chapter I-4, section 3.7). For small angles $\alpha$ and $\beta$, $\sin(\alpha + \beta) \approx \alpha + \beta$, and the lift force becomes:

$$Lift \approx 4 \times \pi \times R \times \rho \times V_O^2 \times (\alpha + \beta)$$

The lift force is proportional to the angle of incidence, or angle of attack, $\alpha$ and to the angle $\beta$ which is a measure of the foil camber. The camber equals $2 \times \Delta y$ (Chapter I-5, section 3.6) and the ratio camber to chord is $\Delta y / (2 \times R_1) = \sin(\beta/2) \approx \beta/2$.

Hence, the lift force per unit width on a circular arc is about:

$$\text{Lift} \approx \pi \times \text{Chord} \times \rho \times V_O^2 \times \left(\alpha + 2 \times \frac{\text{Camber}}{\text{Chord}}\right)$$

for small angles of incidence $\alpha$ with $\alpha$ in radians (Fig. 5.13). In dimensionless form, it yields an estimate of the lift coefficient:

$$C_L \approx 2 \times \pi \times \left(\alpha + 2 \times \frac{\text{Camber}}{\text{Chord}}\right)$$

### Application: lift force on a flat plate

The limiting case of a circular arc without camber ($\beta = 0$) is the flat plate (Fig. 5.E1). The lift force becomes:

$$\text{Lift} = \pi \times L \times \rho \times V_O^2 \times \sin\alpha \qquad\qquad \text{Flow past a flat plate with incidence}$$

where L is the plate length (or chord).

### 3.3    Lift force on a Joukowski airfoil

Considering the flow past a Joukowski airfoil (paragraph 2.3), the lift force on the foil equals:

$$\text{Lift} = 4 \times \pi \times R \times \rho \times V_O^2 \times \sin(\alpha + \beta) \tag{6.14}$$

where $\alpha$ is the angle of incidence and the angle $\beta$ is defined in Figure 6.5. The lift force is positive for the flow sketched in Figures 6.4 and 6.5.

The lift coefficient equals:

$$C_L = 2 \times \pi \times \sin(\alpha + \beta) \tag{6.15}$$

A comparison between the Kutta-Joukowski law and experimental results was presented in Figure 3.3. It shows a fair agreement in lift coefficient predictions for angle of incidence $\alpha$ of up to $10°$.

### Notes

- The airfoil chord equals $4 \times R$.
- The angle $(\alpha + \beta)$ is also called the effective angle of attack.
- Remember, the angle of incidence $\alpha$ is defined herein as the angle between the approaching flow velocity and the airfoil chordline.
- The angle $\beta$ is a measure of the airfoil camber. It is related to the lift for zero angle of incidence:

$$\sin\beta = \frac{C_L(\alpha = 0)}{2 \times \pi}$$

For a symmetrical Joukowski airfoil (Fig. 6.7), $\beta = 0$.
- Physically, the lift on a Joukowski airfoil is zero when the angle of incidence is $\alpha = -\beta$. In other words, $\beta$ is the angle between the chordline and zero lift direction (Mises 1945). Figure 6.8 presents the characteristics of two NACA airfoil profiles. $\beta$ equals $2°$ and $2.5°$ for the NACA2409 and 2415 profiles respectively.

*Figure 6.7*  Water flow past a cavitating Joukowski profile – Symmetrical Joukowski profile, 15% thickness, 5° incidence, V = 8 m/s, cavitation number: 1.5 – Flow from right to left – Note the cavitation bubble cloud on the extrados.

## Discussion

The Joukowski airfoil shape was commonly used in aircraft design during the first half of the 20th century. However, speed and manoeuvrability requirements imply not only good lift characteristics but also low drag and favourable pressure distributions over a broad range of angle of attack. In the particular case of water flows, separation may be associated with cavitation (Fig. 6.7). Further three-dimensional phenomena are not considered in the above analysis but are important in real-fluid flows.

Since the mid-20th century, profiles other than the Joukowski profiles were devised: e.g., the NACA profiles. These were evaluated from experimental data. A family of NACA profiles is determined by its thickness distribution and a midline shape. For the NACA four-digit series, the midline shape and thickness distributions are polynomials with maximum camber, maximum camber location and maximum thickness as specified characteristics. For example, the characteristics of the NACA 2409 and 2415 profiles are shown in Figure 6.8. For the NACA 2415 airfoil, the maximum camber is **2**% of the chord, the maximum camber occurs at **4**0% of the chord and the maximum thickness of the foil is **15**% of the chord. The numbers in **bold** yield 2-4-15, or profile number 2415. The website {http://www.pagendarm.de/trapp/programming/java/profiles/NACA4.html} gives a graphical description of the NACA four-digit series.

A comparison between Figures 3.3 and 6.8 gives some indication of the respective performances of a Joukowski airfoil and NACA profiles. The results are summarised in Table 6.1. Although the NACA profile gives slightly less lift than the Joukowski profile, the drag coefficient is significantly lower for an identical angle of incidence α and stall occurs at a greater angle of attack (Table 6.1). Figure 6.9 shows the V-173 experimental V/STOL aircraft whose basic wing area and planform were shaped with the NACA 0015 section.

(A) NACA profile 2409

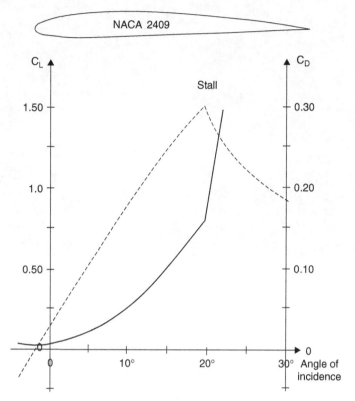

(B) NACA profile 2415

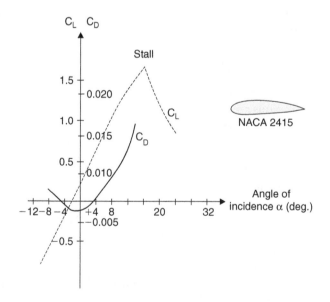

*Figure 6.8*  Drag and lift coefficient for NACA profiles.

*Table 6.1* Comparative performances of Joukowski and NACA airfoil profiles.

| Angle of incidence: | $\alpha = 5°$ | | $\alpha = 10°$ | | |
| --- | --- | --- | --- | --- | --- |
| Airfoil profile | $C_L$ | $C_D$ | $C_L$ | $C_D$ | Stall $\alpha$ |
| Joukowski profile (Fig. 3.3) | 1.2 | 0.04 | 0.75 | 0.003 | 10° |
| NACA 2409 profile (Fig. 6.8B) | 0.5 | 0.01 | 0.85 | 0.055 | 19° |
| NACA 2415 profile (Fig. 6.8A) | 1.5 | 0.085 | 1.2 | 0.011 | 18° |

(A) Chance-Vought V-173 plane at the Futuroscope, Poitiers on 17 Sept. 2008

(B) Maiden flight of the Vought V-173 flown by Test Pilot Boone Guyton on 23 November 1942 (Photograph US Navy)

*Figure 6.9* Chance-Vought V-173 experimental plane – Called Zimmerman "flying panacake" after its designer Charles Horton Zimmerman, this experimental V/STOL aircraft could take off almost vertically with a 45 km/h wind, or in 60 m with zero wind.

**Remarks**

1. NACA stands for National Advisory Council for Aeronautics which was the forerunner of the National Aeronautics and Space Administration (NASA).
2. A comparison between Figures 6.8A and 6.8B illustrate the influence of the profile thickness (also Table 6.1).
3. The drag coefficient is very sensitive to the surface roughness. It may increase up to 300% of its smooth surface value if the surface is rough. Figures 3.3 and 6.8 show some smooth surface data.
4. Another NACA profile designation is the NACA five-digit series. The profile shape are designated as: 22518 where **2** is related to the design lift coefficient ($3/2 \times 0.2 = 0.3$), **25** is linked with the maximum camber location ($1/2 \times 25 = 12.5\%$ chord) and **18** is the section thickness (**18%** chord).

## 3.4   Discussion

A modern application of airfoil is the wings of wind turbines. The aerodynamic lift provides the force responsible for the power yield generated by the turbine, and a high lift-to-drag ratio is selected for rotor blade design. The blade can have a lift-to-drag ratio of about 120 compared to 15 for a commercial airplane. In practice the use of a unique aerofoil shape for the entire wing length would result in an inefficient design. Each section has a different relative air velocity and structural requirement. Typically, at the root, the blade sections have thick profiles, with much thinner sections near the tips (Fig. 6.10).

Modern wind turbines have often 3 blades. Figure 6.10 shows the wings of wind turbines, during construction. The movie Plouarz1.mov presents 4 wind turbines in operation (App. F). For comparison, the movie Moidrey33.mov shows the operation of an old windmill.

*Figure 6.10* Wind turbine at St Aaron (France) on 22 August 2008 and details of the wings.

*Figure 6.10*  Continued.

# Exercises

## 6.1 Drag force

A wing having the lift and drag characteristics of Figure 3.3 has a span of 20.5 m and a chord of 2.1 m. What thrust would the propeller of the plane have to develop in order to overcome the total drag of the wing at a speed of 260 km/h and for an angle of attack of 3°?

## 6.2 Lift and drag forces (1)

Considering an aircraft with two wings, having the lift and drag characteristics of Figure 3.3, and a span of 12.2 m each and a chord of 4.1 m, what thrust would the engine of the plane have to develop in order to overcome the total drag of the wing at a speed of 560 km/h and for an angle of attack of 2°?

For that speed and angle of attack, what is the total lift force? What should be the weight of the aircraft to fly at this speed with this angle of attack?

## 6.3 Lift and drag forces (2)

(a) Plot the ratio of lift to drag forces against the angle of incidence for the data plotted in Figure 3.3. Indicate the angle of best performance.
(b) Plot the ratio of lift to drag forces against the angle of incidence for the data plotted in Figure 6.8. Indicate the angle of best performance.
(c) Discuss the main differences between the two profiles.

## 6.4 Lift and drag forces (3)

As a means of eliminating the wave resistance to a high-speed boat, the hulls are supported well above the water by streamlined struts connected to horizontal hydrofoils. If the hydrofoil has the characteristics plotted in Figure 6.8A, what must be its dimensions to support the load of a 3,200 kg boat at the most favourable lift-drag ratio when the high-speed boat is cruising at 35 knots?

## 6.5 Lift on a circular arc

Using the software FoilSimU III, calculate the lift coefficient of a circular arc with 5% camber and 3% thickness for angles of incidence between 0 and 30°.

The software FoilSimU III was developed by the NASA and can be obtained at {http://www.grc.nasa.gov/www/K-12/UndergradProgs/index.htm}.

# Theorem of Schwarz-Christoffel, free streamlines and applications

**SUMMARY**

The theory of free-streamlines and the theorem of Schwarz-Christoffel are presented. They are applied to flows with separation.

## I   PRESENTATION

In flow situations involving straight boundaries, the application of the theorem of Schwarz-Christoffel and of the theory free-streamlines may provide a technique to solve analytically the flow. A typical example is the flow through an orifice (Fig. 3.1A). Figure 7.1 illustrates some flow situations with separation which may be studied using the method of free streamlines. For example, the flow through a sharp orifice, the flow past a flat plate, and the flow past an object with separation. The movie IMGP0123.mov illustrates the flow past a triangular sharp-crested weir (App. F).

---

**Remarks**

1.   The theory of free-streamline was introduced by Hermann Helmholtz, Gustav Kirchhoff and others.
2.   Hermann Ludwig Ferdinand von Helmholtz (1821–1894) was a German scientist who made basic contributions to physiology, optics, electrodynamics and meteorology.
3.   Gustav Robert Kirchhoff (1824–1887) was a German physicist and mathematician who made notable contributions to spectrum analysis, electricity, the study of light and astronomy, but also to wave motion and viscous fluid flows.

---

### Basic definitions

The theorem of Scharwz-Christoffel introduces the concept of polygon. Herein a polygon is a closed plane figure bounded by straight lines. The term "simple closed polygon" defines a closed polygon such that (a) the polygon boundaries divide the whole z-diagram into two regions: the interior of the polygon and the exterior; and (b) the boundaries may be completely drawn without leaving the boundary. The interior of the polygon is said to be "connected". That is, the path from any point in the interior to another point in the interior may be followed without crossing a boundary. The same may be said from the exterior region.

Figure 7.2 shows simple closed polygons having vertices at infinity. The points at infinite are denoted with the subscript $\infty$ (e.g. $A_\infty$). Dotted arrows indicate the closure of the polygon.

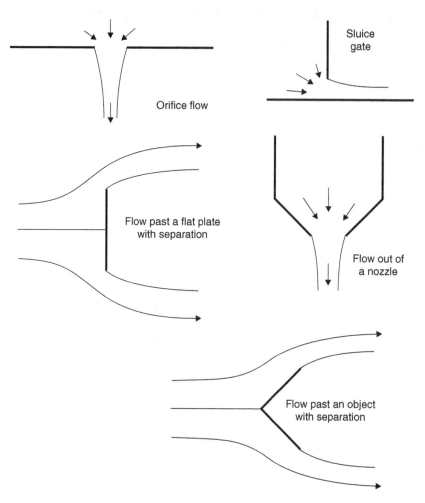

*Figure 7.1* Flows situations with separation.

The interior is in white and the exterior in grey colour. Usually the polygon is drawn in the z-diagram and the interior of the polygon is the flow region.

Two polygons that are often involved in problems requiring successive transformations are the semi-infinite strip and the infinite strip (Fig. 7.2). The semi-infinite strip is a rectangle with two vertices at infinity. The infinite strip a rectangle with two vertices at $+\infty$ and two vertices at $-\infty$.

## Notes

– A polygon is a closed plane figure bounded by straight lines. The word derives from the Greek name *polygonon* for polygon or *polygonos* for polygonal.
– A vertex (plural *vertices*) is a point of a polygon that terminates a line or comprises the intersection of two or more lines.

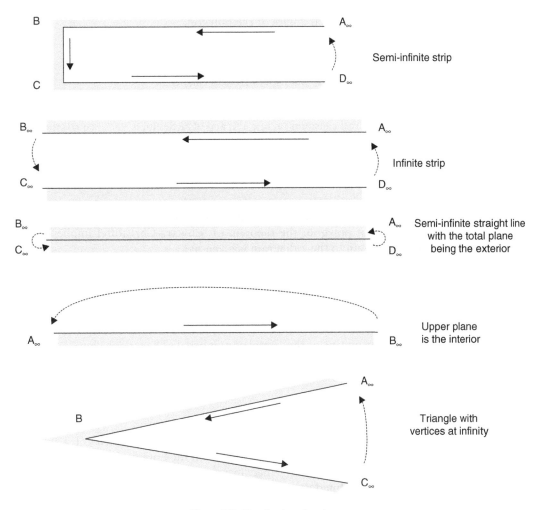

*Figure 7.2*  Simple closed polygons.

## 2  THEOREM OF SCHWARZ-CHRISTOFFEL

### 2.1  Theorem

The Schwarz-Christoffel theorem states that the interior of a simple closed polygon may be mapped ([1]) into the upper half of a plane and the boundaries of the polygon into the real axis (Streeter 1948). The transformation is given by:

$$\frac{dz}{dt} = \frac{A}{(a-t)^{\alpha/\pi} \times (b-t)^{\beta/\pi} \times (c-t)^{\gamma/\pi} \dots} \tag{7.1}$$

---

[1]That is, in the sense of conformal mapping (see Chapters I-5 and I-6).

where $A$ is a complex number in the z-diagram; a, b, c, ... are real constants in ascending order of magnitude; $\alpha$, $\beta$, $\gamma$, ... are the external angles of the polygon.

## Notes

- Hermann Amandus Schwarz (1843–1921) was a German mathematician.
- Elwin Bruno Christoffel (1829–1900) was a German mathematician who worked actively during the second half of the 19th century.
- In Equation (7.1), the number of real constants (a, b, c, ...) and of angles ($\alpha$, $\beta$, $\gamma$, ...) equals the number of vertices of the polygon.
- By simple geometry, the external angles of the polygon satisfy:

$$\alpha + \beta + \gamma + \cdots = 2 \times \pi$$

A polygon with four vertices is sketched in the t-diagram and in the z-diagram respectively in Figures 7.3 and 7.4. For that polygon, the theorem of Schwarz-Christoffel implies that the transformation from the t-diagram to the z-diagram is given by:

$$\frac{dz}{dt} = \frac{A}{(a-t)^{\alpha/\pi} \times (b-t)^{\beta/\pi} \times (c-t)^{\gamma/\pi} \times (d-t)^{\delta/\pi}}$$

For t real and $t < a$, the terms $(a-t)$, $(b-t)$, $(c-t)$ and $(d-t)$ are all reals (Fig. 7.3), and the argument of dz/dt equals the argument of the complex number $A$. The straight line $[t < a]$ in the t-diagram is transformed into a straight line in the z-diagram because

$$dz = k \times A \times dt \quad t < a$$

where k is a real.

The location $t = a$ corresponds to the vertex A. Note that dz/dt is not defined at $t = a$, b, c and d (Eq. (7.1)). The points A, B, C and D must be excluded from the boundaries.

For t real and $a < t < b$, the term $1/(a-t)^{\alpha/\pi}$ becomes complex because $(a-t)$ is negative and $\alpha < \pi$. In the t-diagram, $(a-t)$ is defined as:

$$a - t = r' \times e^{i \times \pi} \quad t > a$$

where $r'$ is the modulus of $(a-t)$. Hence:

$$\frac{1}{(a-t)^{\alpha/\pi}} = \frac{1}{r'^{\alpha/\pi}} \times e^{-i \times \alpha} \quad t > a \tag{7.2}$$

In the z-diagram, the argument of the term $1/(a-t)^{\alpha/\pi}$ is the angle $(-\alpha)$. That is, the straight line $[a < t < b]$, in the t-diagram, transforms into a straight line in the z-diagram and the deflection angle at the point A is equal to $\alpha$ since:

$$dz = k' \times A \times e^{-i \times \alpha} \times dt \quad t < a$$

where $k'$ is a real.

The reasoning may be extended to the segments $[b < t < c]$, $[c < t < d]$ and $[t > d]$ (Fig. 7.3 & 7.4).

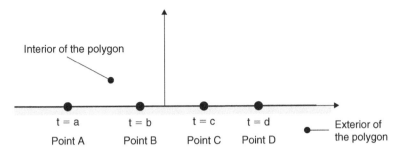

*Figure 7.3*  Polygon in the t-diagram.

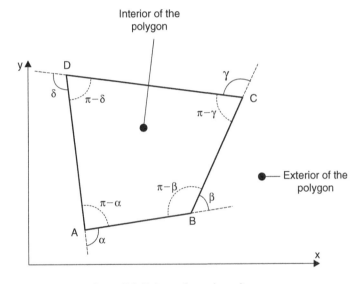

*Figure 7.4*  Polygon from the z-diagram.

The integration of Equation (7.1) yields:

$$z = A \times \left( \int \frac{1}{(a - t)^{\alpha/\pi} \times (b - t)^{\beta/\pi} \times (c - t)^{\gamma/\pi} \cdots} \times dt \right) + B \tag{7.3}$$

where $A$ and $B$ are complex constants. The complex number $A$ affects the scale and orientation of the polygon in the z-diagram while the complex number $B$ determines the location of the polygon with respect to the origin.

---

**Discussion**

1. When the vertex of a polygon corresponds to a point at infinity in the t-diagram, Equation (7.1) becomes independent of that point. Streeter (1948, pp. 162–163) and Vallentine

(1969, p. 195) derived the proof. For example, for a semi-infinite strip (Fig. 7.5), Equation (7.1) becomes:

$$\frac{dz}{dt} = \frac{A}{(b-t)^{\beta/\pi} \times (c-t)^{\gamma/\pi}}$$

The points $A_\infty$ and $D_\infty$ do not contribute to the transformation because they are at infinity.

2. In Equation (7.3), the modulus of the complex number $A$ affects the scale of the polygon and its argument determines the polygon orientation.

3. Practically, three of the real numbers a, b, c, d, ... may be selected arbitrarily while the remaining ones are determined by the shape of the polygon.

## 2.2 Applications

### 2.2.1 Semi-infinite strips

A simple example of the Schwarz-Christoffel transformation is the semi-infinite strip sketched in Figure 7.5 where two vertices are $+\infty$ and two vertices are on the imaginary axis in the z-diagram. The width of the strip is denoted l. The exterior angles or deflection angles are:

$$\alpha = \beta = \gamma = \delta = \frac{\pi}{2}$$

In the t-diagram, the points A, B, C and D are on the real-axis. The points A, B and C are arbitrarily selected such that:

$$a = -\infty$$
$$b = -1$$
$$c = +1$$

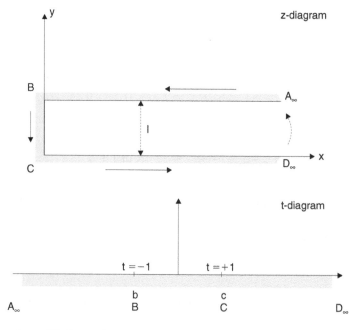

Figure 7.5 Semi-infinite strip mapped in the z-diagram and t-diagram.

The point D must also be at infinity and it yields $d = +\infty$. Equation (7.3) gives:

$$z = A \times \left( \int \frac{1}{(-(1+t))^{1/2} \times (1-t)^{1/2}} \times dt \right) + B \qquad (7.4)$$

It yields:

$$z = A \times \left( \int \frac{1}{\sqrt{t^2 - 1}} \times dt \right) + B$$

The integration gives (see App. D):

$$z = A \times \cosh^{-1}(t) + B$$

The constants $A$ and $B$ are deduced from the boundary conditions. At the point C, $z = 0$ and $t = +1$:

$$0 = 0 + B \quad \text{Point C } (t = +1)$$

Hence $B$ equals zero. At the point B, $z = i \times 1$ and $t = -1$, where $l$ is the strip width (Fig. 7.5). It yields:

$$i \times 1 = A \times \cosh^{-1}(-1) \quad \text{Point B } (t = -1)$$

and hence $A = l/\pi$.

The mapping function is:

$$z = \frac{1}{\pi} \times \cosh^{-1}(t) \qquad (7.5)$$

The interior of the semi-infinite strip covers the whole t-diagram above the real axis as sketched in Figure 7.5.

---

**Remarks**

1. In Equation (7.4), the points $A_\infty$ and $D_\infty$ do not contribute to the transformation because they are at infinity.
2. At the point B, the boundary condition yields:

$$\cosh\left(\frac{i \times 1}{A}\right) = -1$$

and hence:

$$\cos\left(\frac{1}{A}\right) = -1$$

since $\cosh(i \times x) = \cos(x)$ (see App. D). As a result, $1/A = \pi$.

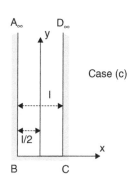

Figure 7.6 Semi-infinite strips mapped in the z-diagram.

---

**Discussion**

Figure 7.5 illustrates the simple case where C is at the origin and B is on the imaginary axis in the z-diagram. Figure 7.6 presents further examples of semi-infinite strips. In Case (a), the mapping function is:

$$z = \frac{1}{\pi} \times \cosh^{-1}(t) + z_C \quad \text{Case (a)}$$

where $z_C$ is the location of the point C in the z-diagram and l is the strip width.

In Case (b) (Fig. 7.6), the strip axis is the real axis in the z-diagram. It is an application of the case (a) for the particular case $z_C = -i \times l/2$. The mapping function is:

$$z = \frac{1}{\pi} \times \cosh^{-1}(t) - i \times \frac{1}{2} \quad \text{Case (b)}$$

In Case (c) (Fig. 7.6), the semi-infinite strip is vertical. The mapping function is:

$$z = \frac{1}{\pi} \times \sin^{-1}(t) \quad \text{Case (c)}$$

---

### 2.2.2   Infinite strips

Considering the infinite strip sketched in Figure 7.7, the width of the strip is l and the vertices $B_\infty$ and $C_\infty$ are on the real axis in the z-diagram. The exterior angles or deflection angles are:

$$\alpha = \beta = \gamma = \delta = \pi/2$$

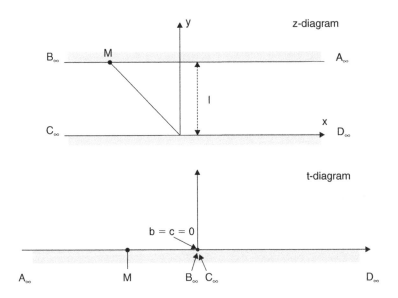

*Figure 7.7* Infinite strip mapped in the z-diagram and t-diagram.

In the t-diagram, the points A, B, C and D must be on the real-axis. The points A, B and C are arbitrarily selected such that:

$$a = -\infty$$

$$b = 0$$

$$c = 0$$

The point D must also be at infinity and it yields $d = +\infty$. After substitution of a, b, c, d, α, β, γ and δ, Equation (7.3) yields:

$$z = A \times \left( \int \frac{1}{t} \times dt \right) + B = A \times \text{Ln}(t) + B \qquad (7.6)$$

The complex constants $A$ and $B$ are deduced from the boundary conditions. At the vertex $D_\infty$, $z = +\infty$ and $t = +\infty$. It yields:

$$+\infty + i \times 0 = A \times \text{Ln}(+\infty) + B \quad \text{Point } D_\infty$$

The condition is satisfied by $B$ being a real. Arbitrarily, let set $B = 0$.
    At the vertex $A_\infty$, $z = +\infty + i \times l$ and $t = -\infty$. It yields:

$$+\infty + i \times l = A \times \text{Ln}(-\infty) = A \times (+\infty + i \times \pi) \quad \text{Point } A_\infty$$

The boundary condition is satisfied for $A = l/\pi$.

The mapping function of the infinite strip sketched in Figure 7.7 is then:

$$z = \frac{1}{\pi} \times Ln(t) \tag{7.7}$$

---

**Remarks**

1.  The constant $B$ may be deduced from a different manner. Considering the point on the boundary $C_\infty - D_\infty$ such that $t = 1$, Equation (7.6) yields:

    $$x = A \times Ln(1) + B = 0 + B$$

    Hence $B$ must be a real constant. If the origin in the z-diagram (i.e. $z = 0$) corresponds to $t = 1$, then $B = 0$.

2.  If the infinite strip is displaced vertically by a distance $i \times Y$ from the position shown in Figure 7.7, the mapping function becomes:

    $$z = \frac{1}{\pi} \times Ln(t) + i \times Y$$

---

*Application: uniform flow in an infinite strip*

An uniform flow in an infinite strip may be modelled by an infinite strip with a source $(+q)$ at $-\infty$ and a sink of equal and opposite strength $(-q)$ located at $+\infty$ (Fig. 7.8). The uniform flow velocity is $V = q/l$ where $l$ is the strip width. In the t-diagram, the corresponding flow pattern is located in the upper-half of the t-diagram with the source located at the origin: i.e., the fluid is injected at the origin with a flow rate $q$ in the half-plane. In the t-diagram, the flow pattern is that of a source of strength $2 \times q$ (Chapter I-4, section 2.2). The complex potential is (see also Chapter I-5, section 2.1):

$$W = -\frac{2 \times q}{2 \times \pi} \times Ln(t)$$

That is:

$$W = -\frac{q}{\pi} \times Ln(t) = \frac{1}{\pi} \times Ln(t)$$

according to Equation (7.7).

The width of the strip equals the absolute value of the discharge.

## 3   THEORY OF FREE STREAMLINES

### 3.1   Presentation

In ideal fluid flows, the fluid acceleration becomes infinite at a sudden corner with zero radius of curvature and the velocity becomes infinite. Such a situation is not physical. In real fluid flows,

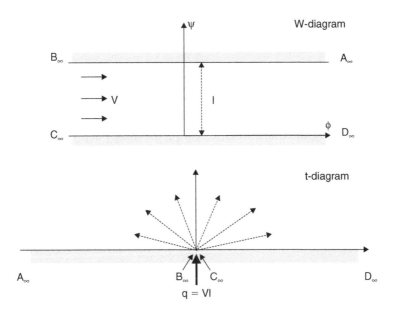

*Figure 7.8* Uniform flow in an infinite strip.

separation takes place. The theory of free-streamlines assumes that the ideal fluid separates from the boundary rather than makes a sharp turn. Separation occurs at those points where the body form makes a sharp change in direction. (This excludes however stagnation points and points located downstream of the first separation.) At separation in a steady flow of ideal fluid, the streamlines leave the body. The dividing streamline is called the free-streamline. The flow region separated from the main stream is called the wake. The (ideal) fluid in the wake is assumed to be at rest in steady flow situations.

If the effects of gravity are ignored ([2]), the pressure in the wake is constant since the fluid is at rest. According to the Bernoulli principle, the velocity along the free-streamline must be constant because the pressure on the streamline is constant itself. That is, the complex velocity has constant modulus along the free streamline.

---

**Discussion**

In real fluid flows, the fluid in the wake is not at rest. The assumption that the (ideal) fluid is at rest in the wake region is inaccurate. As a result, theoretical estimates of drag force in ideal fluid flow conditions often underestimate the real drag force (e.g. paragraph 3.4).

When the wake contains a fluid of lesser density than the main stream flow, the theory of free-streamlines may give results that compare favourably with experiments. A typical example is a water jet discharging into air from an orifice.

---

[2]This will be the case in the following developments.

---

**Remarks**

1.  The term "free streamline" refers to the characteristic streamline issuing from the separation point.
2.  The term bounding streamline refers to a streamline in contact with a boundary upstream of the separation point.

---

In free-streamline problems, the transformations are conformal, but it is a difficult topic that may involved as many as six to eight successive transformations. Usually, it is more convenient to start with the z-diagram. In most problems, four basic transformations are used. A typical example is the orifice flow (section 3.2):

$$\zeta = -\frac{dz}{dW}$$

$$\xi = Ln(\zeta)$$

$$t = \cosh(\xi - \xi_C)$$

$$W = \frac{q}{\pi} \times Ln(t)$$

Straight boundary lines in the z-diagram become radial lines in the $\zeta$-diagram and horizontal lines in the $\xi$-diagram because the velocity direction $\theta_V$ is constant. Free-streamlines in the z-diagram become circular arcs in the z-diagram and vertical lines in the $\xi$-diagram because the velocity modulus is constant along free-streamlines. Next the equation of the free-streamline(s) in the z-diagram must be determined. On the free-streamline, the velocity modulus is constant and the stream function is constant. That is:

$$\xi = Ln(\zeta) = Ln\left(\frac{1}{|V_\infty|}\right) + i \times \theta_V$$

$$\delta W = \delta(\phi + i \times \psi) = \delta\phi = -V \times \delta s$$

The latter equation derives from the definition of the velocity potential $\phi$ (Chapter I-3, section 2.1).

These transformations and the equations of free-streamlines are detailed for the flow through an orifice with separation and further examples.

## 3.2   Flow through an orifice with separation

Considering a two-dimensional orifice at the bottom of a infinitely large tank, the orifice has a length l. Downstream of the orifice, the free jet thickness tends to d and the jet velocity at infinity is $V_\infty$ (Fig. 7.9). By continuity, the flow rate is $q = V_\infty \times d$. At the orifice edges, separation must occur: i.e., points B and C in Figure 7.9. The problem is to find the profile of the free streamline. Along the free-streamline, the velocity is constant and it equals the jet velocity at infinity $V_\infty$. The conformal transformation consists of several successive simple transformations sketched in Figures 7.12 to 7.15.

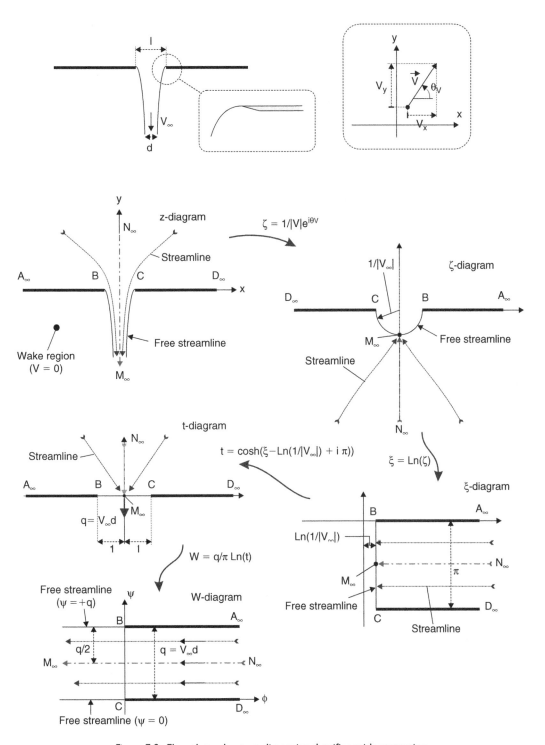

*Figure 7.9*  Flow through a two-dimensional orifice with separation.

Orifice flows were used as water clock, called clepsydra, in ancient Babylon and Egypt ([3]) as well as in parts of Africa and by some North America Indians (Fig. 7.10). They were in use up to the 16-th century. Today the sand glass ([4]) uses the same principle with granular material. Orifices and nozzles are used also as measuring discharges. In his study of orifice flows, J.C. de Borda (1733–1799) made a significant contribution by not only introducing the concept of streamlines but also by developing the "Borda" mouthpiece to measure accurately the orifice flow. A related form of orifice is the sharp-crested weir commonly used for discharge measurement in open channels (e.g. Chanson 1999, pp. 322–323). Figure 7.11 illustrates a triangular sharp-crested weir located at one end of a large tank, and the movie IMGP0123.mov shows its operation (App. F).

When waters flow through a sharp-edged orifice, the jet flow contracts to have its smallest section a small distance downstream of the hole. For a horizontal jet, Bernoulli principle implies that the velocity at vena contracta (i.e. $V_\infty$) equals $\sqrt{2 \times g \times H}$ where H is the reservoir head above orifice centreline (Fig. 7.16). This relationship is called Torricelli theorem, after Evangelista Torricelli (1608–1647) who discovered it in 1643. By continuity, the orifice discharge equals:

$$Q = C_d \times A_O \times \sqrt{2 \times g \times H}$$

where $A_O$ is the orifice cross-section area. The discharge coefficient $C_d$ may be expressed as:

$$C_d = C_c \times C_v$$

where the velocity coefficient $C_v$ account for the energy losses and the contraction coefficient $C_c$ equals $A/A_o$, A being the jet cross-section area at vena contracta. For a water jet discharging horizontally from an infinite reservoir, $C_c$ equals 0.58 and 0.61 for axisymmetrical and two-dimensional jets respectively ([5]). For example, Hunt (1968) for axisymmetrical jets and Mises (1917) for two-dimensional jets. The result for round orifice was proposed first by E. Trefftz. For two-dimensional jets, $C_c = \pi/(\pi + 2)$ (see next section). This result was derived independently by Joukowski and Michell in 1890 (Joukowski 1890, Michell 1890). The value of $C_c$ increases with increasing relative nozzle area when the reservoir is of finite dimensions.

---

**Remarks**

1. Hero of Alexandria was a Greek mathematician (1st century A.D.) working in Alexandria, Egypt. He wrote at least 13 books on mathematics, mechanics and physics. He designed and experimented the first steam engine. His treatise "Pneumatica" described Hero's fountain, siphons, steam-powered engines, a water organ, and hydraulic and mechanical water devices. It influenced directly the waterworks design during the Italian Renaissance. In his book "Dioptra", Hero stated rightly the concept of continuity for incompressible flow: the discharge being equal to the area of the cross-section of the flow times the speed of the flow.

2. Evangelista Torricelli (1608–1647) was an Italian physicist and mathematician who invented the barometer. From 1641, Torricelli worked with the elderly astronomer Galileo

---

[3]Hero of Alexandria (1st century A.D.) wrote a treaty on water clock in four books.
[4]For example, the egg timer.
[5]These results were obtained for sharp-edged orifices.

and was later appointed to succeed him as professor of mathematics at the Florentine Academy.

3.  Richard von Mises (1883–1953) was a Austrian scientist who worked on fluid mechanics, aerodynamics, aeronautics, statistics and probability theory in Germany, Austria, Turkey and USA. During World War I, he flew as test pilot and instructor in the Austro-Hungarian army.

(A) Sketch

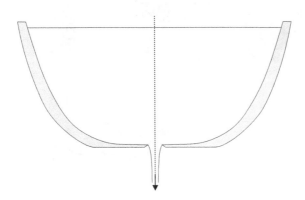

(B) Section shape of a clepsydra with an orifice diameter of 4 mm and an emptying time of 23 hours

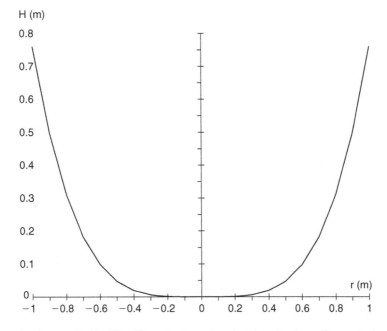

*Figure 7.10* Clepsydra (water clock) – The Clepsydra is equipped with a circular orifice at the bottom of the tank.

*Figure 7.11* Triangular sharp-crested weir overflow.

### Transformation from z-diagram to ζ-diagram

The transformation is defined as:

$$\zeta = -\frac{dz}{dW} \tag{7.8}$$

where W is the complex potential (Chapter I-5). Basically $\zeta$ is the inverse of the complex velocity:

$$\zeta = \frac{-1}{-V_x + i \times V_y} = \frac{1}{|V|} \times e^{i \times \theta_V}$$

where $|V|$ is the modulus of the local velocity and $\theta_V$ is the local velocity direction ([6]), also called the argument of the velocity vector in the z-diagram (Chapter I-5, section 1.3).

In the z-diagram, straight boundaries are streamlines because all solid boundaries are streamlines. In the ζ-diagram, straight solid boundaries become radial lines. In the z-diagram, the velocity modulus and flow direction at the point $A_\infty$ are 0 and 0 respectively. In the ζ-diagram, the point $A_\infty$ is also located at infinity and at $\theta' = 0$. In the z-diagram, the velocity at the point B is that on the free-streamline (i.e. $V_\infty$) and the velocity direction is $\theta_V = 0$. Hence the straight boundary $A_\infty$-B becomes a straight semi-infinite horizontal line in the ζ-diagram (Fig. 7.12). The same reasoning may be applied to the straight boundary $C$-$D_\infty$.

---

[6] That is, $\theta_V$ is the angle of the velocity vector with the horizontal real axis in the z-diagram.

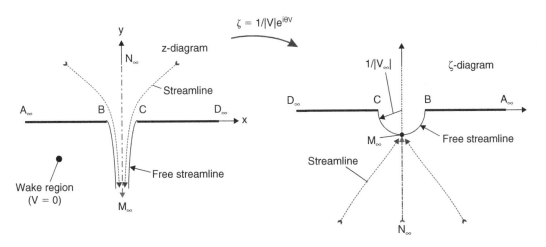

*Figure 7.12*  Flow through a two-dimensional orifice: transformation from z-diagram to ζ-diagram.

*Table 7.1*  Transformation from z-diagram to ζ-diagram for the flow through a two-dimensional orifice.

| Point in the z-diagram (1) | Velocity | | Complex number in ζ-diagram | |
|---|---|---|---|---|
| | Modulus $|V|$ (2) | Direction $\theta_V$ (3) | Real part (4) | Imaginary part (5) |
| $A_\infty$ | 0 | 0 | $+\infty$ | 0 |
| B | $|V_\infty|$ | 0 | $\dfrac{1}{|V_\infty|}$ | 0 |
| C | $|V_\infty|$ | $-\pi$ | $-\dfrac{1}{|V_\infty|}$ | 0 |
| $D_\infty$ | 0 | $-\pi$ | $-\infty$ | 0 |
| $M_\infty$ | $|V_\infty|$ | $-\pi/2$ | 0 | $-\dfrac{1}{|V_\infty|}$ |
| $N_\infty$ | 0 | $-\pi/2$ | 0 | $-\infty$ |

Free-streamlines are lines of constant velocity magnitude. Hence free-streamlines are circular arcs of radii $1/|V|$ in the ζ-diagram. For the flow from a two-dimensional orifice, the circular arc radius is $1/|V_\infty|$ (Fig. 7.12). The streamline $N_\infty-M_\infty$ in the z-diagram becomes a semi-infinite line in the ζ-diagram as sketched in Figure 7.12 (Table 7.1).

## Notes

– The complex velocity w is defined as:

$$w = |V| \times \exp(i \times (\pi - \theta_V))$$

where $\theta_V$ is also called the argument of the velocity vector in the z-diagram (Chapter I-5, section 1.3).

The parameter $\zeta$ equals:

$$\zeta = \frac{-1}{w} = \frac{-1}{|V|} \times \exp(i \times (\theta_V - \pi)) = \frac{1}{|V|} \times \exp(i \times \theta_V)$$

since $\exp(i \times (x - \pi)) = \exp(i \times (x + \pi)) = -\exp(i \times x)$.

- At the orifice edges (points B and C), the flow cannot turn sharply and separation must occur. The fluid leaves the orifice edge in the tangential direction as sketched in Fig. 7.9.
- Calculations are conducted neglecting gravity effects.
- During the mapping of the z-diagram into the $\zeta$-diagram, the upper half of the plane in the z-diagram (i.e. flow upstream of orifice) becomes the lower half of the plane in the $\zeta$-diagram (i.e. beneath the circular arc). For example, see the transformation of the streamlines $N_\infty$–$M_\infty$. Few further streamlines are shown in Figure 7.12. In the $\zeta$-diagram, all the streamlines converge to the point $M_\infty$ because the velocity downstream of orifice and at infinity tends to: $V = V_\infty \times e^{-1 \times \pi/2}$.

### Transformation from $\zeta$-diagram to $\xi$-diagram

The transformation is defined by:

$$\xi = Ln(\zeta) \tag{7.9a}$$

where $\zeta$ is defined by Equation (7–8). It may be rewritten:

$$\xi = Ln\left(\frac{1}{|V|}\right) + i \times \theta_V \tag{7.9b}$$

since $Ln(r \times e^{i \times \theta}) = Ln(r) + i \times \theta$ for $0 \leq \theta < 2 \times \pi$ (See App. D). Basically the real part of $\xi$ is $1/|V|$ and its imaginary part is $\theta_V$ (i.e. local velocity direction).

Straight lines in the z-diagram become radial lines in the $\zeta$-diagram and horizontal lines in the $\xi$-diagram because $\theta_V$ is constant. At the point $A_\infty$, the velocity modulus is zero and the velocity direction is 0 in the z-diagram. The corresponding point in the $\xi$-diagram has an infinite real part and zero imaginary part. The same reasoning may be applied to the points B, C, $D_\infty$, $M_\infty$ and $N_\infty$. Results are summarised in the Table 7.2 and sketched in Figure 7.13.

Free-streamlines in the z-diagram become circular arcs in the z-diagram and vertical lines in the $\xi$-diagram because the velocity modulus is constant along free-streamlines. In the z-diagram, the velocity at the point $M_\infty$ is: $|V| = |V_\infty|$ and $\theta_V = -\pi/2$. In the $\xi$-diagram, the point $M_\infty$ is sketched in Figure 7.13. All streamlines becomes horizontal lines in the $\xi$-diagram.

In the $\xi$-diagram, the flow field becomes a semi-infinite strip.

### Transformation from $\xi$-diagram to t-diagram

The next transformation is based upon the mapping of a semi-infinite strip (section 2.2.1). The transformation is defined by:

$$t = \cosh\left(\xi - Ln\left(\frac{1}{|V_\infty|}\right) + i \times \pi\right) \tag{7.10}$$

where $\xi$ is defined by Equation (7.9):

$$\xi = Ln\left(\frac{1}{|V|} \times e^{i \times \theta_V}\right)$$

*Table 7.2*  Transformation from ζ-diagram to ξ-diagram for the flow through a two-dimensional orifice.

| Point in the z-diagram (1) | Complex number in ξ-diagram | |
|---|---|---|
| | Real part (2) | Imaginary part (3) |
| $A_\infty$ | $+\infty$ | 0 |
| B | $Ln\left(\dfrac{1}{|V_\infty|}\right)$ | 0 |
| C | $Ln\left(\dfrac{1}{|V_\infty|}\right)$ | $-\pi$ |
| $D_\infty$ | $+\infty$ | $-\pi$ |
| $M_\infty$ | $Ln\left(\dfrac{1}{|V_\infty|}\right)$ | $-\pi/2$ |
| $N_\infty$ | $+\infty$ | $-\pi/2$ |

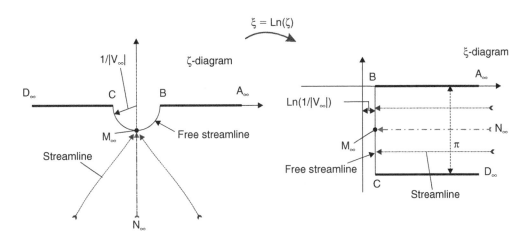

*Figure 7.13*  Flow through a two-dimensional orifice: transformation from ζ-diagram to ξ-diagram.

**Discussion**

Let compare the ξ-diagram in Figure 7.9 with Figure 7.6, Case (a). In the ξ-diagram, the point C is located at:

$$\xi_C = Ln\left(\frac{1}{|V_\infty|}\right) - i \times \pi$$

and the width of the strip is $l = \pi$. Hence the mapping function of the semi-infinite strip is:

$$\xi = \cosh^{-1}(t) + Ln\left(\frac{1}{|V_\infty|}\right) - i \times \pi$$

*Table 7.3*  Transformation from ξ-diagram to t-diagram for the flow through a two-dimensional orifice.

| Point in the z-diagram (1) | Complex number in t-diagram | |
|---|---|---|
| | Real part (2) | Imaginary part (3) |
| $A_\infty$ | $-\infty$ | 0 |
| B | $-1$ | 0 |
| C | $+1$ | 0 |
| $D_\infty$ | $+\infty$ | 0 |
| $M_\infty$ | 0 | 0 |
| $N_\infty$ | 0 | $+\infty$ |

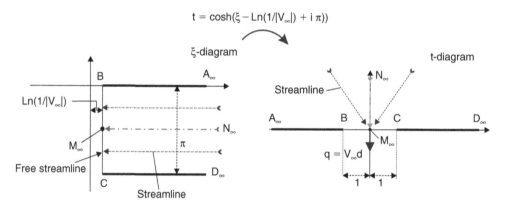

*Figure 7.14*  Flow through a two-dimensional orifice: transformation from ξ-diagram to t-diagram.

Equation (7.10) may be rewritten as:

$$t = \cosh\left(Ln\left(\frac{|V_\infty|}{|V|}\right) + i \times (\theta_V + \pi)\right)$$

At the point $A_\infty$, the velocity modulus $|V|$ and direction $\theta_V$ are respectively: 0 and 0. In the t-diagram, the point $A_\infty$ must be located on the real axis at $-\infty$ because:

$$\cosh(z) = \cosh(x) \times \cos(y) + i \times \sinh(x) \times \sin(y)$$

At the point B, the velocity is $V = |V_\infty| \times e^{i \times 0}$. In the t-diagram, the point B is located on the real-axis at $t = \cosh(i \times \pi) = \cos(\pi) = -1$. The same reasoning may be extended to the points C, $D_\infty$, $M_\infty$ and $N_\infty$. Results are summarised in the Table 7.3 and sketched in Figure 7.14.

**Remarks**

1.  Basic properties of the hyperbolic cosine function cosh are:

    $$\cosh(z) = \cosh(x) \times \cos(y) + i \times \sinh(x) \times \sin(y)$$
    $$\cosh(i \times x) = \cos(x)$$
    $$\cos(x) = \cosh(i \times x)$$

2.  On the free streamlines, the velocity modulus is $|V_\infty|$:

    $$t = \cosh(i \times (\theta_V + \pi)) = \cos(\theta_V + \pi)$$

3.  If the velocity at infinity downstream of the orifice $V_\infty = 1$, Equation (7.10) may be simplified into:

    $$t = \cosh(\xi + i \times \pi)$$

In the t-diagram, the point $M_\infty$ becomes the origin as sketched in Figures 7.9 and 7.14. Note that all streamlines becomes radial lines in the upper half-plane and converging to the origin. In the t-diagram, the flow field becomes that of a half-sink at the origin. The discharge from the half-sink is $q = V_\infty \times d$ where d is the free-jet thickness at infinity downstream of the orifice.

### Transformation from t-diagram to W-diagram

The final transformation from the t-diagram to the W-diagram is basically described in section 2.2.2:

$$W = \frac{q}{\pi} \times Ln(t) \tag{7.11}$$

where $q = V_\infty \times d$ is the flow issued from the orifice.

For the point $A_\infty$ in the z-diagram, $t = -\infty + i \times 0$ (see above) and hence:

$$W(A_\infty) = \frac{q}{\pi} \times (Ln(+\infty) + i \times \pi)$$

It yields that $\phi(A_\infty) = +\infty$ and $\psi(A_\infty) = +q$ since $W = \phi + i \times \psi$. This is sketched in Figure 7.15. For the point B, $t = -1$ and $W(B) = +i \times q$. The same reasoning applies to the points C, $D_\infty$, $M_\infty$ and $N_\infty$. Results are summarised in the Table 7.4 and sketched in Figure 7.15.

In the W-diagram, the flow field is an infinite strip of width q.

**Remarks**

1.  The problem is that of an uniform flow in an infinite strip with uniform flow (section 2.2.2, in this chapter) where the width of the strip is q.
2.  The stream function and velocity potential at the point $M_\infty$ cannot be calculated explicitly because the point $M_\infty$ is a singularity in the t-diagram (i.e. the origin). But the points $N_\infty$ and $M_\infty$ are on the same streamline $\psi = +q/2$ and the flow direction is from $N_\infty$ to $M_\infty$.
3.  A free streamline is the streamline $\psi = 0$. The second free streamline is defined by $\psi = +q$ (Fig. 7.15). The flow rate between the two free streamlines is $q = \Delta\psi$.

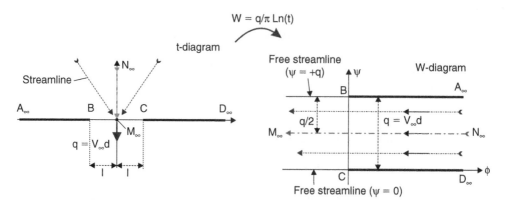

*Figure 7.15* Flow through a two-dimensional orifice: transformation from t-diagram to W-diagram.

*Table 7.4* Transformation from t-diagram to W-diagram for the flow through a two-dimensional orifice.

| Point in the z-diagram (1) | Complex number in W-diagram | |
|---|---|---|
| | Real part $\phi$ (2) | Imaginary part $\psi$ (3) |
| $A_\infty$ | $+\infty$ | $+q$ |
| B | 0 | $+q$ |
| C | 0 | 0 |
| $D_\infty$ | $+\infty$ | 0 |
| $M_\infty$ | $-\infty$ | $+q/2$ |
| $N_\infty$ | $+\infty$ | $+q/2$ |

### Determination of the free streamlines

In the W-diagram, the free streamlines are defined by $\psi = 0$ and $+q$ respectively. The shape of these streamlines in the z-diagram is defined by the conformal transformation.

On the free streamlines, the velocity modulus is a constant: $|V| = |V_\infty|$. Hence:

$$\xi = Ln(\zeta) = Ln\left(\frac{1}{|V_\infty|}\right) + 1 \times \theta_V \tag{7.12}$$

and

$$t = \cosh(i \times (\theta_V + \pi)) = \cos(\theta_V + \pi) = -\cos(\theta_V) \tag{7.13}$$

where $\theta_V$ varies from 0 to $-\pi/2$ along the free streamline B–$M_\infty$ while $-\pi \leq \theta_V \leq -\pi/2$ on the free streamline C–$M_\infty$.

The stream function is also a constant along a streamline. Hence the complex potential satisfies:

$$\delta W = \delta(\phi + i \times \psi) = \delta\phi = -|V_\infty| \times \delta s \tag{7.14}$$

where s is the direction along the streamline. This equation derives from the definition of the velocity potential (Chapter I-3, section 2.1).

The differentiation of Equation (7.11) gives also:

$$\delta W = \frac{q}{\pi} \times \frac{\delta t}{t} \tag{7.15}$$

---

**Discussion: free-jet thickness at infinity**

By definition of the velocity vector angle $\theta_V$, the streamwise coordinate s and the horizontal coordinate x in the z-diagram are related by:

$$\delta x = \cos(\theta_V) \times \delta s$$

Using Equations (7.14) and (7.15), it yields:

$$\delta x = \cos(\theta_V) \times \frac{-q}{\pi \times |V_\infty|} \times \frac{\delta t}{t} = \frac{q}{\pi \times |V_\infty|} \times \delta t$$

Along the free-streamline $B-M_\infty$, the coordinate x increases from $-1/2$ to $-d/2$, while t increases from $-1$ to $0$:

$$\int_{-1/2}^{-d/2} dx = \frac{1-d}{2} = \int_{-1}^{0} \frac{q}{\pi \times |V_\infty|} \times dt = \frac{q}{\pi \times |V_\infty|}$$

Using the continuity equation ($q = |V_\infty| \times d$), this gives the contraction ratio:

$$\frac{d}{1} = \frac{\pi}{\pi + 2} = 0.6110$$

where l is the orifice width and d is the free-jet thickness at infinity (Fig. 7.9).

The result compares well with experimental observations of two-dimensional water jets discharging into atmosphere (see above). Note however that the calculations were conducted neglecting gravity. Chanson et al. (2002) studied two-dimensional jets discharging vertically downwards and their observations showed some influence of the gravity force.

---

### Equation of the free streamlines

On a free-streamline, t equals:

$$t = \cos(\theta_V + \pi) = -\cos(\theta_V) \tag{7.13}$$

and hence:

$$W = \phi + i \times \psi = +\frac{q}{\pi} \times Ln(\cos(\theta_V + \pi)) = \frac{q}{\pi} \times Ln(-\cos(\theta_V))$$

where $\psi = 0$ or $+q$ (Fig. 7.15). On the free streamline B–$M_\infty$, $\theta_V$ varies from 0 to $-\pi/2$ while t ranges from $-1$ to 0. On the free streamline C–$M_\infty$, $\theta_V$ varies from $-\pi$ to $-\pi/2$ while t ranges from $+1$ down to 0.

Along the free-streamline, the definition of the velocity potential gives: $|V_\infty| = -\delta\phi/\delta s$ where s is the streamwise coordinate (Chapter I-3, section 2.1). It yields:

$$\phi = -|V_\infty| \times s$$

where s is measured along the free streamline $\psi = 0$ with $s = 0$ at the stagnation point.

On the free-streamline $\psi = 0$ (C-$M_\infty$), the velocity potential satisfies:

$$\phi = \frac{q}{\pi} \times Ln(-\cos(\theta_V))$$

where $\theta_V$ varies from $-\pi$ to $-\pi/2$ (Fig. 7.9). The combination of the last two equations gives a relationship between the streamwise coordinate s and the velocity vector angle $\theta_V$:

$$s = -\frac{d}{\pi} \times Ln(-\cos(\theta_V)) \tag{7.16}$$

where s is a real positive quantity because $-1 \leq \cos(\theta_V) < 0$ for $\psi = 0$ and $s = 0$ at the point C in the z-diagram.

---

**Remember**

The logarithm of a negative real number t is:

$$Ln(t) = Ln(|t| \times e^{i \times \pi}) = Ln(|t|) + i \times \pi$$

---

By definition, the streamwise coordinate s is tangent to the velocity vector at every point and it satisfies:

$$\delta s = \frac{\delta y}{\sin(\theta_V)} = \frac{\delta x}{\cos(\theta_V)}$$

The differentiation of Equation (7.16) yields also:

$$\delta s = \frac{d}{\pi} \times \tan(\theta_V) \times \delta_V$$

where $-3 \times \pi/2 < \theta_V \leq -\pi$.

The integrations give:

$$x = \frac{d}{\pi} \times \left(\frac{\pi}{2} + \cos(\theta_V)\right) \qquad\qquad \psi = 0$$

$$y = -\frac{d}{\pi} \times \left(Ln\left(-\tan\left(\frac{\theta_V}{2} + \frac{\pi}{4}\right)\right) - \sin(\theta_V)\right) \quad \psi = 0$$

These are the parametric equations in terms of $\theta_V$ describing the free streamline C–$M_\infty$ (i.e. $\psi = 0$). A similar development may be conducted for the streamline B–$M_\infty$ (i.e. $\psi = +q$):

$$x = -\frac{d}{\pi} \times \left( \frac{\pi}{2} + \cos(\theta_V) \right) \qquad\qquad \psi = +q$$

$$y = \frac{d}{\pi} \times \left( Ln\left( \tan\left( \frac{\theta_V}{2} + \frac{\pi}{4} \right) \right) - \sin(\theta_V) \right) \quad \psi = +q$$

### Discussion

In the flow downstream of an orifice with separation, the centreline streamline (i.e. $N_\infty$–$M_\infty$) may be replaced by a solid boundary. The free-streamline then represents the case of a supercritical flow under a sluice gate.

For a horizontal flow downstream of a vertical sluice gate, experimental observations illustrated that the jet contracts to a minimum, called vena contracta, before the developing boundary layer induces some form of flow bulking. Experimental data showed that the vena contracta is located about 1.7 times the orifice height downstream of the orifice (e.g. Montes 1998, pp. 283–284). Such a distance would correspond to $y/l = -1.7$ in Figure 7.9 and the contraction ratio of the free streamline at that location is about $x/(l/2) = 0.305$.

## 3.3  Borda's mouthpiece

### 3.3.1  Presentation

A Borda's mouthpiece is a re-entrant slot in a large tank (Fig. 7.16). It must be long enough such that the issuing jet is not affected by the presence of the walls. Herein the mouthpiece size is

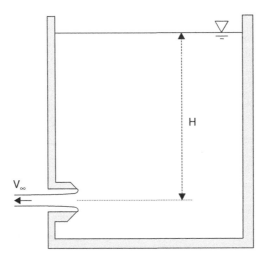

Figure 7.16  Sketch of a Borda mouthpiece in a reservoir.

denoted l. The downstream jet thickness and velocity at infinity are denoted d and $V_\infty$ respectively (Fig. 7.17).

The bounding streamlines $A_\infty$–B and $D_\infty$–C are assumed long enough such that the velocity at $A_\infty$ and $D_\infty$ is zero. The line $M_\infty$–$N_\infty$ is the line of symmetry and the velocity at $N_\infty$ is also zero. Note that the line of symmetry is a streamline. At the point $M_\infty$, the velocity equals $V_\infty$.

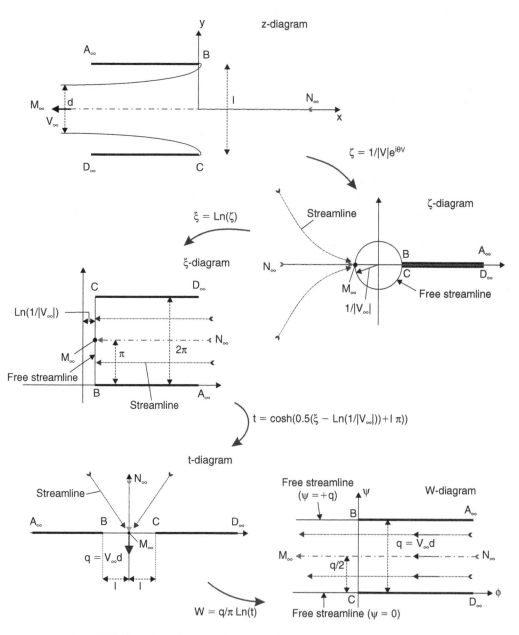

*Figure 7.17* Flow through a two-dimensional Borda mouthpiece with separation.

### Note

The Borda's mouthpiece is a horizontal re-entrant tube in the side of a tank with a length such that the issuing jet is not affected by the presence of the walls (Fig. 7.16). It is named after the French military engineer Jean-Charles de Borda (1733–1799) who investigated the flow through orifices and developed the Borda's mouthpiece.

### 3.3.2  Basic transformations

The basic transformations of the flow through a Borda's mouthpiece with separation are:

$$\zeta = -\frac{dz}{dW} = \frac{1}{|V|} \times e^{i \times \theta_V} \tag{7.17}$$

$$\xi = Ln(\zeta) = Ln\left(\frac{1}{|V|}\right) + i \times \theta_V \tag{7.18}$$

$$t = \cosh\left(\frac{1}{2} \times \left(\xi - Ln\left(\frac{1}{|V_\infty|}\right)\right) + i \times \pi\right) \tag{7.19}$$

$$W = \frac{q}{\pi} \times Ln(t) \tag{7.20}$$

Equation (7.19) is a Schwarz-Christoffel transformation and it may be rewritten:

$$t = \cosh\left(\frac{1}{2} \times \left(Ln\left(\frac{|V|}{|V_\infty|}\right)\right) + i \times \left(\frac{\theta_V}{2} + \pi\right)\right)$$

or

$$t = \cosh\left(\frac{1}{2} \times \left(Ln\left(\frac{|V|}{|V_\infty|}\right)\right)\right) \times \cos\left(\frac{\theta_V}{2} + \pi\right)$$

$$+ i \times \sinh\left(\frac{1}{2} \times \left(Ln\left(\frac{|V|}{|V_\infty|}\right)\right)\right) \times \sin\left(\frac{\theta_V}{2} + \pi\right)$$

At the point $A_\infty$ in the z-diagram, the velocity modulus and direction are respectively $|V| = 0$ and $\theta_V = 0$. As a result $t(A_\infty) = (+\infty) \times e^{i \times \pi}$, and $\phi = +\infty$ and $\psi = +q$. The same reasoning may be applied to the points B, C, $D_\infty$, $M_\infty$ and $N_\infty$. The transformations are sketched in Figure 7.17 and summarised in the Table 7.5.

### Remarks

1.  There are a few important differences between the conformal transformations of the orifice flow and of the Borda's mouthpiece flow. First the velocity direction $\theta_V$ is equal to 0 at $A_\infty$ and B, while it is taken as $2 \times \pi$ at C and $D_\infty$. Second the Schwarz-Christoffel transformation (Eq. (7.19)) is different because it has different boundary conditions (Streeter 1948). The constants in Equation (7.19) were selected to satisfy the boundary conditions (section 2.2.1, in this chapter).

2.   The point $M_\infty$ is a singularity in the t-diagram. The stream function at that point is deduced from the value of the stream function at the point $N_\infty$.
3.   In the t-diagram, the flow rate at the origin (Point $M_\infty$) is $q = V_\infty \times d$ where d is the free jet thickness.

On a free streamline, the velocity modulus is constant and equal to $|V_\infty|$, and $t = \cos(\theta_V/2 + \pi)$. It yields:

$$\delta s = \frac{\delta y}{\sin(\theta_V)} = \frac{\delta x}{\cos(\theta_V)} = \frac{d}{\pi} \times \frac{\delta t}{t} = \frac{d}{2 \times \pi} \times \frac{\sin(\theta_V/2)}{\cos(\theta_V/2)} \times \delta\theta_V$$

The integration gives the equations of the free-streamline. Complete calculations demonstrate that the coefficient of contraction d/l equals 0.5.

### 3.3.3   Discussion

The flow through a Borda's mouthpiece is the limiting case of the flow through an orifice.

For a two-dimensional water jet issued from an orifice, the contraction coefficient is a function of the relative gate opening l/B where B is the pipe height and l is the orifice opening, and of the nozzle geometry (Fig. 7.18). Ideal-fluid flow calculations are summarised in Table 7.6. The cases $\delta = 90°$ and $180°$ correspond respectively to the orifice flow (section 3.2) and to the Borda's mouthpiece (section 3.3).

## 3.4   Flow normal to a flat plate with separation

In Chapter I-5, the flow past a plate perpendicular to the flow direction was investigated in absence of separation. For large Reynolds numbers, separation is observed and the theory of free streamlines provides a more realistic solution (Fig. 7.19).

At the point A, in the z-diagram, the velocity modulus is that on the free streamline: $|V| = |V_\infty|$, while the velocity direction is $+\pi$. At the plate edges (points A and B), the flow cannot turn sharply

Table 7.5  Characteristics of the basic transformations for the flow through a two-dimensional Borda mouthpiece with separation.

| Points | $A_\infty$ | B | C | $D_\infty$ | $M_\infty$ | $N_\infty$ |
|---|---|---|---|---|---|---|
| Velocity modulus $|V|$ | 0 | $|V_\infty|$ | $|V_\infty|$ | 0 | $|V_\infty|$ | 0 |
| Velocity direction $\theta_V$ | 0 | 0 | $2 \times \pi$ | $2 \times \pi$ | $\pi$ | $\pi$ |
| Real part of $\zeta$ | $+\infty$ | $\dfrac{l}{|V_\infty|}$ | $\dfrac{l}{|V_\infty|}$ | $+\infty$ | $\dfrac{-l}{|V_\infty|}$ | $-\infty$ |
| Imaginary part of $\zeta$ | 0 | 0 | 0 | 0 | 0 | 0 |
| Real part of $\zeta$ | $+\infty$ | $Ln\left(\dfrac{l}{|V_\infty|}\right)$ | $Ln\left(\dfrac{l}{|V_\infty|}\right)$ | $+\infty$ | $Ln\left(\dfrac{l}{|V_\infty|}\right)$ | $+\infty$ |
| Imaginary part of $\xi$ | 0 | 0 | $2 \times \pi$ | $2 \times \pi$ | $\pi$ | $\pi$ |
| Real part of t | $-\infty$ | $-l$ | $+l$ | $+\infty$ | 0 | 0 |
| Imaginary part of t | 0 | 0 | 0 | 0 | 0 | $+\infty$ |
| Velocity potential $\phi$ | $+\infty$ | 0 | 0 | $+\infty$ | 0 | $+\infty$ |
| Stream function $\psi$ | $+q$ | $+q$ | 0 | 0 | $+q/2$ | $+q/2$ |

and separation must occur. The fluid leaves the edges in the tangential direction as sketched in Fig. 7.18. That is, $\theta_V = \pi$ and 0 at points A and B respectively.

On the free streamline, the velocity is constant and it equals the fluid velocity at infinity. Downstream of the plate and outside of the wake, the velocity modulus and direction tend respectively to $|V| = |V_\infty|$ and $\theta_V = -\pi/2$ at infinity. In the wake region, the pressure is assumed to be the ambient pressure and the velocity is zero.

Note the stagnation point S on the upstream face of the plate and the two separation points A and B. On the streamline through the stagnation point S, the stream function is taken as $\psi = 0$. It can be shown that the velocity potential equals zero at stagnation.

The basic transformations are:

$$\zeta = -\frac{dz}{dW} = \frac{1}{|V|} \times e^{i \times \theta_V} \tag{7.21}$$

$$\xi = Ln(\zeta) = Ln\left(\frac{1}{|V|}\right) + i \times \theta_V \tag{7.22}$$

$$t = \cosh\left(\frac{1}{2} \times \left(\xi - Ln\left(\frac{1}{|V_\infty|}\right)\right) + i \times \pi\right) \tag{7.23}$$

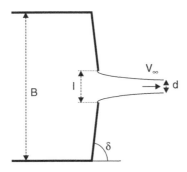

Figure 7.18   Definition sketch of a two-dimensional orifice flow.

Table 7.6   Contraction coefficient of a two-dimensional Borda mouthpiece.

| Relative gate opening l/B (1) | Coefficient of contraction d/l | | | | | Remarks (6) |
|---|---|---|---|---|---|---|
| | $\delta = 45°$ (2) | $\delta = 90°$ Two-dimensional orifice (3) | | $\delta = 135°$ (4) | $\delta = 180°$ Borda's mouthpiece (5) | |
| 1 | 1.0 | 1.0 | | 1.0 | 1.0 | Two-dimensional pipe flow discharging into air. |
| 0.8 | 0.789 | 0.722 | | 0.698 | 0.691 | |
| 0.6 | 0.758 | 0.662 | | 0.620 | 0.613 | |
| 0.4 | 0.749 | 0.631 | | 0.580 | 0.564 | |
| 0.2 | 0.747 | 0.616 | | 0.555 | 0.528 | |
| 0.1 | 0.747 | 0.612 | | 0.546 | 0.513 | |
| 0 | 0.746 | 0.611 | | 0.537 | 0.500 | Small orifice flow. |

Notes: calculations by Von Mises in Rouse (1946); $\delta$ is defined in Figure 7.18.

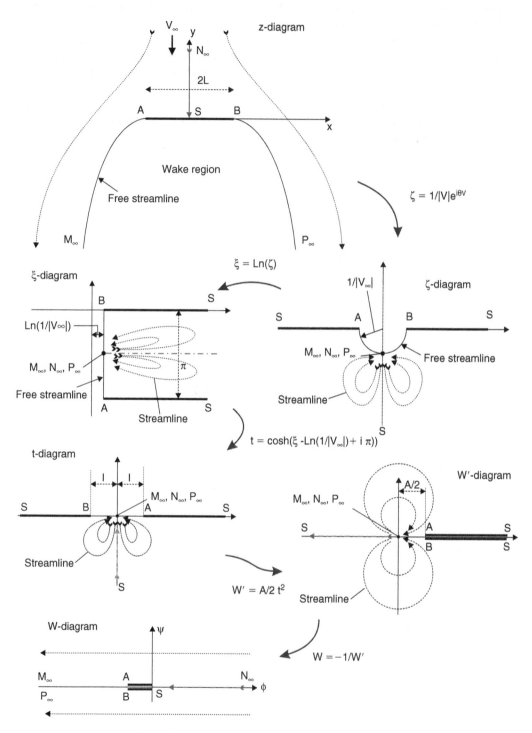

*Figure 7.19* Flow normal to a flat plate with separation.

$$W' = A \times \frac{t^2}{2} \tag{7.24}$$

$$W = -\frac{1}{W'} \tag{7.25}$$

The flow pattern is sketched in Figure 7.19. There is one more transformation than in the previous applications. The transformation from the t-diagram to the W'-diagram (Eq. (7.24)) is a Schwarz-Christoffel transformation for a semi-infinite strip of zero thickness (section 2.2.1). In the t-diagram, the real constants a, b and c are arbitrarily selected such that: $a = -\infty$, $b = 0$, $c = +\infty$, and the external angles of the polygon are $\beta = -\pi$ and $\gamma = 0$. The Schwarz-Christoffel transformation is:

$$W' = A \times \int \frac{dt}{t^{-1}} + B \tag{7.24a}$$

which gives:

$$W' = A \times \frac{t^2}{2}$$

assuming $B = 0$ for $t = 0$: i.e., at the points $N_\infty$, $M_\infty$ and $P_\infty$. It can be demonstrated that $W'$ must be real along the plate ($-1 < t < +1$) and that $A$ must be a real positive constant (Streeter 1948, p. 178, Vallentine 1969, p. 224). The constant $A$ is a function of the dimensions of the plate:

$$A = \frac{4 + \pi}{L \times |V_\infty|} \tag{7.26}$$

where L is the half-breadth of the plate (Fig. 7.19).

The basic transformations are summarised in Table 7.7 for the characteristic points A, B, S, $N_\infty$, $M_\infty$ and $P_\infty$ and sketched in Figure 7.19. Note the singularity of the stagnation point.

In Figure 7.19, some streamlines are sketched. They have a pattern somehow similar to a doublet flow pattern with the origin at the points $M_\infty$, $N_\infty$ and $P_\infty$. In the W'-diagram, the streamline pattern is exactly that of a doublet.

---

**Application**

Demonstrate that the constant $A$ (Eq. (7.26)) is a function of the plate size and free stream velocity only.

*Solution*

On the plate, let consider the streamline $\psi = 0$ between the stagnation point S and the plate edge A. At stagnation (point S), the velocity modulus and direction are 0 and $+\pi$ respectively, while $|V| = |V_\infty|$ and $\theta_V = +\pi$ at the plate edge (point A). Between the points S and A, the velocity vector is: $V = |V| \times e^{i \times \pi}$. That is, t is real and varies from $+\infty$ down to $+1$:

$$t = \cosh\left( \text{Ln}\left( \frac{|V_\infty|}{|V|} \right) \right) \qquad \text{along streamline S-A}$$

It gives:

$$t = \frac{1}{2} \times \left( \exp\left( \mathrm{Ln}\left( \frac{|V_\infty|}{|V|} \right) \right) + \exp\left( -\mathrm{Ln}\left( \frac{|V_\infty|}{|V|} \right) \right) \right) = \frac{1}{2} \times \left( \frac{|V_\infty|}{|V|} + \frac{1}{\frac{|V_\infty|}{|V|}} \right)$$

along streamline S-A

and hence:

$$\frac{|V_\infty|}{|V|} = t + \sqrt{t^2 - 1}$$

along streamline S-A

which satisfies $|V| = 0$ at the stagnation point S ($t = +\infty$).
   By definition of the velocity potential (Chapters I-2 and I-3):

$$\frac{\delta \phi}{\delta s} = -|V| = -\frac{\delta \phi}{\delta x}$$

along streamline S-A

since $\delta x = -\delta s$ along the streamline S-A.
   On the streamline S-A, the stream function is zero and the complex potential satisfies $W = \phi$:

$$W = \phi = -\frac{1}{W'} = -\frac{2}{A} \times \frac{1}{t^2}$$

along streamline S-A

Integrating dx in the z-diagram between S and A:

$$\int_S^A dx = -L = \int_{t=+\infty}^{t=+1} \frac{dx}{d\phi} \times \frac{d\phi}{dt} \times dt$$

along streamline S-A

where L is the half-breadth of the plate (Fig. 7.19). Since $\delta x/\delta \phi = 1/|V|$ and

$$\frac{d\phi}{dt} = \frac{4}{A \times t^3}$$

the integration between S and A gives:

$$\int_S^A dx = -L = \frac{4}{A \times |V_\infty|} \times \int_{t=+\infty}^{t=+1} \frac{t + \sqrt{t^2 - 1}}{t^3} \times dt$$

along streamline S-A

The right-handside integral term may be integrated analytically (e.g. Spiegel 1968). The exact solution yields:

$$A = \frac{4 + \pi}{L \times |V_\infty|}$$

Table 7.7 Characteristics of the basic transformations for the flow normal to a flat plate with separation.

| Point | A | B | $M_\infty$ | $P_\infty$ | $N_\infty$ | Stagnation point S |
|---|---|---|---|---|---|---|
| Velocity modulus $|V|$ | $|V_\infty|$ | $|V_\infty|$ | $|V_\infty|$ | $|V_\infty|$ | $|V_\infty|$ | 0 |
| Velocity direction $\theta_V$ | $-\pi$ | 0 | $-\pi/2$ | $-\pi/2$ | $-\pi/2$ | $+\pi/2$ ([1]) <br> 0 ([2]) <br> $+\pi$ ([3]) |
| Real part of $\zeta$ | $\dfrac{-1}{|V_\infty|}$ | $\dfrac{+1}{|V_\infty|}$ | 0 | 0 | 0 | – |
| Imaginary part of $\zeta$ | 0 | 0 | $\dfrac{-1}{|V_\infty|}$ | $\dfrac{-1}{|V_\infty|}$ | $\dfrac{-1}{|V_\infty|}$ | – |
| Real part of $\xi$ | $Ln\left(\dfrac{1}{|V_\infty|}\right)$ | $Ln\left(\dfrac{1}{|V_\infty|}\right)$ | $Ln\left(\dfrac{1}{|V_\infty|}\right)$ | $Ln\left(\dfrac{1}{|V_\infty|}\right)$ | $Ln\left(\dfrac{1}{|V_\infty|}\right)$ | $+\infty$ |
| Imaginary part of $\xi$ | $-\pi$ | 0 | $-\pi/2$ | $-\pi/2$ | $-\pi/2$ | – |
| Real part of t | $+1$ | $-1$ | 0 | 0 | 0 | 0 ([1]) <br> $-\infty$ ([2]) <br> $+\infty$ ([3]) |
| Imaginary part of t | 0 | 0 | 0 | 0 | 0 | $-\infty$ ([1]) <br> 0 ([2]) <br> 0 ([3]) |
| Real part of W′ | $+\dfrac{A}{2}$ | $+\dfrac{A}{2}$ | 0 | 0 | 0 | $-\infty$ ([1]) <br> $-\infty$ ([2]) <br> $+\infty$ ([3]) |
| Imaginary part of W′ | 0 | 0 | 0 | 0 | 0 | 0 ([1]) <br> 0 ([2]) <br> 0 ([3]) |
| Velocity potential $\phi$ | $-\dfrac{2}{A}$ | $-\dfrac{2}{A}$ | $-\infty$ | $-\infty$ | $+\infty$ | 0 |
| Stream function $\psi$ | 0 | 0 | 0 | 0 | 0 | 0 |

Notes: ([1]): along $N_\infty$–S; ([2]): along S–B; ([3]): along S-A; A: real constant: $A = (4 + \pi)/(L \times |V_\infty|)$

## Equation of the free streamlines

On a free-streamline, the velocity modulus is $|V_\infty|$ and the stream function is $\psi = 0$. The complex potential W equals:

$$W = \phi = -\frac{2}{A \times t^2}$$

where $t = -\cos(\theta_V)$ and $A$ is a real constant (Eq. (7.26)). The definition of the velocity potential gives:

$$|V_\infty| = -\frac{\delta\phi}{\delta s}$$

and

$$\phi = -|V_\infty| \times s$$

where s is measured along the free streamline ($\psi = 0$) with $s = 0$ at the plate edge. It yields:

$$\delta s = \frac{4}{A \times |V_\infty|} \times \frac{\sin(\theta_V)}{\cos^3(\theta_V)} \times \delta\theta_V$$

Integrating and replacing $A$ by its expression:

$$s = \frac{4 \times L}{4 + \pi} \times \frac{1}{\cos^2(\theta_V)}$$

where $\theta_V$ varied between 0 and $-\pi/2$ along the free streamline B–$P_\infty$ and between $-\pi$ and $-\pi/2$ along the free streamline A–$M_\infty$.

By definition, the streamwise coordinate s is tangent to the velocity vector at every point and it satisfies:

$$\delta s = \frac{\delta y}{\sin(\theta_V)} = \frac{\delta x}{\cos(\theta_V)}$$

Hence:

$$\delta x = \frac{4}{A \times |V_\infty|} \times \frac{\sin(\theta_V)}{\cos^2(\theta_V)} \times \delta\theta_V$$

$$\delta y = \frac{4}{A \times |V_\infty|} \times \frac{\sin^2(\theta_V)}{\cos^3(\theta_V)} \times \delta\theta_V$$

On the free-streamline B–$P_\infty$, the integrations yield:

$$x = \frac{4 \times L}{4 + \pi} \times \left( \frac{\pi}{4} + \frac{1}{\cos(\theta_V)} \right) \qquad\qquad \text{free streamline B–}P_\infty$$

$$y = \frac{2 \times L}{4 + \pi} \times \left( \tan(\theta_V) - \text{Ln}\left( \tan\left( \frac{\theta_V}{2} + \frac{\pi}{4} \right) \right) \right) \qquad\qquad \text{free streamline B–}P_\infty$$

assuming the z-diagram origin at the stagnation point (Fig. 7.19). These parametric equations in terms of $\theta_V$ describe the free streamline B–$P_\infty$ for $\theta_V$ varying between 0 and $-\pi/2$.

A similar development may be conducted for the streamline A–$M_\infty$:

$$x = -\frac{4 \times L}{4 + \pi} \times \left( \frac{\pi}{4} + \frac{1}{\cos(\theta_V)} \right) \qquad\qquad \text{free streamline A–}M_\infty$$

$$y = \frac{2 \times L}{4 + \pi} \times \left( \tan(\theta_V) + \text{Ln}\left( -\tan\left( \frac{\theta_V}{2} + \frac{\pi}{4} \right) \right) \right) \qquad\qquad \text{free streamline A–}M_\infty$$

The free streamlines A–$M_\infty$ and B–$P_\infty$ define the wake region. They are plotted in Figure 7.20 for the flow pattern defined in Figure 7.19.

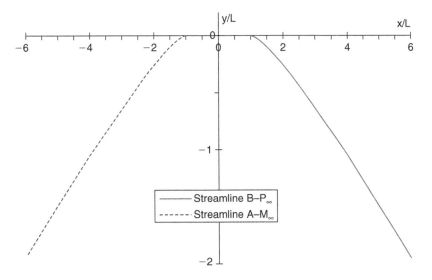

*Figure 7.20*  Free-streamlines for the flow normal to a flat plate with separation.

### Drag force on the plate

The Bernoulli equation states that:

$$\frac{|V|^2}{2} + \frac{P}{\rho} = \frac{|V_\infty|^2}{2} + \frac{P_o}{\rho} \tag{7.27}$$

along a streamline for a steady flow motion neglecting gravity effects (Chapter I-2, section 4.3). $P_o$ is the ambient static pressure far away upstream of the plate. It is also the pressure in the wake region. It may be rewritten:

$$P - P_o = \rho \times \frac{|V_\infty|^2}{2} \times \left(1 - \left(\frac{|V|}{|V_\infty|}\right)^2\right)$$

The drag force exerted on the plate equals:

$$\text{Drag} = \int_{-L}^{+L} (P - P_o) \times dx$$

That is:

$$\text{Drag} = 2 \times \int_{0}^{+L} \frac{1}{2} \times \rho \times |V_\infty|^2 \times \left(1 - \frac{|V|^2}{|V_\infty|^2}\right) \times dx$$

Between the points S and B, the velocity modulus |V| increases from 0 to $|V_\infty|$ while the velocity direction $\theta_V$ is zero. That is:

$$t = -\cosh\left(\text{Ln}\left(\frac{|V_\infty|}{|V|}\right)\right) \qquad\qquad \text{between points S and B}$$

Hence:

$$-t = \frac{1}{2} \times \left( \frac{|V_\infty|}{|V|} + \frac{|V|}{|V_\infty|} \right)$$

It can be shown that:

$$-2 \times \sqrt{t^2 - 1} = \frac{1}{2} \times \left( \frac{|V_\infty|}{|V|} + \frac{|V|}{|V_\infty|} \right)$$

Between S and B, the stream function y equals zero and the complex potential satisfies:

$$\delta W = \delta\phi = -|V| \times \delta x = -\frac{4}{A \times t^3} \times \delta t$$

Replacing into the expression of the drag force, it yields:

$$\text{Drag} = -8 \times \frac{\rho \times |V_\infty|}{A} \times \int\limits_{-\infty}^{-1} \frac{\sqrt{t^2 - 1}}{t^3} \times dt$$

where $A = (4 + \pi)/(L \times |V_\infty|)$. It may be rewritten:

$$\text{Drag} = C_D \times (2 \times L) \times \rho \times \frac{V_\infty^2}{2} \tag{7.28}$$

where the drag coefficient $C_D$ equals:

$$C_D = \frac{2 \times \pi}{\pi + 4} = 0.8798 \tag{7.29}$$

---

**Remark**

The integration of the drag force has an exact solution since:

$$\int\limits_{-\infty}^{-1} \frac{\sqrt{t^2 - 1}}{t^3} \times dt = -\frac{\sqrt{t^2 - 1}}{2 * t^2} + \frac{1}{2} \times \cos^{-1}\left( \frac{1}{|t|} \right)$$

---

*Discussion*

The above analysis for an ideal fluid flow assumes implicitly that the wake is at rest and that the pressure in the wake region equals the pressure $P_o$ in the undisturbed stream. This approximation may apply for a ventilated cavity when a stream of water flows past a flat plate and the space between the separation lines is filled with air [7]. If the wake is filled with a fluid of density

---

[7] or water vapour in the case of high-velocity flows and cavitating flows.

Table 7.8 Drag coefficient on a blunt flat plate normal to the flow (real fluid flow observations) at large Reynolds numbers (Re > 1 × 10³).

| B/(2 × L) (1) | $C_D$ (2) | Re (3) | Remarks (4) |
|---|---|---|---|
| +∞ | 1.90 | > 1 × 10³ | Two-dimensional flows |
| 20 | 1.50 | | |
| 5 | 1.20 | | |
| 1 | 1.16 | | |

comparable to the main stream, flow recirculation develops behind the plate and the pressure behind the plate is less than that of the undisturbed stream, yielding greater drag on the plate.

Considering a real fluid flow past a two-dimensional plate, experiments show that the drag coefficient is $C_D = 1.90$ for Re > 1 × 10³, where the drag coefficient is defined as:

$$\text{Drag} = C_D \times A \times \rho \times \frac{V_\infty^2}{2}$$

where A is the projected area of the body. The result is valid for two dimensional flows in absence of ventilation. Note that it is more than twice the calculated drag coefficient (Eq. (7.29)). Plates of finite widths have a lower drag because three-dimensional flows occur at the ends. For a plate of length 2 × L (Fig. 7.9) and of width B [8], the measured drag coefficient is listed in Table 7.8.

For a circular disk, the observed drag coefficient is about $C_D = 1.12$ for Re > 1 × 10³.

---

**Remarks**

1.  Experimental results showed that the drag coefficient is basically independent of the Reynolds number for Re > 1 × 10³.
2.  For the flow past a blunt body (e.g. flat plate), the drag coefficient is defined using the projected area. For example, $A = \pi \times R^2$ for a circular cylinder of radius R aligned with the flow direction, and $A = 2 \times R \times L$ for a cylinder of length L normal to the flow direction. But, in the study of flows past airfoils, the area A is defined as the chord times the width of the foil (Chapter I-6).

---

## 4    CONCLUSION

The theory of free streamlines associated with the theorem of Schwarz-Christoffel is a power technique to solve analytically ideal fluid flow with separation. Basic applications include the flow through an orifice and the flow past a normal plate.

However it must be emphasised that it is a difficult technique involving several successive transformations. In the wake region, the velocity is assumed zero. While the approximation may be suitable when the wake contains a fluid of lesser density than the main stream, it yields often lower drag force estimates. Further gravity effects were ignored in the above developments.

---

[8]in the direction normal to the x-y plane.

# Exercises

7.1   For the flow through an orifice discharging from a infinitely large reservoir, draw the free streamlines for $l = 0.070$ m and $V_\infty = 6.1$ m/s. (Plot the free streamlines on a graph paper using a 1:1 scale.) What is the coefficient of contraction at a distance $|y|/l = 1$ beneath the orifice.
  *Neglect gravity effects.*

7.2   For the flow through a Borda's mouthpiece, calculate the equations of the free streamlines. For $l = 0.07$ m and a head above orifice $H = 1.3$ m, plot the free streamlines on a graph paper using a 1:1 scale. Show that the coefficient of contraction is 0.5.
  *Neglect gravity effects.*

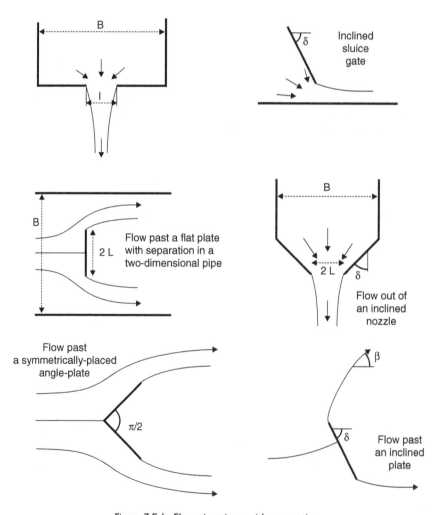

*Figure 7.E.1* Flow situations with separation.

7.3 For the water flow through an orifice discharging from a finite width reservoir into air (Fig. E7.1), calculate the equations of the free streamlines. Show that the coefficient of contraction is 0.644 for $B/l = 2$. If the pressure $P_o$ in the reservoir is known, develop the relationship between the flow rate, the pressure $P_o$ and the orifice dimensions.

   *Neglect gravity effects.*

## Solution

The continuity equation and Bernoulli principle are written between upstream and far downstream of the orifice:

$$q = V_o \times B = V_\infty \times d \qquad \text{Continuity equation}$$

$$\frac{|V_o|^2}{2} + \frac{P_o}{\rho} = \frac{|V_\infty|^2}{2} + \frac{P_{atm}}{\rho} \qquad \text{Bernoulli principle}$$

where $P_{atm}$ is the atmospheric pressure downstream of the orifice. After transformation, it yields:

$$q = \frac{C_c}{\sqrt{1 - \frac{d_c^2}{B^2}}} \times 1 \times \sqrt{2 \times \frac{P_o - P_{atm}}{\rho}}$$

where $C_c = d/l$.

7.4 For the flow beneath an inclined sluice gate (Fig. E7.1), calculate the relationship between the contraction coefficient and the sluice angle $\delta$. Show that the contraction coefficient is 0.746 for $\delta = \pi/4$.

   *Neglect gravity effects. Assume and infinitely long gate.*

7.5 For the flow past a normal plate in a pipe (Fig. E7.1), sketch the W-diagram showing the locations of the transformed boundaries.

   *Neglect gravity effects.*

7.6 For the flow out of an inclined nozzle (Fig. E7.1), sketch the W-diagram showing the locations of the transformed boundaries. Calculate the relationship between the contraction coefficient, the relative orifice width $l/B$ and the nozzle angle $\delta$.

   *Neglect gravity effects.*

## Solution

See Vallentine (1969, pp. 217–220). (Beware of sign convention differences.)

7.7 For the flow past a symmetrically place angle plate (Fig. E7.1), sketch the W-diagram showing the locations of the transformed boundaries. Determine the transformation equations.

   *Neglect gravity effects.*

## Solution

See Vallentine (1969, p. 229). (Beware of sign convention differences.)

7.8 For the flow past an inclined plate (Fig. E7.1), sketch the W-diagram showing the locations of the transformed boundaries. Determine the transformation equations. Express the relationship between the upper free streamline angle at infinity $\beta$ as a function of the plate angle $\delta$.

   *Neglect gravity effects.*

## Solution

Vallentine (1969, pp. 229–230). (Beware of sign convention differences.)

# Real fluid flows: Theory and applications

Dust storm over China on 7 April 2001 (Courtesy of NASA earth Observatory) – At ground level, the dust storm blocked enough sunlight to reduce visibility to roughly 20 m

# Real fluid flows: Theory and applications

# Introduction

## 1 INTRODUCTION

In Nature, most practical applications are associated with turbulent flows. These include river flows, wind flows around land forms (e.g. hills, mountains), ocean waves. In engineering applications, many fluid flows are also turbulent: e.g., spillway flows, wind flows between buildings in a city. The fluid involved may be gas or liquid. In external flows, some relative motion occurs between an object and a large mass of fluid: e.g., the wind flow around a paraglider wing (Fig. 1.1 & movie Para10.mov, App. F). Internal flows include pipeline systems, and irrigation

Figure 1.1 Para-glider suspended below a hollow fabric wing above Pingtung county (Taiwan) on 16 December 2008.

*Figure 1.2* La Grande Arche, La Défense (France) (Courtesy of VF communication, La Grande Arche) – Looking West at sunset.

water channel. The flow may be steady or unsteady: e.g., a plane at take off. While many applications involve turbulent flows, laminar flows occur in medical engineering (e.g. blood flows), slow-moving landslides and lubrication.

Many civil engineering structures are affected by hydro- and aero-dynamics. In some cases, the fluid structure interactions may lead to failures. While the failure of the Tacoma Narrows suspension bridge on 7 November 1940 was well illustrated, other failures are documented. On 18 March 1998, several transmission towers were damaged in the Brisbane Valley. Most damage occurred near Harlin, Colinton and Kangaroo Creek. It is believed that the towers collapsed during gusty, tornado-like winds with wind gusts in excess of 200 km/h. The replacement towers were designed to withstand winds up to 210 km/h. Another example of aerodynamic failure is the La Grande Arche building in La Défense, a western suburb of Paris (France). Completed in 1989, the building is 110.9 m high, 112 m long and 106.9 m wide. Shortly after completion, very strong winds were felt at the feet of the Arch. This was caused by natural strong winds experienced at La Défense associated by a Venturi effect induced by the arch shape (Fig. 1.2). As a result, a 2,300 m² surface area membrane (PTFE-coated glass fibre fabric) was installed at the foot of the arch to reduce the wind speed at pedestrian level. The membrane is clearly seen in Figure 1.2. An example of successful integration of fluid structure interactions is the Pont de Normandie suspension bridge

*Figure 1.3* Viaduc de Millau (France) on 1 May 2009 (Courtesy of Mr and Mrs Chanson).

on the Seine river. Completed in 1995, the bridge's total length is 2,141 m and the central span is 856 m long. The estuary of the Seine river is well-known for strong Westerly winds and the bridge was designed to sustain wind speeds of up to 300 km/h. The bridge deck was streamlined to reduce the wind drag and its design was tested in wind tunnel. A further example is the Millau bridge (or Millau viaduct) in Southern France (Fig. 1.3). Completed in 2004, the bridge is 342 m tall and 2,460 m long. It was designed to sustain storm wind speeds up to 250 km/h. The final design was tested in both wind and water tunnels. In the water tunnel, the physical model was built at a 1:3,000 scale.

---

**Remark**

Several movies of real fluid flows are included in Appendix F. These include smoke dispersion in the atmospheric boundary layer, turbulence in rivers and spillways, turbulent water jets an wave motion.

---

## 2   LAMINAR AND TURBULENT FLOWS

Let us consider a particular situation such as the pipe flow shown in Figure 1.4 and movie P1010317.mov (App. F). A low-velocity flow motion is characterised by fluid particles moving along smooth paths in laminas or layers, with one layer gliding smoothly over an adjacent layer: i.e., the *laminar flow regime*.

With increasing flow velocity, there is a critical velocity above which the flow motion becomes characterised by an unpredictable behaviour, strong mixing properties and a broad spectrum of length scales: i.e., the *turbulent flow regime*. In his classical experiment, Osborne Reynolds (1842–1912) illustrated the strong mixing properties of turbulent flows with some rapid mixing of dye (Reynolds 1883). This is seen in Figure 1.4 showing the original Reynolds experiment (Fig. 1.4A) and a modified Reynolds experiment (Fig. 1.4B) in the Gordon McKay Hydraulics Laboratory at the University of Queensland. In turbulent flows, the fluid particles move in very irregular paths, causing an exchange of momentum from one portion of the fluid to another as shown

(A) Gravure of the original experimental apparatus of Osborne Reynolds with flow motion from left to right

(B) Photographs of dye injection for Reynolds numbers between 349 (Top) to 2,320 (Bottom) – Flow direction from right to left

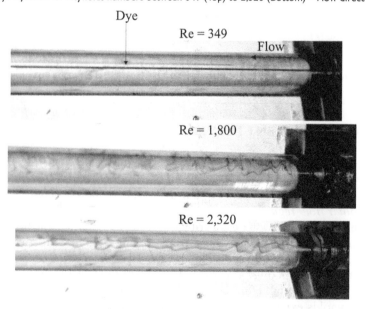

*Figure 1.4* Dye dispersion in laminar and turbulent flows.

in Figure 1.4B (Bottom) where dye is rapidly dispersed in the turbulent flow regime. The fluid particles move in very irregular paths in turbulent flows, causing an exchange of momentum from one portion of the fluid to another. Turbulent flows have a great mixing potential and they involve a wide range of eddy length scales. For example, in natural streams, the flow is turbulent and strong turbulent mixing occurs.

In industrial pipes flows, laminar flows are observed for Reynolds numbers less than 1,000 to 3,000 where the Reynolds number is defined as $Re = \rho \times V \times D_H/\mu$, $\rho$ and $\mu$ are the fluid density and dynamic viscosity respectively, V is the flow velocity or cross-sectional averaged velocity, and $D_H$ is the equivalent pipe diameter. Turbulent flows occur for Reynolds numbers greater than 5,000 to 10,000 typically.

---

**Remarks**

1. Osborne Reynolds (1842–1912) was a British physicist and mathematician who expressed first the Reynolds number (Reynolds 1883) and later the Reynolds stress or turbulent shear stress. Jackson presented some pertinent insight into the contribution of Osborne Reynolds (Jackson 1995, Jackson and Launder 2007).
2. The Reynolds number Re characterises the ratio of inertial force to viscous force. In the Moody diagram, and the Colebrook-White formula for turbulent flows, the Reynolds number is defined in terms of the equivalent pipe diameter $D_H$. This definition is used in both pipe flows and open channel flows (Henderson 1966, Chanson 2004).
3. The hydraulic diameter $D_H$ or equivalent pipe diameter is defined as:

$$D_H = 4 \times \frac{A}{P_w}$$

   where A is the flow cross-section area and $P_w$ is the wetted perimeter. For a circular pipe, the equivalent pipe diameter equals the pipe diameter: $D_H = D$ where D is the internal pipe diameter.
4. The movie P1010317.mov starts by showing a laminar motion in a circular pipe (App. F). The dye flows as a ribbon and there is no dispersion. The discharge is increased. When the flow becomes turbulent, the dye is rapidly mixed across the entire flow cross-section.

---

## 3  SHEAR STRESS

The term *shear flow* characterises a fluid motion with a velocity gradient in a direction normal to the mean flow direction: e.g., a boundary layer flow along a flat plate (Fig. 1.5). In a shear flow, some momentum (i.e. momentum per unit volume $= \rho \times V$) is transferred from the region of high velocity to that of low-velocity. The fluid tends to resist the shear associated with the transfer of momentum, and the shear stress is proportional to the rate of transfer of momentum.

In laminar flows ([1]), Newton's law of viscosity relates the shear stress to the rate of angular deformation:

$$\tau = \mu \times \frac{\partial v}{\partial y} \tag{1.1}$$

---

[1] That is, for a Reynolds number less than 1,000 to 3,000 in industrial pipes.

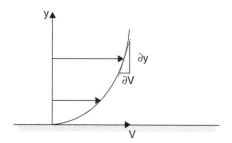

*Figure 1.5* Sketch of a shear flow.

where $\tau$ is the shear stress between two adjacent laminae of fluid, $\mu$ is the dynamic viscosity of the flowing fluid, v is the local fluid velocity and y is the direction normal to the flow direction (Fig. 1.5).

**Note**

The velocity gradient $\partial v/\partial y$ is related to the angular deformation. It is also called the strain rate.

For large shear stresses, the fluid cannot sustain the viscous shear stress and some turbulence spots develop. After the apparition of turbulence spots, the turbulence expands rapidly to the entire shear flow. The apparent shear stress in turbulent flows is expressed as:

$$\tau = \rho \times (v + v_T) \times \frac{\partial v}{\partial y} \tag{1.2}$$

where $v$ is the fluid kinematic viscosity (i.e. $v = \mu/\rho$) and $v_T$ is called the *momentum exchange coefficient* (or "eddy viscosity") in the turbulent flow. The momentum exchange coefficient $v_T$ is a factor depending upon the flow motion.

Practically, $v_T \gg v$ in turbulent flows and Equation (1.2) is often approximated as:

$$\tau = \rho \times v_T \times \frac{\partial v}{\partial y} \tag{1.3}$$

In turbulent flows, the momentum exchange coefficient $v_T$ and the shear rate $\partial v/\partial y$ may be related by introducing the concept of mixing length:

$$v_T = l_m^2 \times \frac{\partial v}{\partial y} \tag{1.4}$$

The mixing length $l_m$ is the characteristic distance travelled by a particle of fluid before its momentum is changed by the new environment.

**Notation**

Herein the notation V characterises the cross-sectional averaged velocity. The symbol v defines the local velocity. The subscript x and y may be introduced to characterise the x- and y-velocity components where x is aligned with the main flow direction and y is the distance normal to the solid boundary.

> **Remarks**
>
> 1.  The concept of momentum exchange coefficient was first introduced by the Frenchman J.V. Boussinesq (1877, 1896). The momentum exchange coefficient is sometimes called "eddy viscosity" or "turbulent viscosity".
> 2.  Joseph Valentin Boussinesq (1842–1929) was a French hydrodynamicist and Professor at the Sorbonne University (Paris) (Bois 2007). His treatise "Essai sur la théorie des eaux courantes" (1877) remains an outstanding contribution in hydraulic engineering literature.
> 3.  The mixing length theory is a turbulence theory developed by L. Prandtl (Prandtl 1925).
> 4.  Ludwig Prandtl (1875–1953) was a German physicist and aerodynamicist who introduced the concept of boundary layer (Prandtl 1904). He was Professor at the University of Göttingen.

At the boundary ($y = 0$, Fig. 1.5), the fluid shear stress equals the boundary shear stress $\tau_O$. The *boundary shear stress* $\tau_O$ represents the shear force per unit area transferred between the fluid and the boundary.

The *shear velocity* $V_*$ is defined as:

$$V_* = \sqrt{\frac{\tau_O}{\rho}} \tag{1.5}$$

where $\rho$ is the density of the flowing fluid. The shear velocity is a measure of shear stress and velocity gradient near the boundary.

### Application: the Couette flow

A simple shear flow is the steady flow between two parallel plates moving at different velocities and called a Couette flow. The Couette flow is characterised by a constant shear stress distribution.

Let us consider the two-dimensional flow between a fixed plate and a moving plate (Fig. 1.6). In the laminar flow regime, the shear stress $\tau$ is a constant, and the integration of Equation (1.2) yields a linear velocity profile. The boundary conditions are $v(y = 0) = 0$ and $v(y = D) = V_O$ where D is the space between the plate and $V_O$ is the speed of the moving boundary (Fig. 1.6). That, the velocity distribution of a laminar Couette flow is

$$\frac{v}{V_O} = \frac{y}{D} \tag{1.6}$$

With increasing Reynolds numbers, some turbulence develops. The velocity profile exhibits a S-shape at high Reynolds numbers although the shear stress distribution remains constant between the plates and is equal to that at the wall $\tau_o$. Two-dimensional turbulent Couette flows are observed for $\rho \times V_O \times D/\mu > 3\,E+3$.

**Remarks**

1. Maurice Marie Alfred Couette (1858–1943) was a French scientist who measured experimentally the viscosity of fluids with a rotating viscometer (Couette 1890).
2. In a plane Couette flow, the shear stress distribution is uniform:

$$\tau(y) = \tau_O = \text{constant} \quad 0 \leq y/D \leq 1$$

3. Recent investigations in the plane Couette flow include Tsukahara et al. (2010).
4. A rotating viscometer consists of two co-axial cylinders rotating in opposite direction. It is used to measure the viscosity of the fluid placed in the space between the cylinders. In a steady state, the torque transmitted from one cylinder to another is proportional to the fluid viscosity and relative angular velocity.

### Discussion: turbulent Couette flow

Turbulent Couette flows are experienced for $\rho \times V_O \times D/\mu > 3\,E{+}3$. In the Couette flow, the shear stress distribution is uniform and the momentum exchange coefficient may be estimated by a parabolic shape:

$$v_T = K \times V_* \times y \times \left(1 - \frac{y}{D}\right)$$

where $K$ is the von Karman constant ($K = 0.40$), $V_*$ is the shear velocity, D is the distance between plates and y is the distance normal to the plates with $y = 0$ at one plate.
    The velocity distribution between two smooth plates follows:

$$\frac{v}{V_*} = \frac{\frac{V_O}{2}}{V_*} + \frac{1}{K} \times Ln\left(\frac{\frac{y}{D}}{1 - \frac{y}{D}}\right) - 0.41 \times \left(1 - \frac{2 \times y}{D}\right)$$

with:

$$\frac{\frac{V_O}{2}}{V_*} = \frac{1}{K} \times Ln\left(\frac{V_* \times \frac{D}{2}}{v}\right) + 7.1$$

    The result was found to be in good agreement with experiments (Schlichting and Gersten 2000). It is sketched in Figure 1.6.

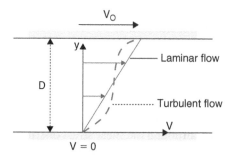

*Figure 1.6*  Sketch of a steady Couette flow.

# Exercises

## Exercise No. 1

Let us consider the following fluid flows:

| Fluid | Density kg/m³ | Dynamic viscosity Pa.s | Channel dimensions | V Laminar-Turbulent m/s |
|---|---|---|---|---|
| Water | 998.2 | 1.005 E−3 | Ø = 0.15 m | |
| Air | 1.2 | 1.7 E−5 | Wind tunnel (3 m × 2 m) | |
| Bentonite suspension (mud) | 1100 | 0.15 | Rectangular open channel (0.35 m wide × 0.10 m depth) | |
| Blood | 1050 | 4 E−3 | Ø = 2.2 mm | |

In each case, calculate the critical range of flow velocities above which the flow becomes turbulent.

### Solution

| Fluid | $D_H$ Hydraulic diameter | V |
|---|---|---|
| Water | 0.15 m | 1 to 7 cm/s |
| Air | 2.4 m | 0.6 to 6 cm/s |
| Bentonite suspension (mud) | 0.25 m | 0.5 to 5 m/s |
| Blood | 2.2 mm | 1.7 to 17 m/s |

## Exercise No. 2

(a)  Considering a plane Couette flow, the gap between the plates is 1 mm. One plate is fixed and the other is moving at 10 cm/s. If the measured shear stress is 1.34 Pa, calculate the fluid viscosity?

(b)  For the same experiment and fluid, what the maximum plate speed to ensure a laminar Couette flow motion.

*The fluid density is 960 kg/m³.*

### Solution

(a)  $\mu = 0.013$ Pa.s
(b)  $V_O = 41$ m/s

## Exercise No. 3

A rotating viscometer consists of two 0.6 m long co-axial cylinders with Ø = 0.3 m and 0.32 m respectively. (The gap between the cylinders is 1 cm.) The external cylinder is fixed and the inner cylinder is rotating at 2.5 rpm.

(a)   Calculate the shear stress on the cylinder walls.
(b)   What is the shear stress in the fluid at 5 mm from the walls?
(c)   Calculate the power required to drive the rotating viscometer.

*The fluid is a SAE40 oil (density: 871 kg/m³, viscosity: 0.6 Pa.s).*
   *Remember, the power equals the product of the angular velocity time the torque between the fluid and the moving cylinder.*

### Solution

(a)   $\tau_0 = 2.43$ Pa
(b)   $\tau = 2.43$ Pa (Remember: the Couette flow is characterised by a constant shear stress distribution)
(c)   0.06 W (Remember: the power equals the force times the velocity, or the torque times the angular velocity)

## Exercise No. 4

A blood solution is tested in a cylindrical Coutte viscosimeter. The apparatus is 0.100 m high. The inner, rotating cylinder has an outer diameter of 40.0 mm and the outer (fixed) cylinder has an inner diameter of 40.4 mm. The rotation speed is 6.1 rpm.

(a)   Calculate the shear stress on the outer cylinder wall.
(b)   What is the shear stress in the fluid at 0.2 mm from the walls?
(c)   Sketch the velocity profile between the cylinders.
(d)   Calculate the torque on the inner cylinder.
(e)   Calculate the power required to drive the viscometer.

*The blood density is 1051 kg/m³ and its viscosity is 4.1 E−3 Pa.s.*

### Solution

(a)   The shear stress distribution is uniform between the two cylinders. Assuming a quasi-two-dimensional Couette flow:

$$\tau = \tau_O = \mu \times V/D = 0.262 \, Pa$$

where $V = R \times \omega = 0.0128$ m/s and $D = 0.2$ mm.
(d)   Torque $= 6.6$ E−5 N/m
(e)   Power $= 4.2$ E−5 W

## Exercise No. 5

Considering a plane flow between two plates, one plate is at rest while the other is moving at speed $V_o = 0.75$ m/s. The gap between the plates is 5 mm and the fluid is a viscous oil ($\rho = 1{,}050$ kg/m³, $\mu = 0.011$ Pa.s). Calculate the shear stress on the plate at rest and the flow rate per unit width between the plates.

### Solution

The Reynolds number is: $\rho \times V_o \times D/\mu = 360$ (laminar flow motion)

$$\tau_o = \mu \times V_o/D = 1.65 \, Pa$$

$$q = \int_0^D V \times dy = 1.875 \times 10^{-3} \, m^2/s$$

# An introduction to turbulence

## SUMMARY

In this chapter, the basic characteristics of turbulence are introduced. Some turbulent flow properties are described and discussed. Some basic equations are developed.

## I   PRESENTATION

Turbulence describes a flow motion characterised by an unpredictable, pseudo-random behaviour, some very-strong mixing properties and a broad spectrum of time and length scales (Lesieur 1994). Bradshaw (1971, p. 17) added: "turbulence is a three-dimensional time-dependent motion in which vortex stretching causes velocity fluctuations to spread to all wavelengths between a minimum determined by viscous forces and a maximum determined by the boundary conditions of the flow".

Figure 2.1 illustrates some instantaneous velocity measurements in turbulent flows. Figure 2.1A presents a 2-minutes record of the longitudinal velocity component in a developing boundary layer. Figure 2.1B illustrates the fluctuations of the longitudinal velocity component in a small estuary for about 17 minutes during the ebb tide. Both data sets show some quasi-random fluctuations of the longitudinal velocity around the mean. They highlight further the rapid fluctuations in turbulent velocities.

Appendix F includes several movies illustrating turbulence in geophysical flows: e.g., movie 'Developing shear layer and large vortical structures' in a large estuary, movie 'Smoke plume in the atmosphere' in the atmospheric boundary layer, movie 'Turbulence and free-surface vortices in an inundated urban environment' and movies 'Turbulence and free-surface vortices during a major flood' in a river during a major flood. All the movies highlight the mixing induced by the flow turbulence as well as a wide range of turbulent structure dimensions.

**Remarks**

1. The turbulence may be analysed in terms of the statistical properties of the velocity components. However the turbulence scales are of interest in addition to the turbulence intensity and statistical moments of the turbulent velocity fluctuations. Turbulent scales may be obtained from either spatial correlation using two measurement devices or from the auto-correlation calculated from the time signal measured by a single measurement system.
2. *Homogeneous turbulence* is statistically independent of the position in space.
3. *Isotropic turbulence* is statistically independent of the direction: e.g., $v'_x = v'_y = v'_z$ where $v'$ is the standard deviation of the velocity component V.

4.  A flow is statistically *stationary* if the time averaged velocity $\bar{v}$, and the time averaged pressure $\bar{P}$, are independent of the time t at mid-point. That is:

$$\bar{v} = \frac{1}{2 \times T} \times \int_{-T+t}^{t+T} v(t') \times dt' = \text{constant}$$

$$\bar{P} = \frac{1}{2 \times T} \times \int_{-T+t}^{t+T} P(t') \times dt' = \text{constant}$$

(A) Developing boundary layer flow in a laboratory open channel – Data: Koch and Chanson (2005), d = 0.08 m, δ = 0.056 m, y/δ = 0.31, $V_O$ = 1.12 m/s, sampling rate: 25 Hz

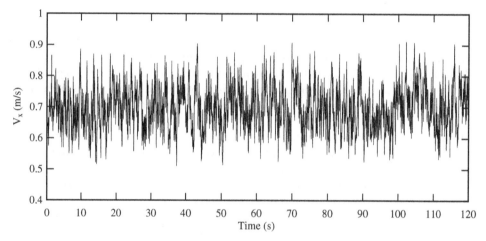

(B) Turbulence measurements in a small estuary – Data: Trevethan et al. (2006), sampling point at 0.2 m above the bed, mid-ebb tide, sampling rate: 25 Hz

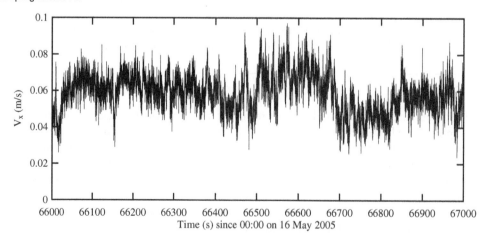

*Figure 2.1* Instantaneous longitudinal velocity measurements.

## 2   TURBULENCE CHARACTERISTICS

### 2.1   Presentation

In a steady turbulent flow, the instantaneous velocity is usually decomposed into a time-averaged component and a turbulent fluctuation. If the instantaneous velocity is V, $\bar{v}$ denotes the time-averaged velocity and v is the instantaneous velocity fluctuation defined as:

$$V = v - \bar{v} \tag{2.1}$$

In a three dimensional flow, the Cartesian velocity components are denoted $V_x$, $V_y$, and $V_z$, where x is the streamwise direction, y is the direction normal to the boundary (e.g. vertical direction in a horizontal channel) and z is the transverse direction.

---

**Discussion**

When the flow is unsteady, a time average is not meaningful because the long-term trend and the fast turbulent fluctuations must be processed separately. The turbulence in unsteady flows may be analysed by two techniques: ensemble-averaging and variable interval time averaging.

If the unsteady turbulent flow is gradually-varied with some distinct long-term and short-term fluctuation frequencies, $\bar{v}$ is a low-pass filtered component, sometimes called variable-interval time average or VITA (Piquet 1999, p. 28). The cutoff frequency $1/(2 \times T)$ must be selected such that the averaging time $2 \times T$ is greater than the characteristic period of fluctuations, and small with respect to the characteristic period for the time-evolution of the mean properties:

$$\frac{|\bar{v}_i(t + T) - \bar{v}_i(t - T)|}{\bar{v}_i(t)} \ll 1$$

In a transient highly unsteady flow, the experiments have to be repeated many times. Each time, the instantaneous velocity is recorded at a point at a particular instant. The average of all these instantaneous measurements, namely the ensemble average, is the appropriate mean velocity at the point at that particular instant (Bradshaw 1971, pp. 9–10). The deviation of the instantaneous velocity from the ensemble average is the turbulent velocity fluctuation.

---

### 2.2   Turbulent (Reynolds) stresses

In a turbulent flow, the flux of the x-momentum in the y-direction induces an additional shear stress in the x-direction acting on the surface element normal to the y-direction. This additional stress is called the Reynolds stress or turbulent stress. It is denoted $\rho \times v_x \times v_y$, or more generally $\rho \times v_i \times v_j$ where $i, j = x, y, z$. The Reynolds stress $\rho \times v_i \times v_j$ characterises the additional shear stress on the faces $dx_i\, dx_j$ of an elementary control volume (dx, dy, dz), where $x_i, x_j = x, y, z$.

The Reynolds stresses are a transport effect resulting from turbulent motion induced by velocity fluctuations with its subsequent increase of momentum exchange and of mixing (Piquet 1999, p. 47). Turbulent transport is a property of the flow itself. It is not a property of the fluid. The *Reynolds stress tensor* include the normal stresses ($\rho \times v_i \times v_i$) and the tangential stresses ($\rho \times v_i \times v_j$) with $i \neq j$ (Hinze 1975, p. 30). The tangential stresses are usually negative. There is however no fundamental difference between the normal stresses and tangential stresses (Bradshaw 1971, p. 24). For example, $(v_x + v_y)/\sqrt{2}$ is the component of the velocity fluctuation along a line in the xy-plane at 45 degrees to the x-axis; hence its mean square $(v_x^2 + v_y^2 + 2 \times v_x \times v_y)/2$ is the component of the normal stress (over the density) in this direction, although it is a combination of normal and tangential stresses in the x- and y-axes.

---

## Remarks

1.  The Reynolds stresses are a fundamental concept in turbulent flows. They are called also turbulent stresses.
2.  The turbulent stresses are defined in terms of the instantaneous velocity fluctuation: $v = V - \bar{v}$.
3.  At the solid boundary surface, the tangential stress $\rho \times v_x \times v_y$ tends to the boundary shear stress $\tau_O$.
4.  The Reynolds stress tensor is symmetrical since:

$$\rho \times v_i \times v_j = \rho \times v_j \times v_i$$

5.  In open channel flows, the skewness and kurtosis of the Reynolds stress fluctuations are large (Nezu and Nakagawa 1993, p. 181, Trevethan et al. 2006, 2007).

---

### Discussion

A Reynolds stress may be expressed in a dimensionless form as a coefficient of correlation. The normalised cross-correlation coefficient is related to the Reynolds stress as:

$$R_{ij} = \frac{\rho \times \overline{v_i \times v_j}}{\rho \times \sqrt{\overline{v_i^2}} \times \sqrt{\overline{v_j^2}}} \tag{2.2}$$

where $R_{ij}$ is the normalised cross-correlation coefficient function, i, j = x, y, z, and the time lag $\tau$ is zero $(R_{ij} = R_{ij}(\tau = 0))$. The mean product $\overline{v_i \times v_j}$ is called the *co-variance* of $v_i$ and $v_j$. Equation (2.2) is a dimensionless turbulent stress.

A triple correlation characterises the transport of turbulent shear stress by velocity fluctuations (Bradshaw 1976, p. 27). The triple correlation is defined as:

$$R_{ijk} = \frac{\overline{v_i \times v_j \times v_k}}{\sqrt{\overline{v_i^2}} \times \sqrt{\overline{v_j^2}} \times \sqrt{\overline{v_k^2}}} \tag{2.3}$$

Physically, the gradient of the triple correlation $R_{xxx}$ contributes to the streamwise diffusive flux of the streamwise kinetic energy $\overline{v_x^2}$ (Tachie 2001). $\overline{v_x \times v_x \times v_y}$ represents the transport of $\overline{v_x \times v_y}$ in the x direction, while $\overline{v_x \times v_y \times v_y}$ describes the turbulent transport of $\overline{v_x \times v_y}$ in the y direction. More generally, $R_{ijk}$ characterises in dimensionless form the transport of $\overline{v_j \times v_k}$ in the i direction, which is equal to the transport of $\overline{v_i \times v_j}$ in the k direction.

---

**Remarks**

1. The term $\sqrt{\overline{v_i^2}}$ is the standard deviation of the velocity component $v_i$:

$$\sqrt{\overline{v_i^2}} = \sqrt{\overline{v_i \times v_i}}$$

It is also denoted $\sqrt{v_i'^2}$, and sometimes $v_i'$.

2. The co-variance is defined as:

$$\overline{v_i \times v_j} = \frac{1}{2 \times T} \times \int\limits_{-T+t}^{t+T} v_i(t') \times v_j(t') \times dt'$$

at a time t.

3. In discussing the Reynolds stress equations, Piquet (1999) introduced the triple correlation tensor as the diffusion tensor, or transport of Reynolds stresses. The diffusion tensor is built with three contributions: the first (i.e. the triple correlation itself) is the turbulent diffusion of kinetic energy in the direction $x_i$, the second is so-called pressure diffusion effect, while the last term is almost equal to the molecular transport of a quantity K (Piquet 1999, pp. 59–60).

---

## 2.3  Turbulence statistics

In the studies of turbulence, the measured statistics include usually: (a) the spatial distribution of Reynolds stresses and of the turbulent kinetic energy per unit volume

$$0.5 \times \rho \times (\overline{v_x^2} + \overline{v_y^2} + \overline{v_z^2})$$

which equals half of the sum of the normal Reynolds stresses, (b) the rates at which turbulent kinetic energy, or individual Reynolds stresses, are produced, destroyed, or transported from one point in space to another, (c) the contribution of different sizes of eddy to the Reynolds stresses, and (d) the contribution of different sizes of eddy to the rates mentioned in (b) and to rate at which energy or Reynolds stresses are transferred from one range of eddy size to another (Bradshaw 1971, pp. 22–23).

### Note

The turbulence structure discussed in terms of dimensionless turbulent properties (e.g. correlation shapes, ratios like $\overline{v_y}/\overline{v_x}$), as opposed to absolute values like integral scales or intensities (Bradshaw 1971, p. xvii).

### *Discussion*

Although turbulence is a pseudo-random process, hence quasi-Gaussian, "the small departures from a Gaussian probability distribution are the most interesting features of the turbulence: for instance, the triple products like $\overline{v_x \times v_x \times v_y}$, which would be zero in a Gaussian process, are connected with energy transfer by the turbulence" (Bradshaw 1971, p. 22).

The skewness $Sk_i$ and kurtosis $Ku_i$ give some information regarding the temporal distribution of the velocity fluctuation around its mean value, where the subscript i refers to the velocity component: $i = x, y, z$. A non-zero skewness indicate some degree of temporal asymmetry of the turbulent fluctuation: e.g., acceleration versus deceleration, sweep versus ejection. The skewness retains some sign information and it can be used to extract basic information without ambiguity. A kurtosis larger than zero is associated with a peaky signal: e.g., produced by intermittent turbulent events.

---

**Remarks**

1. The Gaussian distribution is also called the normal distribution.
2. For a Gaussian (random) distribution, the skewness is zero: $Sk = 0$. A positive skewness indicates a preponderance of smaller data relative to the mean and a long right-sided tail. Conversely, a negative skewness indicates that the probability distribution function of the data has a long left-sided tail.
3. A common kurtosis definition is the Fisher kurtosis, also called excess kurtosis, and defined as:

$$Ku_i = \frac{\overline{v_i^4}}{(\overline{v_i^2})^2} - 3$$

For a Gaussian (random) distribution, $Ku = 0$. A positive kurtosis indicates a velocity probability density function more "peaky" than a Gaussian distribution.

---

Figure 2.2 presents the probability distribution function of some turbulent velocity data collected in a developing boundary layer. For the 100 s sampling duration, the time-averaged longitudinal velocity was 10.14 m/s, and the maximum and minimum recorded velocities were 14.55 and 4.14 m/s respectively. The probability distribution is compared with a Gaussian distribution in Figure 2.2, and the standard deviation, skewness and kurtosis are listed in the figure caption.

## 2.4 Eulerian auto-correlation and turbulent time scales

Let us define the normalised auto-correlation function $R_{ii}(\tau)$ of the i-velocity fluctuations:

$$R_{ii}(\tau) = R(v_i(t), v_i(t - \tau)) = \frac{\overline{v_i(t) \times v_i(t + \tau)}}{\overline{v_i^2}} \tag{2.4}$$

for a single-point measurement, where $\tau$ is the time lag, V equals: $V = v - \overline{v}$, $\overline{v}$ is the time average velocity component, v is the instantaneous velocity, $i = x, y, z$, and R is the co-variance function.

Figure 2.3 shows a typical auto-correlation function. The auto-correlation function is unity for a zero time lag ($\tau = 0$) and it is between $-1$ and $+1$ for $\tau \neq 0$. Figure 2.3B presents an experimental data in a small estuary.

| Velocity property | Time-average (m/s) | Standard deviation (m/s) | Skewness | Kurtosis |
|---|---|---|---|---|
| | 10.14 | 1.311 | −0.7575 | +0.1168 |

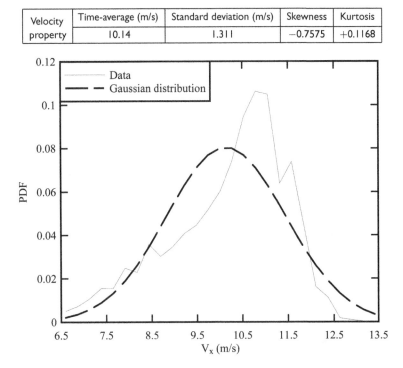

*Figure 2.2*  Probability distribution function of the longitudinal velocity $V_x$ in a developing boundary layer above a rough plate – Wind tunnel data, $x = 0.98$ m, $y = 0.025$ m, sampling rate: 400 Hz, sampling duration: 100 s, hot-wire anemometry.

---

**Remarks**

1. The co-variance $\overline{v_i(t) \times v_j(t + \tau)}$ is defined as:

$$\overline{v_i(t) \times v_j(t + \tau)} = \frac{1}{2 \times T} \times \int_{-T+t}^{t+T} v_i(t') \times v_j(t' + \tau) \times dt'$$

where $\tau$ is the time lag.

2. The normalised auto-correlation coefficient $R_{ii}$ ranges between $-1$ and $+1$.

---

The *dissipation time scale* can estimated from the curvature of the auto-correlation function at the origin $\tau = 0$ (Fig. 2.3). It is also called the Taylor micro scale. In the neighbourhood of $\tau = 0$, the shape of the auto-correlation function is about:

$$R_{xx}(\tau) \approx 1 - \left(\frac{\tau}{\tau_E}\right)^2 \qquad \tau \approx 0 \tag{2.5}$$

(A) Definition sketch of a turbulent velocity auto-correlation function

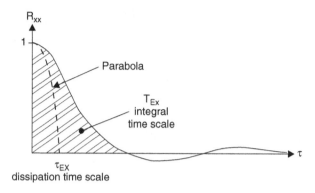

(B) Velocity auto-correlation functions of the longitudinal velocity $v_x$ in a developing boundary layer above a rough plate – Wind tunnel data, $x = 0.3$ m, $y = 0.025$ m, sampling rate: 400 Hz, sampling duration: 100 s, hot-wire anemometry

| Velocity property | Time-average (m/s) | Standard deviation (m/s) | Skewness | Kurtosis |
|---|---|---|---|---|
| | 10.20 | 1.337 | −1.176 | +1.585 |

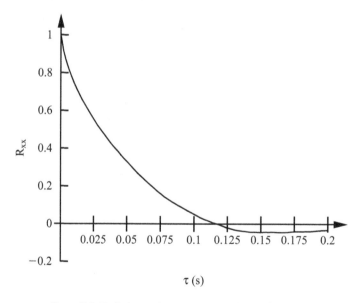

*Figure 2.3* Turbulent velocity auto-correlation function.

where $\tau_E$ is the Eulerian dissipation time scale. $\tau_E$ is a measure of the most rapid changes that occur in the fluctuations of $v_x(t)$ (Hinze 1975, p. 45). Indeed it is reasonable to expect a relationship between $\tau_E$ and the highest frequencies occurring in the fluctuations. $\tau_E$ is viewed as the smallest energetic time scale.

**Discussion**

Using a Taylor series expansion of the auto-correlation function, the dissipation time scale may also be expressed as:

$$\tau_E = \sqrt{2 \times \frac{\overline{v_i^2}}{\left(\frac{\partial v_i}{\partial t}\right)^2}}$$

Note that there is a dissipation time scale for each velocity component: i = x, y, z.

Several researchers used the above result to obtain a measured time scale $\tau_E^{(\delta t)}$ based upon the time increment $\delta t$:

$$\tau_E^{(\delta t)} = \sqrt{2 \times \frac{\overline{v_i^2}}{\left(\frac{\delta \overline{v_i}}{\delta t}\right)^2}}$$

where $1/\delta t$ is the sampling rate. The calculation is repeated for several time increments $\delta t$, and the limit of $\tau_E^{(\delta t)}$ for $\delta t \to 0$ is the dissipation time scale $\tau_E$. Practically, the relationship between $\tau_E^{(\delta t)}$ and $\delta t$ may be approximated by a quadratic interpolation function:

$$\tau_E^{(\delta t)} = \tau_E + a_1 \times \delta t + a_2 \times \delta t^2$$

The above method is deemed a more accurate way to estimate the dissipation time scale $\tau_E$ (Hallback et al. 1989, Fransson et al. 2005, Koch and Chanson 2005).

The *Eulerian integral time scale* is defined as:

$$T_E = \int_{\tau=0}^{+\infty} R_{ii}(\tau)\, d\tau \tag{2.6}$$

where $R_{ii}(\tau)$ is the normalised auto-correlation function of the turbulent velocity fluctuation $v_i$. The integral time scale $T_E$ is a rough measure of the longest connection in the turbulent behaviour of $v_i(\tau)$. It is also called macro time scale. It may be related to the space integral scale $\Lambda_f$ by:

$$\Lambda_f = \overline{V} \times T_E$$

The integral scale $\Lambda_f$ is a measure of the longest connection (or correlation distance) between the velocities at two points of the flow field (Hinze 1975, p. 43).

Often an upper bound to the integral time scale $T_E$ calculation is introduced:

$$T_E = \int_{\tau=0}^{t(R_{ii}(\tau)=0)} R_{ii}(\tau) \times d\tau \tag{2.7}$$

Figure 2.3A illustrates the definition of the integral time scale.

## 3   MEAN FLOW EQUATIONS

### 3.1   Presentation

In a turbulent flow, the velocity and pressure may be divided into a time-average component plus a turbulent fluctuation component. That is:

$$V = \bar{v} + v \tag{2.8}$$

where V is the instantaneous velocity, $\bar{v}$ is the time averaged velocity, and v is the instantaneous velocity fluctuation. Similarly for the pressure:

$$P = \bar{p} + p \tag{2.9}$$

The minuscule refers to the fluctuating parameter and the overbar refers to the time-averaged quantity. The time-average of a parameter V at a time t is defined as:

$$\bar{v} = \frac{1}{2 \times T} \times \int_{t-T}^{t+T} v \times dt \tag{2.10}$$

where T is significantly longer than the characteristic turbulence time scale.

In a steady flow, the limits of the integration must be large such that the time-average becomes independent of the limits. In an unsteady flow, the limits must be large in comparison with the turbulent time scales but small compared to the time scale of the flow motion (Bradshaw 1971, Liggett 1994). The term "unsteady flow" is used in the sense that the time-average velocities vary with time. Figure 2.4 illustrates the differences between steady and unsteady flows. Figure 2.4A shows a steady flow data set, and Figure 2.4B presents some unsteady flow data. In both graphs, the time-average velocity is shown. The data set shown in Figure 2.4B corresponds to some turbulent velocity measurements beneath an undular tidal bore. Figure 2.5 illustrates such a phenomenon in Nature.

---

**Remarks**

1.   In steady turbulent flows, the sampling duration does influence the results because the turbulence characteristics may be biased with small sample numbers. Some basic turbulence studies showed the needs for larger sample sizes: e.g., 60,000 to 90,000 data samples per sampling location (Karlsson and Johansson 1986, Krogstad et al. 2005, Chanson et al. 2007).

2.   In presence of some slow-frequency flow fluctuations, a triple decomposition of the instantaneous velocity data may be performed. The instantaneous velocity time-series may be represented as a superposition of three components:

$$V = <V> + [V] + v$$

where V is the instantaneous velocity, <V> is the mean velocity contribution, [V] is the slow fluctuating component of the velocity and v corresponds to the turbulent motion. The technique was previously applied to periodic turbulent flows and riverine flows with large coherent structures (Hussain and Reynolds 1972, Fox et al. 2005, Brown and Chanson 2013).

(A) Steady developing boundary layer flow: $y/d = 0.72$, $y/\delta \approx 1$, $d = 0.079$ m and $V_O = 1.12$ m/s

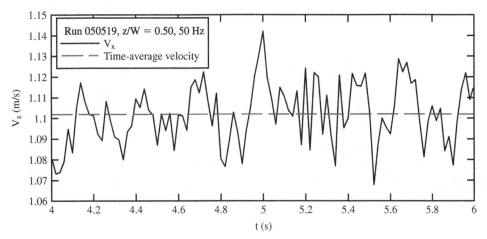

(B) Unsteady flow beneath an undular bore (Fr = 1.4) at $y = 0.057$ m on the channel centreline

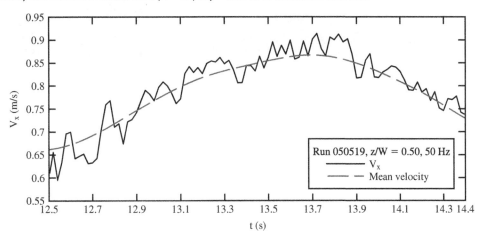

*Figure 2.4* Turbulence velocity fluctuations in steady and unsteady flows: longitudinal velocity component $v_x$ in an open channel flow (Data: Koch and Chanson 2005).

## Notes: time-average properties

– By definition, the time-average of a fluctuating parameter is zero. That is:

$$\frac{1}{2 \times T} \times \int_{t-T}^{t+T} v \times dt = 0$$

– Several properties of the time-average are:

$$\overline{v + p} = \overline{v} + \overline{p}$$

*Figure 2.5*  Undular tidal bore of the Dordogne River at St Pardon (France) on 23 September 2010 – Bore front propagation from left to right.

$$\overline{\frac{\partial v}{\partial x}} = \frac{\partial \overline{v}}{\partial x}$$

$$\overline{\int v \times dx} = \int \overline{v} \times dx$$

– Further properties of the fluctuating component are:

$$\overline{v} = 0$$

$$\overline{v \times v} \geq 0$$

– Note that:

$$\overline{v \times p} = 0$$

only and only if v and p are independent and un-correlated.

---

**Discussion: tidal bores**

A tidal bore is a series of waves propagating upstream as the tidal flow turns to rising (Fig. 2.5). It forms during spring tide conditions when the tidal range exceeds 4 to 6 m and the flood tide is confined to a narrow funnelled estuary (Chanson 2011). A tidal bore occurs only during the flood tide under spring tidal ranges and low freshwater flow conditions. Its passage is very rapid: e.g., from a few seconds to a few minutes at most. It is so easy to miss ... A classical example is the Qiantang River tidal bore that has been documented for nearly 3,000 years; but Marco Polo who visited the city of Hangzhou on the Qiantang River never mentioned its bore called the "Silver Dragon".

There are some related geophysical, as well as man-made, processes. These include the storm-surge-induced bores in the Bay of Bengal, the tsunami-induced bores and the positive surges in canals. In the Bay of Bengal, the development of a storm surge during the early flood tide with spring tidal conditions may yield a rapid rise in water level generating a bore front. The wind shear amplifies the tidal range and the phenomenon has been observed in Bangladesh where the storm events are called locally 'tidal bores' (Sommer and Mosley 1972). Another related process is the tsunami-induced bore (Shuto 1985). In shallow bay and river systems, the tsunami wave is somehow similar to a tidal bore and the tsunami-induced bore may propagate far upstream. During the 26 December 2004 and 10 March 2011 tsunami catastrophes, a number tsunami-induced bores were observed in shallow-water bays and river mouths in Malaysia, Thailand, Sri Lanka and Japan.

## 3.2 Continuity and momentum equations

For an incompressible turbulent flow motion, the fluid density is constant and does not need to be divided into time-average and fluctuation quantities. The differential form of the continuity equation is:

$$\frac{\partial v_x}{\partial x} + \frac{\partial v_y}{\partial y} + \frac{\partial v_z}{\partial z} = 0 \tag{2.11}$$

It may be transformed by introducing the velocity time-average and fluctuation quantities:

$$\frac{\partial}{\partial x}(\bar{v}_x + v_x) + \frac{\partial}{\partial y}(\bar{v}_y + v_y) + \frac{\partial}{\partial z}(\bar{v}_z + v_z) = 0 \tag{2.11b}$$

Let us time-average the continuity equation (Eq. (2.11)). Since the time-average of a fluctuating quantity is zero, the continuity equation of the time-average velocity is the same as the continuity equation of the instantaneous velocity:

$$\frac{\partial \bar{v}_x}{\partial x} + \frac{\partial \bar{v}_y}{\partial y} + \frac{\partial \bar{v}_z}{\partial z} = 0 \tag{2.12}$$

Note that the fluctuating velocities are also continuous:

$$\frac{\partial v_x}{\partial x} + \frac{\partial v_y}{\partial y} + \frac{\partial v_z}{\partial z} = 0 \tag{2.13}$$

For a steady flow, the Navier-Stokes equations are:

$$\rho \times \left( v_x \times \frac{\partial v_x}{\partial x} + v_y \times \frac{\partial v_x}{\partial y} + v_z \times \frac{\partial v_x}{\partial z} \right) = -\rho \times g \times \frac{\partial z_O}{\partial x} - \frac{\partial p}{\partial x} + \sum_j \mu \times \frac{\partial^2 v_x}{\partial x_j \, \partial x_j} \tag{2.14a}$$

$$\rho \times \left( v_x \times \frac{\partial v_y}{\partial x} + v_y \times \frac{\partial v_y}{\partial y} + v_z \times \frac{\partial v_y}{\partial z} \right) = -\rho \times g \times \frac{\partial z_O}{\partial y} - \frac{\partial p}{\partial y} + \sum_j \mu \times \frac{\partial^2 v_y}{\partial x_j \, \partial x_j} \qquad (2.14b)$$

$$\rho \times \left( v_x \times \frac{\partial v_z}{\partial x} + v_y \times \frac{\partial v_z}{\partial y} + v_z \times \frac{\partial v_z}{\partial z} \right) = -\rho \times g \times \frac{\partial z_O}{\partial z} - \frac{\partial p}{\partial z} + \sum_j \mu \times \frac{\partial^2 v_z}{\partial x_j \, \partial x_j} \qquad (2.14c)$$

where $j = x$, $y$, $z$, and $z_O$ is the vertical elevation. For the sake of simplicity, the Navier-Stokes equations (Eq. (2.14)) may be written as a series of three equations:

$$\sum_j \rho \times v_j \times \frac{\partial v_i}{\partial x_j} = -\rho \times g \times \frac{\partial z_O}{\partial x_i} - \frac{\partial p}{\partial x_i} + \sum_j \mu \times \frac{\partial^2 v_i}{\partial x_j \, \partial x_j} \qquad (2.15)$$

for $x_i$, $x_j$, $x_k = x$, $y$, $z$.

**Notes**

- The Navier-Stokes equations form a vector equation:

$$\frac{D(\rho \times \vec{v})}{Dt} = \rho \times \overrightarrow{\text{grad } g \times z_O} - \overrightarrow{\text{grad } P} + \vec{f}_{visc}$$

where $\vec{f}_{visc}$ is the resultant of the viscous forces on the control volume.
- In the Navier-Stokes equation, the velocity is the instantaneous velocity.
- The equation was first derived by Henri Navier in 1822 and Siméon Denis Poisson in 1829 by an entirely different method. They were derived in a manner similar as above by Adhémar Jean Claude Saint-Venant in 1843 and George Gabriel Stokes in 1845.
- Henri Navier (1785–1835) was a French engineer who primarily designed bridges but also extended Euler's equations of motion.
- Siméon Denis Poisson (1781–1840) was a French mathematician and scientist. He developed the theory of elasticity, a theory of electricity and a theory of magnetism.
- The Frenchman Adhémar Jean Claude Barré de Saint-Venant (1797–1886) developed the equations of motion of a fluid particle in terms of the shear and normal forces exerted on it. He is remembered for his contribution to unsteady open channel flow: i.e., the Saint-Venant equations.
- George Gabriel Stokes (1819–1903), British mathematician and physicist, is known for his research in hydrodynamics and a study of elasticity.

In Equation (2.15), the velocity and pressure are the instantaneous velocity and pressure respectively. They may be replaced in terms of their respective time-average and fluctuation using Equations (2.8) and (2.9).

Then, let us "time-average" the Navier-Stokes equations (2.15). The time-averaged Navier-Stokes equations may be simplified since the time-average of a fluctuating parameter is zero. The result yields:

$$\sum_j \rho \times \overline{v_j} \times \frac{\partial \overline{v_i}}{\partial x_j} = -\rho \times g \times \frac{\partial z_O}{\partial x_i} - \frac{\partial \overline{P}}{\partial x_i} + \sum_j \frac{\partial}{\partial x_j} \left( \mu \times \frac{\partial \overline{v_i}}{\partial x_j} - \rho \times \overline{v_i \times v_j} \right) \qquad (2.16)$$

Equation (2.16) is called the *Reynolds Averaged Navier-Stokes* equation (RANS). In the time-averaged Navier-Stokes equation, the last term is a time-averaged turbulent stress. In this form, the Navier-Stokes equation requires 6 additional terms: i.e., the three normal stresses and three tangential stresses (since the Reynolds stress tensor is symmetrical). We now have more unknowns than basic equations because the Reynolds stresses are unknowns. This situation is known as the *closure problem*.

---

**Remarks**

1. The Reynolds averaged Navier-Stokes equation (RANS) (Eq. (2.16)) is also called the Reynolds-Navier-Stokes equation.
2. The closure problem is directly linked with the introduction of the time-average and fluctuating velocities in the Navier-Stokes equation.
3. Traditionally, researchers try to solve the closure problem by "guessing" some relationship, also called constitutive relationship, between the Reynolds stresses and time-average velocities. The type of closure relationship has some significant consequences on the mathematical structure of the problem.

---

Measurements in turbulent flows showed that the turbulent stresses are larger than the turbulent viscous stresses ($\mu \times \partial v/\partial y$). In a channel, the viscous stress is much smaller than the Reynolds stress across most of the flow. In the vicinity of the wall, however, the viscous stress predominates in the "viscous sublayer" (see section in Turbulent boundary layers).

---

**Discussion**

Let us consider the shear flow next to a solid boundary: i.e., a turbulent boundary layer. Very close to the wall, the viscous shear is predominant and the boundary shear stress equals:

$$\tau_O = \mu \times \left(\frac{\partial v_x}{\partial y}\right)_{y=0} \approx \mu \times \frac{v_x}{y} \qquad\qquad \text{very close to the wall}$$

Introducing the shear velocity $V_* = \sqrt{\tau_0/\rho}$, the above equation gives the velocity field in the viscous sublayer immediately next to the wall:

$$\frac{v_x}{V_*} = \frac{\rho \times V_*}{\mu} \times y$$

This result is valid in the viscous layer of a turbulent boundary layer, also called inner wall region (see section on Turbulent boundary layers, Chap. II-4).

---

## 3.3  Application

Some turbulence closure models are based upon the concept of "eddy viscosity" which links the turbulent shear stress to the velocity gradient:

$$\rho \times \overline{v_i \times v_j} = \rho \times \nu_T \times \frac{\partial \overline{v_j}}{\partial x_i} \tag{2.17}$$

where $\nu_T$ is the momentum exchange coefficient or "eddy viscosity". The "eddy viscosity" is not a fluid property, but a characteristic of the turbulent flow motion itself.

---

**Remark**

The concept of "eddy viscosity" was first introduced by Joseph Valentin Boussinesq (1877, 1896) who was a French hydrodynamicist and Professor at the Sorbonne University (Paris) (Bois 2007).

---

A well-known turbulence model is the mixing length model which relates the "eddy viscosity" to the velocity gradient:

$$v_T = l_m^2 \times \frac{\partial v_x}{\partial y} \tag{2.18}$$

where y is the distance normal to the wall, and $l_m$ is the mixing length. $l_m$ is a characteristic distance travelled by a particle of fluid before its momentum is changed by the new environment.

Considering a turbulent flow bounded by a solid boundary (e.g. boundary layer flow, Fig. 1.6), the mixing length may be assumed by be about:

$$l_m = K \times y$$

where $K$ is the von Karman constant ($K = 0.4$).

At the wall, the boundary shear stress equals:

$$\tau_O = \rho \times v_T \times \left(\frac{\partial v_x}{\partial y}\right)_{y=0} \approx K^2 \times y^2 \times \left(\left(\frac{\partial v_x}{\partial y}\right)_{y=0}\right)^2 \qquad \text{very close to the wall}$$

Introducing the shear velocity $V_*$, it yields next to the wall:

$$\frac{\partial v_x}{V_*} = \frac{1}{K} \times \frac{\partial y}{y}$$

The integration gives the logarithmic law in the wall region:

$$\frac{v_x}{V_*} = \frac{1}{K} \times Ln\left(\frac{\rho \times V_* \times y}{\mu}\right) + D_1$$

where $D_1$ is an integration constant. The above result is the well-known law of wall for a developing turbulent boundary layer and it is valid for $y/\delta < 0.1$ to $0.15$ where $\delta$ is the boundary layer thickness. It will be discussed in the section Turbulent boundary layers (Chapter II-4).

---

**Remark**

1.  The mixing length theory is a turbulence model developed by Ludwig Prandtl (Prandtl 1925).
2.  Although the mixing length theory is more than 80 years old, it is an useful model based upon some simple physically-based concept.

---

# Exercises

## Exercise No. 1

Turbulent velocity measurements were conducted in Eprapah Creek at 0.2 m above the creek bed. The data were recorded mid-estuary on 16 May 2005.

| Time (s) | $V_x$ (cm/s) | Time (s) | $V_x$ (cm/s) | Time (s) | $V_x$ (cm/s) | Time (s) | $V_x$ (cm/s) |
|---|---|---|---|---|---|---|---|
| 66030.98 | 0.0548 | 66033.34 | 0.0566 | 66035.82 | 0.0537 | 66038.42 | 0.0583 |
| 66031.02 | 0.0559 | 66033.38 | 0.0576 | 66035.86 | 0.0554 | 66038.46 | 0.0554 |
| 66031.06 | 0.0532 | 66033.42 | 0.054 | 66035.9 | 0.0593 | 66038.5 | 0.059 |
| 66031.1 | 0.0588 | 66033.46 | 0.0578 | 66035.94 | 0.0548 | 66038.54 | 0.0559 |
| 66031.14 | 0.0588 | 66033.5 | 0.0546 | 66035.98 | 0.0616 | 66038.58 | 0.0506 |
| 66031.18 | 0.0568 | 66033.54 | 0.0593 | 66036.02 | 0.0544 | 66038.62 | 0.0629 |
| 66031.22 | 0.0585 | 66033.58 | 0.06 | 66036.06 | 0.051 | 66038.66 | 0.0591 |
| 66031.26 | 0.057 | 66033.62 | 0.0546 | 66036.1 | 0.0542 | 66038.7 | 0.0534 |
| 66031.3 | 0.0592 | 66033.66 | 0.061 | 66036.14 | 0.0575 | 66038.74 | 0.062 |
| 66031.34 | 0.0596 | 66033.7 | 0.0608 | 66036.18 | 0.0496 | 66038.78 | 0.0591 |
| 66031.38 | 0.0602 | 66033.74 | 0.0615 | 66036.22 | 0.0517 | 66038.82 | 0.0549 |
| 66031.42 | 0.0603 | 66033.78 | 0.0616 | 66036.26 | 0.0538 | 66038.86 | 0.0642 |
| 66031.46 | 0.0618 | 66033.82 | 0.064 | 66036.3 | 0.0547 | 66038.9 | 0.0649 |
| 66031.5 | 0.0622 | 66033.86 | 0.0563 | 66036.34 | 0.0506 | 66038.94 | 0.0653 |
| 66031.54 | 0.0643 | 66033.9 | 0.0606 | 66036.38 | 0.0487 | 66038.98 | 0.0687 |
| 66031.58 | 0.0597 | 66033.94 | 0.0615 | 66036.42 | 0.0535 | 66039.02 | 0.0606 |
| 66031.62 | 0.0646 | 66033.98 | 0.06 | 66036.46 | 0.0516 | 66039.06 | 0.0639 |
| 66031.66 | 0.0641 | 66034.02 | 0.0595 | 66036.5 | 0.0513 | 66039.1 | 0.0609 |
| 66031.7 | 0.059 | 66034.06 | 0.0589 | 66036.54 | 0.0494 | 66039.14 | 0.0662 |
| 66031.74 | 0.0612 | 66034.1 | 0.0633 | 66036.58 | 0.0497 | 66039.18 | 0.0708 |
| 66031.78 | 0.0589 | 66034.14 | 0.0589 | 66036.62 | 0.0538 | 66039.22 | 0.0702 |
| 66031.82 | 0.0563 | 66034.18 | 0.0578 | 66036.66 | 0.0503 | 66039.26 | 0.0689 |
| 66031.86 | 0.0549 | 66034.22 | 0.0544 | 66036.7 | 0.0523 | 66039.3 | 0.0617 |
| 66031.9 | 0.0582 | 66034.26 | 0.0548 | 66036.74 | 0.0554 | 66039.34 | 0.0639 |
| 66031.94 | 0.0546 | 66034.3 | 0.0638 | 66036.78 | 0.0567 | 66039.38 | 0.0666 |
| 66031.98 | 0.0567 | 66034.34 | 0.0632 | 66036.82 | 0.057 | 66039.42 | 0.0616 |
| 66032.02 | 0.0551 | 66034.38 | 0.0615 | 66036.86 | 0.0601 | 66039.46 | 0.064 |
| 66032.06 | 0.0542 | 66034.42 | 0.0626 | 66036.9 | 0.0563 | 66039.5 | 0.0686 |
| 66032.1 | 0.0505 | 66034.46 | 0.0637 | 66036.94 | 0.0588 | 66039.54 | 0.0649 |
| 66032.14 | 0.0517 | 66034.5 | 0.0609 | 66036.98 | 0.0584 | 66039.58 | 0.0618 |
| 66032.18 | 0.0505 | 66034.54 | 0.0677 | 66037.02 | 0.0525 | 66039.62 | 0.0634 |
| 66032.22 | 0.0541 | 66034.58 | 0.0671 | 66037.06 | 0.0596 | 66039.66 | 0.0659 |
| 66032.26 | 0.0506 | 66034.62 | 0.0684 | 66037.1 | 0.0556 | 66039.7 | 0.0668 |
| 66032.3 | 0.0541 | 66034.66 | 0.0651 | 66037.14 | 0.0543 | 66039.74 | 0.0578 |
| 66032.34 | 0.0492 | 66034.7 | 0.0664 | 66037.18 | 0.0584 | 66039.78 | 0.0537 |
| 66032.38 | 0.0515 | 66034.74 | 0.0668 | 66037.22 | 0.056 | 66039.82 | 0.0614 |
| 66032.42 | 0.0478 | 66034.78 | 0.0687 | 66037.26 | 0.0552 | 66039.86 | 0.0609 |
| 66032.46 | 0.0568 | 66034.82 | 0.0648 | 66037.3 | 0.0575 | 66039.9 | 0.0647 |
| 66032.5 | 0.0511 | 66034.86 | 0.0706 | 66037.34 | 0.0555 | 66039.94 | 0.0666 |

(Continued)

| Time (s) | $V_x$ (cm/s) | Time (s) | $V_x$ (cm/s) | Time (s) | $V_x$ (cm/s) | Time (s) | $V_x$ (cm/s) |
|----------|----------|----------|----------|----------|----------|----------|----------|
| 66032.54 | 0.0531 | 66034.9 | 0.0618 | 66037.38 | 0.0559 | 66039.98 | 0.0656 |
| 66032.58 | 0.0535 | 66034.94 | 0.0629 | 66037.42 | 0.0576 | 66040.02 | 0.0643 |
| 66032.62 | 0.0581 | 66034.98 | 0.0677 | 66037.46 | 0.0603 | 66040.06 | 0.0645 |
| 66032.66 | 0.0507 | 66035.02 | 0.0673 | 66037.5 | 0.0599 | 66040.1 | 0.0659 |
| 66032.7 | 0.0549 | 66035.06 | 0.066 | 66037.54 | 0.062 | 66040.14 | 0.0564 |
| 66032.74 | 0.0533 | 66035.1 | 0.0658 | 66037.58 | 0.0579 | 66040.18 | 0.0566 |
| 66032.78 | 0.0449 | 66035.14 | 0.067 | 66037.62 | 0.052 | 66040.22 | 0.0643 |
| 66032.82 | 0.0572 | 66035.18 | 0.0667 | 66037.66 | 0.0558 | 66040.26 | 0.0637 |
| 66032.86 | 0.051 | 66035.22 | 0.0607 | 66037.7 | 0.0515 | 66040.3 | 0.0653 |
| 66032.9 | 0.0599 | 66035.26 | 0.0633 | 66037.74 | 0.0499 | 66040.34 | 0.0626 |
| 66032.94 | 0.0568 | 66035.3 | 0.0669 | 66037.78 | 0.0551 | 66040.38 | 0.064 |
| 66032.98 | 0.0579 | 66035.34 | 0.0617 | 66037.82 | 0.0521 | 66040.42 | 0.0626 |
| 66033.02 | 0.0507 | 66035.38 | 0.0591 | 66037.86 | 0.0508 | 66040.46 | 0.0674 |
| 66033.06 | 0.0521 | 66035.42 | 0.0617 | 66037.9 | 0.0584 | 66040.5 | 0.0638 |
| 66033.1 | 0.0554 | 66035.46 | 0.0616 | 66037.94 | 0.0503 | 66040.54 | 0.0652 |
| 66033.14 | 0.0558 | 66035.5 | 0.0551 | 66037.98 | 0.0558 | 66040.58 | 0.0648 |
| 66033.18 | 0.0581 | 66035.54 | 0.0593 | 66038.02 | 0.0521 | 66040.62 | 0.064 |
| 66033.22 | 0.0579 | 66035.58 | 0.0537 | 66038.06 | 0.0579 | 66040.66 | 0.0642 |
| 66033.26 | 0.0551 | 66035.62 | 0.0597 | 66038.1 | 0.0551 | 66040.7 | 0.0624 |
| 66033.3 | 0.0594 | 66035.66 | 0.0628 | 66038.14 | 0.054 | 66040.74 | 0.0669 |
|  |  | 66035.7 | 0.0583 | 66038.18 | 0.0569 | 66040.78 | 0.0636 |
|  |  | 66035.74 | 0.0583 | 66038.22 | 0.0576 | 66040.82 | 0.0705 |
|  |  | 66035.78 | 0.0554 | 66038.26 | 0.0567 | 66040.86 | 0.065 |
|  |  |  |  | 66038.3 | 0.0535 | 66040.9 | 0.0706 |
|  |  |  |  | 66038.34 | 0.0581 | 66040.94 | 0.0691 |
|  |  |  |  | 66038.38 | 0.0591 |  |  |

(a)  For the following 10 s record, plot the instantaneous velocity as a function of time.
(b)  Calculate the time-average, standard deviation, skewness and kurtosis of the longitudinal velocity.
(c)  Calculate the integral time scale and the dissipation time scale.
(d)  calculate and plot the probability distribution function of the longitudinal velocity.

*For the calculation of the dissipation time scale, compare the results obtained using a parabolic approximation of the auto-correlation function and the method of Hallback et al. (1989). The data were collected by Trevethan et al. (2006).*

### Solution

$V_{avg}$  = 0.059 m/s      time-averaged
$V_{med}$  = 0.059 m/s      median value
$V_{std}$  = 0.00534 m/s    standard deviation
Skew = 0.0812              skewness
Kurt  = −0.669             kurtosis

$T_E$    = 0.3985 s         integral time-scale
$\tau_E$    = **−0.00332 s**    dissipation time-scale (**meaningless !**)

### Exercise No. 2

The following data set was recorded in a steady open channel flow at 0.057 m above the bed.

| Time (s) | $V_x$ (cm/s) | Time (s) | $V_x$ (cm/s) | Time (s) | $V_x$ (cm/s) | Time (s) | $V_x$ (cm/s) |
|---|---|---|---|---|---|---|---|
| 4.00 | 108.1 | 4.5 | 108.7 | 5.00 | 114.2 | 5.5 | 110.2 |
| 4.02 | 107.3 | 4.52 | 110.2 | 5.02 | 111.9 | 5.52 | 106.7 |
| 4.04 | 107.4 | 4.54 | 109.3 | 5.04 | 110.9 | 5.54 | 108.71 |
| 4.06 | 107.9 | 4.56 | 110.2 | 5.06 | 109.7 | 5.56 | 109.9 |
| 4.08 | 109.4 | 4.58 | 108.4 | 5.08 | 111.4 | 5.58 | 109.5 |
| 4.1 | 108.3 | 4.6 | 110.1 | 5.1 | 111.1 | 5.6 | 109.2 |
| 4.12 | 110.5 | 4.62 | 110.1 | 5.12 | 110.4 | 5.62 | 110.6 |
| 4.14 | 111.7 | 4.64 | 109.4 | 5.14 | 110.0 | 5.64 | 112.8 |
| 4.16 | 110.8 | 4.66 | 111.5 | 5.16 | 111.3 | 5.66 | 112.2 |
| 4.18 | 110.2 | 4.68 | 111.8 | 5.18 | 108.7 | 5.68 | 112.6 |
| 4.2 | 110.1 | 4.7 | 111.2 | 5.2 | 112.4 | 5.7 | 111.6 |
| 4.22 | 109.2 | 4.72 | 112.2 | 5.22 | 108.4 | 5.72 | 111.8 |
| 4.24 | 108.9 | 4.74 | 110.8 | 5.24 | 112.2 | 5.74 | 109.8 |
| 4.26 | 110.8 | 4.76 | 109.6 | 5.26 | 111.9 | 5.76 | 108.1 |
| 4.28 | 109.8 | 4.78 | 111.1 | 5.28 | 109.2 | 5.78 | 109.8 |
| 4.3 | 109.1 | 4.8 | 108.0 | 5.3 | 111.1 | 5.8 | 110.5 |
| 4.32 | 108.9 | 4.82 | 107.6 | 5.32 | 109.1 | 5.82 | 109.8 |
| 4.34 | 108.0 | 4.84 | 108.9 | 5.34 | 107.6 | 5.84 | 110.4 |
| 4.36 | 109.3 | 4.86 | 110.2 | 5.36 | 112.0 | 5.86 | 108.4 |
| 4.38 | 109.6 | 4.88 | 109.3 | 5.38 | 109.5 | 5.88 | 109.1 |
| 4.4 | 110.9 | 4.9 | 107.8 | 5.4 | 109.9 | 5.9 | 107.7 |
| 4.42 | 110.5 | 4.92 | 109.4 | 5.42 | 112.1 | 5.92 | 109.8 |
| 4.44 | 111.4 | 4.94 | 110.5 | 5.44 | 111.5 | 5.94 | 111.4 |
| 4.46 | 110.4 | 4.96 | 112.0 | 5.46 | 111.5 | 5.96 | 112.1 |
| 4.48 | 110.1 | 4.98 | 113.0 | 5.48 | 112.1 | 5.98 | 110.9 |

(a)   For the data set, plot the instantaneous velocity as a function of time.
(b)   Calculate the time-average and standard deviation of the longitudinal velocity. What is the turbulence intensity defined as $Tu = \sqrt{\overline{v^2}}/\overline{V}$?
(c)   Calculate the integral time scale and the dissipation time scale.

## Solution

$V_{avg}$ = 110.14 cm/s   time-averaged
$V_{med}$ — 110.10 cm/s   median value
$V_{std}$ = 1.489 cm/s   standard deviation
Skew = 0.0296   skewness
Kurt = −0.419   kurtosis
Tu = 1.35%   turbulence intensity
$T_E$ = 0.028 s   integral time-scale
$\tau_E$ = 0.0028 s   dissipation time-scale

## Exercise No. 3

The steady open channel flow situation, analysed in Exercise 2, is suddenly affected by the passage of a tidal bore. Both the instantaneous and time-averaged velocity data are listed below. The data was recorded at 0.057 m above the bed. (Note that the flow is unsteady and the "time-averaged velocity" is a variable time average.)

## Instantaneous longitudinal velocity

| Time (s) | $V_x$ (cm/s) | Time (s) | $V_x$ (cm/s) | Time (s) | $V_x$ (cm/s) | Time (s) | $V_x$ (cm/s) |
|---|---|---|---|---|---|---|---|
| 12 | 67.5026 | 12.5 | 60.0794 | 13 | 79.6718 | 13.5 | 86.1862 |
| 12.02 | 69.7268 | 12.52 | 65.5508 | 13.02 | 80.8693 | 13.52 | 89.5882 |
| 12.04 | 64.5371 | 12.54 | 59.4951 | 13.04 | 79.989 | 13.54 | 86.3987 |
| 12.06 | 66.7903 | 12.56 | 63.3117 | 13.06 | 78.3692 | 13.56 | 88.8158 |
| 12.08 | 66.3739 | 12.58 | 69.5743 | 13.08 | 76.1492 | 13.58 | 86.9115 |
| 12.1 | 66.5177 | 12.6 | 69.9598 | 13.1 | 77.1707 | 13.6 | 89.9447 |
| 12.12 | 69.6151 | 12.62 | 64.1634 | 13.12 | 82.6693 | 13.62 | 85.8272 |
| 12.14 | 67.807 | 12.64 | 64.6971 | 13.14 | 84.1862 | 13.64 | 86.1953 |
| 12.16 | 66.1124 | 12.66 | 65.1664 | 13.16 | 82.9088 | 13.66 | 89.7416 |
| 12.18 | 66.3661 | 12.68 | 63.1063 | 13.18 | 83.9595 | 13.68 | 86.5472 |
| 12.2 | 67.3087 | 12.7 | 63.2548 | 13.2 | 82.643 | 13.7 | 90.1726 |
| 12.22 | 60.6616 | 12.72 | 64.2278 | 13.22 | 84.8607 | 13.72 | 91.3913 |
| 12.24 | 61.6533 | 12.74 | 70.6123 | 13.24 | 85.4088 | 13.74 | 88.1052 |
| 12.26 | 62.1375 | 12.76 | 75.9225 | 13.26 | 85.1977 | 13.76 | 86.5433 |
| 12.28 | 61.298 | 12.78 | 76.8025 | 13.28 | 86.2025 | 13.78 | 90.6055 |
| 12.3 | 67.3146 | 12.8 | 71.0797 | 13.3 | 84.7838 | 13.8 | 90.0265 |
| 12.32 | 68.2022 | 12.82 | 71.7941 | 13.32 | 85.5499 | 13.82 | 91.2321 |
| 12.34 | 68.7461 | 12.84 | 67.3993 | 13.34 | 83.6273 | 13.84 | 89.5507 |
| 12.36 | 63.1782 | 12.86 | 72.2021 | 13.36 | 80.6722 | 13.86 | 90.2612 |
| 12.38 | 67.9926 | 12.88 | 72.7026 | 13.38 | 80.7049 | 13.88 | 86.9007 |
| 12.4 | 70.3569 | 12.9 | 74.1936 | 13.4 | 84.4192 | 13.9 | 81.6948 |
| 12.42 | 66.5494 | 12.92 | 76.7423 | 13.42 | 84.3725 | 13.92 | 81.8264 |
| 12.44 | 65.036 | 12.94 | 79.0459 | 13.44 | 83.2162 | 13.94 | 85.5667 |
| 12.46 | 60.316 | 12.96 | 78.1772 | 13.46 | 86.4487 | 13.96 | 87.0048 |
| 12.48 | 58.4519 | 12.98 | 76.7767 | 13.48 | 83.8549 | 13.98 | 81.9291 |

## "Time-averaged velocity" (variable time average)

| Time (s) | $V_x$ (cm/s) | Time (s) | $V_x$ (cm/s) | Time (s) | $V_x$ (cm/s) | Time (s) | $V_x$ (cm/s) |
|---|---|---|---|---|---|---|---|
| 12 | 68.59311 | 12.5 | 66.13795 | 13 | 77.34149 | 13.5 | 85.44806 |
| 12.02 | 68.44279 | 12.52 | 66.27083 | 13.02 | 77.86089 | 13.52 | 85.65064 |
| 12.04 | 68.27196 | 12.54 | 66.4389 | 13.04 | 78.36785 | 13.54 | 85.84535 |
| 12.06 | 68.08396 | 12.56 | 66.64473 | 13.06 | 78.8607 | 13.56 | 86.02884 |
| 12.08 | 67.88306 | 12.58 | 66.89004 | 13.08 | 79.33774 | 13.58 | 86.19742 |
| 12.1 | 67.6742 | 12.6 | 67.17561 | 13.1 | 79.79732 | 13.6 | 86.34727 |
| 12.12 | 67.4626 | 12.62 | 67.50117 | 13.12 | 80.23791 | 13.62 | 86.47462 |
| 12.14 | 67.25348 | 12.64 | 67.86545 | 13.14 | 80.65817 | 13.64 | 86.57596 |
| 12.16 | 67.0517 | 12.66 | 68.26624 | 13.16 | 81.05708 | 13.66 | 86.64817 |
| 12.18 | 66.86144 | 12.68 | 68.70055 | 13.18 | 81.43394 | 13.68 | 86.68868 |
| 12.2 | 66.68604 | 12.7 | 69.1648 | 13.2 | 81.78855 | 13.7 | 86.69551 |
| 12.22 | 66.5278 | 12.72 | 69.65499 | 13.22 | 82.12116 | 13.72 | 86.66734 |
| 12.24 | 66.38803 | 12.74 | 70.16697 | 13.24 | 82.43256 | 13.74 | 86.60349 |
| 12.26 | 66.26709 | 12.76 | 70.69659 | 13.26 | 82.72404 | 13.76 | 86.50391 |
| 12.28 | 66.16457 | 12.78 | 71.23991 | 13.28 | 82.9974 | 13.78 | 86.36909 |
| 12.3 | 66.07955 | 12.8 | 71.79326 | 13.3 | 83.2548 | 13.8 | 86.2 |
| 12.32 | 66.01086 | 12.82 | 72.35341 | 13.32 | 83.49874 | 13.82 | 85.99802 |
| 12.34 | 65.95744 | 12.84 | 72.91752 | 13.34 | 83.73186 | 13.84 | 85.76487 |

(Continued)

| Time (s) | $V_x$ (cm/s) | Time (s) | $V_x$ (cm/s) | Time (s) | $V_x$ (cm/s) | Time (s) | $V_x$ (cm/s) |
|---|---|---|---|---|---|---|---|
| 12.36 | 65.91856 | 12.86 | 73.48317 | 13.36 | 83.95681 | 13.86 | 85.50253 |
| 12.38 | 65.8941 | 12.88 | 74.04832 | 13.38 | 84.17605 | 13.88 | 85.21317 |
| 12.4 | 65.88468 | 12.9 | 74.61121 | 13.4 | 84.39172 | 13.9 | 84.89915 |
| 12.42 | 65.89173 | 12.92 | 75.17034 | 13.42 | 84.60543 | 13.92 | 84.56292 |
| 12.44 | 65.91749 | 12.94 | 75.72429 | 13.44 | 84.81812 | 13.94 | 84.20699 |
| 12.46 | 65.96483 | 12.96 | 76.27171 | 13.46 | 85.02999 | 13.96 | 83.83385 |
| 12.48 | 66.03713 | 12.98 | 76.81126 | 13.48 | 85.24045 | 13.98 | 83.44594 |

(a)   For the data set, plot both the instantaneous velocities as a function of time.
(b)   Plot the variable time average velocity as a function of time.
(c)   Calculate the standard deviation and turbulence intensity defined as $Tu = \sqrt{\overline{v^2}}/\overline{V}$ of the longitudinal velocity.
(d)   Calculate the integral time scale.
(e)   Compare the results with the steady flow conditions (Exercise 2).

*All the velocities are positive in the downstream direction. Data obtained by Koch and Chanson (2005).*

## Solution

$V_{avg} = 76.29$ cm/s   time-averaged
$V_{std} = 3.113$ cm/s   standard deviation
$Skew = -0.459$   skewness
$Kurt = -0.5176$   kurtosis
$Tu = 4.1$ %   turbulence intensity
$T_E = 0.024$ s   integral time-scale

A comparison with the steady flow results obtained in the same channel prior to the arrival of the tidal bore shows that the passage of the undular bore is associated with higher turbulence levels (3 times higher) and smaller integral turbulent time scale.

## Exercise No. 4

For the following data set:

(a)   Plot the data, and
(b)   Calculate the time-averaged and standard deviation of the normal and tangential stresses.

| Time (s) | $V_x$ (cm/s) | $V_z$ (cm/s) | Time (s) | $V_x$ (cm/s) | $V_z$ (cm/s) | Time (s) | $V_x$ (cm/s) | $V_z$ (cm/s) |
|---|---|---|---|---|---|---|---|---|
| 0.02 | −1.54 | 3.19 | 0.82 | −2.46 | 3.37 | 1.62 | −2.14 | 2.92 |
| 0.06 | −1.97 | 3.16 | 0.86 | −2.27 | 3.01 | 1.66 | −2.47 | 2.88 |
| 0.1 | −1.95 | 3.28 | 0.9 | −3.45 | 2.62 | 1.7 | −2.35 | 3.45 |
| 0.14 | −2.28 | 3.06 | 0.94 | −2.82 | 2.62 | 1.74 | −1.77 | 2.87 |
| 0.18 | −1.83 | 3.12 | 0.98 | −2.13 | 2.42 | 1.78 | −1.8 | 2.75 |
| 0.22 | −2.3 | 3.16 | 1.02 | −2.47 | 3.01 | 1.82 | −2.38 | 2.86 |
| 0.26 | −1.5 | 2.75 | 1.06 | −2.6 | 3.27 | 1.86 | −2.76 | 2.88 |
| 0.3 | −1.92 | 3.71 | 1.1 | −2.21 | 2.68 | 1.9 | −1.68 | 3.09 |
| 0.34 | −2.49 | 3.55 | 1.14 | −2.62 | 2.56 | 1.94 | −2.36 | 3.05 |

*(Continued)*

| Time (s) | $V_x$ (cm/s) | $V_z$ (cm/s) | Time (s) | $V_x$ (cm/s) | $V_z$ (cm/s) | Time (s) | $V_x$ (cm/s) | $V_z$ (cm/s) |
|---|---|---|---|---|---|---|---|---|
| 0.38 | −2.67 | 2.77 | 1.18 | −2.63 | 2.93 | 1.98 | −2.7 | 2.92 |
| 0.42 | −2.41 | 3.08 | 1.22 | −4.61 | 2.9 | 2.02 | −2.71 | 3.11 |
| 0.46 | −2.83 | 3.04 | 1.26 | −2.72 | 3.03 | 2.06 | −2.21 | 3.1 |
| 0.5 | −2.21 | 2.92 | 1.3 | −2.29 | 2.79 | 2.1 | −2.58 | 2.86 |
| 0.54 | −2 | 2.83 | 1.34 | −2.84 | 2.4 | 2.14 | −2.6 | 3.61 |
| 0.58 | −1.7 | 2.97 | 1.38 | −2.8 | 2.55 | 2.18 | −1.61 | 2.87 |
| 0.62 | −4.05 | 3.32 | 1.42 | −2.39 | 2.61 | 2.22 | −2.04 | 3.29 |
| 0.66 | −2.59 | 2.99 | 1.46 | −3.12 | 1.95 | 2.26 | −1.76 | 3.49 |
| 0.7 | −2.68 | 2.84 | 1.5 | −2.41 | 3.14 | 2.3 | −1.64 | 3.76 |
| 0.74 | −2.41 | 3.41 | 1.54 | −1.88 | 3.29 | 2.34 | −1.72 | 3.32 |
| 0.78 | −2.24 | 3.09 | 1.58 | −2.08 | 2.61 | 2.38 | −2.14 | 2.93 |

*The data set was collected in Eprapah Creek estuary on 4 April 2004 (Chanson 2003). The water density was about 1015 kg/m³ (for brackish waters). $V_x$ is positive upstream and the transverse velocity $V_z$ is positive towards the right bank.*

### Solution

| | $\rho \times v_x^2$ Pa | $\rho \times v_x^2$ Pa | $\rho \times v_x \times v_z$ Pa |
|---|---|---|---|
| Time-averaged: | 0.0306 | 0.01102 | 0.00450 |
| Standard deviation: | 0.07586 | 0.01849 | 0.0185 |

## Exercise No. 5

Considering a turbulent Couette flow between two parallel plates (Fig. 1.7), one plate moving at speed $V_1$ and the other moving at speed $V_2$, express the velocity distribution between the plate. Assume that the "eddy viscosity" may be estimated by a parabolic distribution:

$$v_T = K \times V^* \times y \times \left(1 - \frac{y}{D}\right)$$

where $K$ is the von Karman constant ($K = 0.40$) and D is the distance between plates (Chapter II-1).

### Solution

For $V_1 = 0$ and $V_2 = V_o$, the solution is developed in Chapter II-1. It may be extended easily for $V_1 \neq 0$.

## Exercise No. 6

In a steady turbulent flow between two flat plates (Fig. E2.1), find a suitable Prandtl mixing length distribution. Derive the velocity profile. Discuss the result.

### Solution

Let us define y the distance from the channel centreline. The mixing length may be taken proportional the distance from the plate:

$$l_m = K \times \left(\frac{D}{2} - y\right) \quad \text{for } y \geq 0$$

(A) Definition sketch

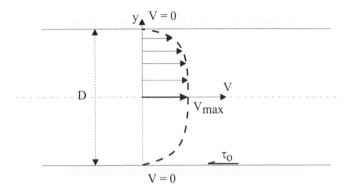

(B) Dimensionless velocity and mixing length distribution

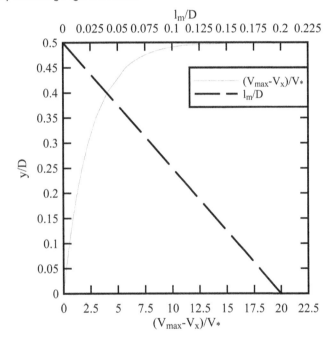

*Figure E2.1*  Turbulent flow between plates.

Replacing into Equation (2.18),

$$\frac{V_{max} - V_x}{V_*} = \frac{1}{K} \times Ln\left(\frac{1}{1 - 2 \times \frac{y}{D}}\right) \quad \text{for } y \geq 0$$

where $V_*$ is the shear velocity ($V_* = (\tau/\rho)^{1/2}$) and $V_{max}$ is the centreline velocity. The results are presented in Figure E2.1.

Note that, on the channel centreline ($y = 0$), the velocity derivative is not continuous.

# Boundary layer theory – application to laminar boundary layer flows

**SUMMARY**

In this chapter, the basics of boundary layers are reviewed. The basic definitions and the momentum integral equation development are valid for both laminar and turbulent flows. The boundary layer flow properties are described and their application is discussed in the context of laminar boundary layers. Turbulent boundary layers will be discussed in the next chapter.

## 1 INTRODUCTION

A *boundary layer* is defined as the flow region next to a solid boundary where the flow field is affected by the presence of the boundary (Fig. 3.1). The concept was originally introduced by Ludwig Prandtl (1904). In the boundary layer, some momentum is gained by the flow region next to the boundary. Momentum is transferred from the main stream (or free stream) and contributes to the boundary layer growth. At the boundary itself, the velocity is zero. This is called the no-slip condition.

---

**Remarks**

1. The basic reference in boundary layer flows is the text by Schlichting (1960, 1979) (also Schlichting and Gersten 2000). On laminar boundary layer flows, solid references include Schlichting (1979) and Liggett (1994).
2. Hermann Schlichting (1907–1982) was a Ph.D. student under the supervision of Ludwig Prandtl and became later a professor at the Technical University of Braunschweig (Germany).
3. Ludwig Prandtl (1875–1953) was a German physicist and aerodynamicist who introduced the concept of boundary layer (Prandtl 1904).
4. Meier (2004) presented a historical perspective on Prandtl's scientific life and contribution.

---

At each cross-section located at a distance x from the inception of a boundary layer, the boundary layer may be characterised by some key parameters. A basic characteristic is the boundary layer *thickness* δ which is defined in terms of 99% of the free-stream velocity $V_O$:

$$\delta = y(v_x = 0.99 \times V_O) \tag{3.1}$$

where $V_x$ is the longitudinal velocity at a distance y measured perpendicular to the boundary. The free-stream velocity is the ideal fluid flow velocity outside of the boundary layer where the effects of boundary friction are nil (Fig. 3.1).

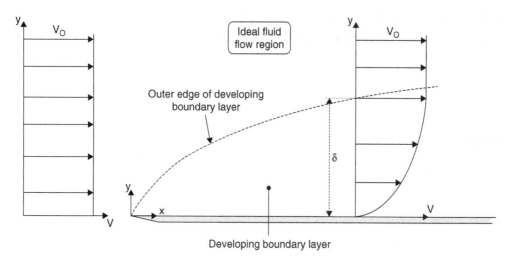

*Figure 3.1*  Sketch of a developing laminar boundary layer along a flat plate.

Another important parameter is the *displacement thickness* $\delta_1$ defined as:

$$\delta_1 = \int_0^\delta \left(1 - \frac{v_x}{V_O}\right) \times dy \tag{3.2}$$

For the ideal fluid flow outside of the boundary layer, the presence of a boundary layer region is equivalent to an apparent displacement (or "thickening") of the plate. The amount of "apparent" displacement is the displacement thickness $\delta_1$.

Similarly, some "displacement thickness" may be defined in terms of the momentum transport, or momentum flux, and of the energy flux: i.e., the momentum thickness and the energy thickness respectively. The *momentum thickness* $\delta_2$ is defined as:

$$\delta_2 = \int_0^\delta \frac{v_x}{V_O} \times \left(1 - \frac{v_x}{V_O}\right) \times dy \tag{3.3}$$

The *energy thickness* $\delta_3$ is defined as:

$$\delta_3 = \int_0^\delta \frac{v_x}{V_O} \times \left(1 - \left(\frac{v_x}{V_O}\right)^2\right) \times dy \tag{3.4}$$

---

**Remarks**

1.  Although the boundary layer thickness $\delta$ is (arbitrarily) defined in terms of 99% of the free-stream velocity $V_O$, the real extent of the effects of boundary friction on the flow is probably about 1.5 to 2 times $\delta$.
2.  The energy thickness is also called the energy dissipation thickness. It represents some form of kinetic energy loss due to viscous effects.
3.  The above definitions of boundary layer thickness, displacement thickness, momentum thickness and energy thickness are valid for both laminar and turbulent boundary layers.

## 2   BASIC EQUATIONS OF THE LAMINAR BOUNDARY LAYER

### 2.1   Presentation

Considering a laminar boundary layer along a smooth flat plate, and for an incompressible fluid, the Navier-Stokes equation may be simplified. It may be shown that the pressure variations $\partial P/\partial y$ in the boundary layer are small (Schlichting 1979, Liggett 1994). Simply the pressure field is driven by the ideal-fluid flow above the developing boundary layer. After simplifications, the continuity and momentum equations yield:

$$\frac{\partial v_x}{\partial x} + \frac{\partial v_y}{\partial y} = 0 \tag{3.5}$$

$$\frac{\partial v_x}{\partial t} + v_x \times \frac{\partial v_x}{\partial x} + v_y \frac{\partial v_x}{\partial y} = -\frac{1}{\rho} \times \frac{\partial P}{\partial x} - g \times \frac{\partial z_O}{\partial x} + \frac{\mu}{\rho} \times \frac{\partial^2 v_x}{\partial y^2} \tag{3.6}$$

where P is the pressure, $z_O$ is the elevation positive upwards, $\mu$ and $\rho$ are the fluid dynamic viscosity and density respectively. In the ideal fluid flow region, above the boundary layer, the flow properties satisfy:

$$\frac{1}{\rho} \times \frac{\partial P}{\partial y} + g \times \frac{\partial z_O}{\partial y} = 0 \tag{3.7}$$

Equation (3.7) is the differential form of the Bernoulli equation for the ideal fluid flow, with $\partial V_O/\partial x = 0$ for a flat plate.

Usually the boundary conditions are: $V_x(y=0)=0$, $V_y(y=0)=0$, and $V_x(y=+\infty)=V_O$. The first condition is the no-slip condition at the boundary. The second one states that there is no mass flux through a solid boundary. The last condition expresses that the longitudinal velocity tends to the free-stream velocity $V_O$ outside of the boundary layer (Fig. 3.1).

---

### Discussion

The free-stream velocity $V_O$ is basically the ideal-fluid flow velocity outside of the boundary layer region. Above the boundary layer, the flow may be treated as an ideal fluid flow (see Part I).

---

### 2.2   Developing boundary layer on a flat plate

A simple solution of the laminar boundary layer is that for the developing flow over of flat plate in absence of pressure gradient. Figure 3.1 illustrates such a situation. The pressure gradient is zero everywhere (i.e. $\partial P/\partial x = \partial P/\partial y = 0$), the plate is assumed horizontal ($\partial z_O/\partial x = 0$) and the flow is steady ($\partial V_x/\partial t = 0$). The boundary layer equations become:

$$\frac{\partial v_x}{\partial x} + \frac{\partial v_y}{\partial y} = 0 \tag{3.8}$$

$$v_x \times \frac{\partial v_x}{\partial x} + v_y \times \frac{\partial v_x}{\partial y} = \frac{\mu}{\rho} \times \frac{\partial^2 v_x}{\partial y^2} \tag{3.9}$$

with the following boundary conditions: $V_x(y=0)=0$, $V_y(y=0)=0$, and $V_x(y=+\infty)=V_O$.

The developing boundary layer is laminar for $Re_x < 3\,E{+}5$ where $Re_x$ is the Reynolds number defined in terms of the distance from the upstream edge of the plate:

$$Re_x = \rho \times \frac{V_O \times x}{\mu} \tag{3.10}$$

Note however that the critical value for $Re_x$ for the flow to become turbulent is affected by the inflow conditions and the free-stream turbulence, and by the plate roughness. The above criterion ($Re_x < 3\,E{+}5$) is given for a smooth plate with a sharp leading edge (Schlichting and Gersten 2000, p. 420).

---

**Remember**

Importantly, Equations (3.6) to (3.9) were developed for laminar flows. They should not be applied to turbulent flows.

---

**Notes**

– For a smooth flat plate with a sharp leading edge, the flow remains laminar for:

$$Re_x = \rho \times \frac{V_O \times x}{\mu} < 3 \times 10^5$$

However a laminar flow motion can be maintained for $Re_x$ as high as $3 \times 7$ by some appropriate profile shaping and streamlining. NACA developed some laminar-flow airfoils that are used in the design of subsonic aircraft, and tested some supersonic laminar flow control using a suction system. Figure 3.2 illustrates a test aircraft: a major portion of the left wing (on the right of the photograph) was equipped with a suction system which drew turbulent boundary-layer air through millions of tiny laser-drilled holes in a titanium "glove" fitted to the upper left wing.
– The section shapes of laminar flow NACA profiles are designated as:

$$66_2 - 115$$

where **6** is the series designation, the next digit **6** describes the distance of the minimum pressure area in tens of percent of chord, the subscript **2** is related to the maximum lift coefficient for favourable pressure gradient (**0.2**), **1** gives the design lift coefficient (**0.1**) and **15** is the section thickness (**15%**).
– The National Advisory Committee for Aeronautics (NACA) was the predecessor of NASA.

## 2.3  Blasius equation

The solution of Equations (3.8) and (3.9) is called the Blasius equation. Liggett (1994) presented an elegant development. Since the flow is assumed incompressible, the stream function $\psi$ does exist because it satisfies the continuity equation (Eq. (3.8)). The momentum equation (Eq. (3.9)) may be written in terms of the stream function:

$$\frac{\partial \psi}{\partial y} \times \frac{\partial^2 \psi}{\partial x\,\partial y} - \frac{\partial \psi}{\partial x} \times \frac{\partial^2 \psi}{\partial y^2} = \frac{\mu}{\rho} \times \frac{\partial^3 \psi}{\partial y^3} \tag{3.11}$$

The boundary conditions are: $\partial\psi/\partial x = 0$ and $\partial\psi/\partial y = 0$ at the plate ($y = 0$), and $\partial\psi/\partial y = -V_O$ for $y = +\infty$. That is, the stream function is a constant at the solid boundary, and $\psi$ is a linear function of y far away from the plate.

*Figure 3.2*   F-16XL supersonic laminar flow control research aircraft (Courtesy of NASA Dryden Flight Research Center, Photo number EC96-43811-2) – The project demonstrated that laminar airflow could be achieved over a major portion of a upper left wing at supersonic speeds by use of a suction system which drew turbulent boundary-layer air through millions of tiny laser-drilled holes.

---

**Remember**

The stream function is defined for steady and unsteady *incompressible* flows because it does satisfy the continuity equation. For a two-dimension flow, the relationship between the stream function and the velocity is:

$$v_x = -\frac{\partial \psi}{\partial y}$$

$$v_y = \frac{\partial \psi}{\partial x}$$

The stream function has the dimensions $L^2 T^{-1}$ (i.e. m²/s).

Any solid boundary is a streamline since there no flow across. Along a streamline, the stream function is a constant.

---

Let us introduce the dimensionless variables $\eta$ and $\Lambda$:

$$\eta = \frac{y}{x} \times \sqrt{Re_x}$$

$$\Lambda = -\frac{\psi}{\left(\frac{\mu}{\rho}\right)^2} \times \sqrt{Re_x}$$

in which $\Lambda$ is a dimensionless stream function that is a function of the dimensionless co-ordinate $\eta$. Equation (3.11) may be rewritten into a differential equation in terms of the dimensionless stream function $\Lambda$:

$$\frac{\partial^3 \Lambda}{\partial \eta^3} + \frac{\Lambda}{2} \times \frac{\partial^2 \Lambda}{\partial \eta^2} = 0 \tag{3.12}$$

Equation (3.12) is the *Blasius equation*. The boundary conditions are: $\Lambda = 0$ and $\partial \Lambda/\partial \eta = 0$ at the plate ($\eta = 0$), and $\partial \Lambda/\partial \eta = 1$ for $\eta = +\infty$. If $\Lambda$ is only a function of $\eta$, the differentiation of $\Lambda$ with respect to $\eta$ becomes the first derivative (i.e. $\partial \Lambda/\partial \eta = \Lambda'$) and so on for the second and third differentiation (i.e. $\partial^2 \Lambda/\partial \eta^2 = \Lambda''$ & $\partial^3 \Lambda/\partial \eta^3 = \Lambda'''$).

The Blasius equation may be expressed as:

$$\Lambda''' + \frac{\Lambda}{2} \times \Lambda'' = 0 \tag{3.12b}$$

Equation (3.12b) may be integrated as a power series. The complete solution is shown in Figure 3.3 and tabulated in Table 3.1.

---

**Remarks**

1. The Blasius equation was derived by Blasius (1907,1908) as part of his doctoral thesis at the University of Göttingen under the supervision of Ludwig Prandtl (1875–1953). The problem was considered the first historical application of the boundary layer theory of L. Prandtl (Liggett 1994, Meier 2004)).

2. Paul Richard Heinrich Blasius (1883–1970) was a German fluid mechanician. Commonly called Heinrich Blasius, he lectured at the University of Hamburg from 1912 to 1950. His career in Germany spanned over the period of two World Wars, and his greatest contributions included the Blasius equation and the Blasius friction factor formula for smooth turbulent pipe flows (Hager 2003).

3. The dimensionless stream function $\Lambda$ is assumed to be a power series:

$$\Lambda = a_0 + a_1 \times \eta + \frac{a_2}{2!} \times \eta^2 + \frac{a_3}{3!} \times \eta^3 + \cdots$$

where 3! is the factorial of 3: $3! = 3 \times 2 \times 1$.

4. The Blasius equation was extended (e.g. Rouse 1959, pp. 325–328). Some applications may include boundary layer flows with a temperature-dependant property (e.g. Fang et al. 2006).

5. Practically, the solution of the Blasius equation yields $\Lambda' \approx 1$ and $V_x/V_O \approx 1$ for $\eta > 6$.

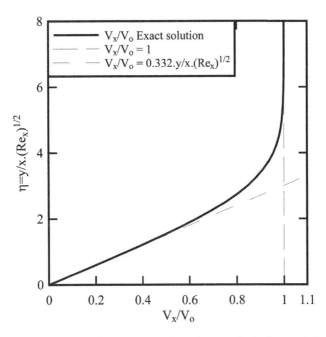

*Figure 3.3* Dimensionless velocity profile in a laminar boundary profile (solution of the Blasius equation).

*Table 3.1* Solution of the Blasius equation.

| $\eta$ (1) | $\Lambda$ (2) | $\Lambda'$ $V_x/V_O$ (3) | $\eta$ (1) | $\Lambda$ (2) | $\Lambda'$ $V_x/V_O$ (3) |
|---|---|---|---|---|---|
| 0 | 0 | 0 | 3.25 | 1.61 | 0.883 |
| 0.1 | 0.002 | 0.033 | 3.5 | 1.84 | 0.913 |
| 0.2 | 0.007 | 0.066 | 3.75 | 2.07 | 0.937 |
| 0.3 | 0.015 | 0.100 | 4 | 2.31 | 0.955 |
| 0.4 | 0.027 | 0.133 | 4.25 | 2.55 | 0.969 |
| 0.5 | 0.041 | 0.166 | 4.5 | 2.79 | 0.979 |
| 0.6 | 0.060 | 0.199 | 4.75 | 3.04 | 0.987 |
| 0.7 | 0.081 | 0.232 | 5 | 3.28 | 0.991 |
| 0.8 | 0.106 | 0.265 | 5.5 | 3.78 | 0.997 |
| 0.9 | 0.134 | 0.297 | 6 | 4.28 | 0.999 |
| 1 | 0.166 | 0.330 | 6.5 | 4.78 | 1 |
| 1.25 | 0.258 | 0.409 | 7 | 5.28 | 1 |
| 1.5 | 0.370 | 0.487 | 7.5 | 5.78 | 1 |
| 1.75 | 0.501 | 0.560 | 8 | 6.28 | 1 |
| 2 | 0.650 | 0.630 | 8.5 | 6.78 | 1 |
| 2.25 | 0.816 | 0.693 | 9 | 7.28 | 1 |
| 2.5 | 0.996 | 0.751 | 9.5 | 7.78 | 1 |
| 2.75 | 1.191 | 0.802 | 10 | 8.28 | 1 |
| 3 | 1.40 | 0.846 | | | |

**Discussion**

The solution of the Blasius equation has two asymptotes in terms of the velocity distribution (Fig. 3.3). These are:

$$\frac{v_x}{V_O} = 0.332 \times \frac{y}{x} \times \sqrt{Re_x} \quad \text{for} \ \frac{y}{x} \times \sqrt{Re_x} < 1$$

$$\frac{v_x}{V_O} = 1 \quad \text{for} \ \frac{y}{x} \times \sqrt{Re_x} > 6$$

Both asymptotes are compared with the Blasius solution in Figure 3.3.

For $y/x \times \sqrt{Re_x} > 0.15$, the velocity distribution may be grossly approximated by:

$$\frac{v_x}{V_O} \approx \frac{1}{1 + 9.719 \times \exp(-1.396 \times \frac{y}{x} \times \sqrt{Re_x})} \quad \text{for} \ \frac{y}{x} \times \sqrt{Re_x} > 0.15$$

The above approximation is an empirical correlation that matches closely the analytical solution, but next to the wall.

## 2.4  Applications

Using the Blasius equation solution (Table 3.1), the solutions of the boundary layer characteristics are:

$$\delta = 4.91 \times \frac{x}{\sqrt{Re_x}} \tag{3.13}$$

$$\delta_1 = 1.72 \times \frac{x}{\sqrt{Re_x}} \tag{3.14}$$

$$\delta_2 = 0.664 \times \frac{x}{\sqrt{Re_x}} \tag{3.15}$$

At a distance x from the plate leading edge, the local boundary shear stress is

$$\tau_O = \mu \times \left( \frac{\partial v_x}{\partial y} \right)_{y=0}$$

The boundary shear stress on the plate is hence:

$$\tau_O = \Lambda''(0) \times \sqrt{\frac{V_O^3 \times \mu \times \rho}{x}} = 0.332 \times \sqrt{\frac{V_O^3 \times \mu \times \rho}{x}} \tag{3.16}$$

**Remember**

In laminar flows, Newton's law of viscosity relates the shear stress to the rate of angular deformation, or strain rate:

$$\tau = \mu \times \frac{\partial v_x}{\partial y}$$

where $\tau$ is the shear stress between two adjacent laminas of fluid, $\mu$ is the dynamic viscosity of the flowing fluid and y is the direction normal to the flow direction.

The total shear force per unit width is:

$$\int_{x=0}^{L} \tau_O \times dx$$

where L is the plate length. The shear force represents the friction force on the plate per unit width. The solution of the Blasius equation yields a total shear force per unit width:

$$\int_{x=0}^{L} \tau_O \times dx = \frac{1.328}{2} \times \rho \times \sqrt{\frac{\mu}{\rho} \times L \times V_O^3} \tag{3.17}$$

In dimensionless form, the shear force per unit width is called the overall skin friction coefficient and it equals:

$$\frac{\int_{x=0}^{L} \tau_O \times dx}{\frac{1}{2} \times \rho \times V_O^2 \times L} = \frac{1.328}{\sqrt{Re_L}} \tag{3.18}$$

where $Re_L$ is the Reynolds number defined in terms of the plate length: $Re_L = \rho \times V_O \times L/\mu$.

The application of the Blasius equation shows that the boundary layer thickness $\delta$ increases with the square-root of the distance x from the leading plate: $\delta \propto \sqrt{x}$ (Eq. (3.13)). As the boundary layer develops, the velocity gradient $\partial V_x / \partial y$ decreases with increasing distance x and Equation (3.16) shows: $\tau_O \propto 1/\sqrt{x}$. Note further that the boundary shear stress is proportional to $V_O^{1.5}$.

---

**Remark**

The solution of the Blasius equation is not necessarily valid at the plate leading edge. At x = 0 and at the wall (y = 0), there is a discontinuity of the fluid velocity from $V_x = V_O$ for x < 0 to $V_x = 0$ for x > 0.

---

The Blasius equation is seen as an important contribution to fluid mechanics. It demonstrated the validity of the boundary layer concept for a laminar flow. Practical applications include blood flows, slow geophysical flows (e.g. mud flows), lubrication and other forms of industrial flows.

## 2.5  Discussion

The solution of the Blasius equation was developed assuming that (1) the free-stream velocity $V_O$ is a constant and that (2) the velocity profiles were self-similar: $V_x/V_O \sim F(y/\delta)$. This approach may be extended to flow situations in which the ideal fluid flow velocity outside of the boundary layer follows:

$$V_O \propto x^m \tag{3.19}$$

where the exponent m is a constant. For m = 0, the solution is the Blasius equation.

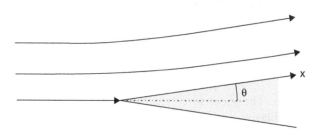

*Figure 3.4* Ideal fluid flow past a wedge.

Considering the flow past a wedge (Fig. 3.4), the exponent m satisfies:

$$m = \frac{\theta}{\pi - \theta} \tag{3.20}$$

for $\theta < \pi$. For a flat plate, $\theta = 0$. Liggett (1994) developed the basic differential equation, called the Falkner-Skan equation, and the associated boundary conditions.

## 3   THE MOMENTUM INTEGRAL EQUATION

In boundary layer studies, an important principle is the momentum integral equation which is derived from the integration of the Navier-Stokes equation to a large control volume encompassing a vertical slice of both the boundary layer and ideal fluid flow regions (Fig. 3.5). This fundamental equation is valid for both laminar and turbulent boundary layers. Herein it is developed in the general case (section 3.1) and later applied to laminar boundary layers (sections 3.2 and 3.3). The applications to turbulent boundary layers will be discussed in the next chapter (Chap. II-4, section 2.2).

### 3.1   Basic equations

Let us consider the boundary layer developing along the flat plate and apply the equation of conservation of momentum in its integral form to the control volume sketched in Figure 3.5. The integration of the momentum principle (Eq. (3.6)) yields:

$$\int_{y=0}^{+\infty} \left( v_x \times \frac{\partial v_x}{\partial x} + v_y \times \frac{\partial v_x}{\partial y} \right) \times dy = \int_{y=0}^{+\infty} \left( -\frac{1}{\rho} \times \frac{\partial P}{\partial x} - g \times \frac{\partial z_O}{\partial x} + \frac{\mu}{\rho} \times \frac{\partial^2 v_x}{\partial y^2} \right) \times dy \tag{3.21}$$

In Equation (3.21), the last term in the right handside is the boundary shear which opposes the flow motion. Using the differential form of the Bernoulli equation for the ideal fluid flow, the above equation may be transformed as:

$$\int_{y=0}^{+\infty} \left( v_x \times \frac{\partial v_x}{\partial x} + v_y \times \frac{\partial v_x}{\partial y} - V_O \times \frac{\partial V_O}{\partial x} \right) \times dy = -\frac{\tau_O}{\rho} \tag{3.21b}$$

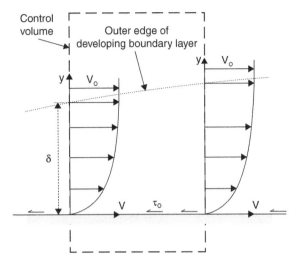

*Figure 3.5* Definition sketch of a developing boundary layer and the selection of the control volume for the application of the momentum integral method.

where $\tau_O$ is the boundary shear stress. The negative sign reflects that the shear force opposes the fluid motion.

---

**Remember**

1.  The differential form of the Bernoulli equation derives from the Navier-Stokes equation (Liggett 1993, Chanson 1999). Considering the flow is along a *streamline*, assuming that the fluid is *frictionless*, and for a *steady* and *incompressible* flow, the Navier Stokes equation along any streamline becomes:

    $$\rho \times V \times dV + \rho \times g \times dz_O + dP = 0$$

    where V is the velocity magnitude along the streamline in the ideal fluid flow region: i.e., $V = V_O$. The above equation may be re-arranged as:

    $$V_O \times \frac{\partial V_O}{\partial x} = -g \times \frac{\partial z_O}{\partial x} - \frac{1}{\rho} \times \frac{\partial P}{\partial x}$$

    where x is the direction along the streamline.

2.  For a laminar flow, the boundary shear stress equals the strain rate at the boundary $(y = 0)$:

    $$\tau_0 = \mu \times \left( \frac{\partial v_x}{\partial y} \right)_{y=0}$$

The continuity equation (Eq. (3.5)) gives:

$$v_y = -\int_0^y \frac{\partial v_x}{\partial x} \times dy \tag{3.22}$$

Let us consider the second term in Equation (3.21b). Using Equation (3.22), it gives:

$$\int_{y=0}^{+\infty} v_y \times \frac{\partial v_x}{\partial y} \times dy = -\int_{y=0}^{+\infty} \left( \int_0^y \frac{\partial v_x}{\partial x} \times dy \right) \times \frac{\partial v_x}{\partial y} \times dy \tag{3.23}$$

The integration by parts yields:

$$\int_{y=0}^{+\infty} v_y \times \frac{\partial v_x}{\partial y} \times dy = -\left( v_x \times \int_0^y \frac{\partial v_x}{\partial x} \times dy \right)_{y=0}^{y=+\infty} + \int_{y=0}^{+\infty} v_x \times \frac{\partial v_x}{\partial x} \times dy \tag{3.24}$$

with $V_x(y=0)=0$ and $V_x(y=+\infty)=V_O$.

After substitution of Equation (3.24) into Equation (3.21b), it yields:

$$\int_{y=0}^{+\infty} \left( 2 \times v_x \times \frac{\partial v_x}{\partial x} - V_O \times \frac{\partial v_x}{\partial x} - V_O \times \frac{\partial V_O}{\partial x} \right) \times dy = -\frac{\tau_o}{\rho} \tag{3.25}$$

After some re-arrangement, the above equation becomes:

$$\frac{\partial}{\partial x} \left( \int_{y=0}^{+\infty} v_x \times (V_O - v_x) \times dy \right) + \frac{\partial V_O}{\partial x} \times \int_{y=0}^{+\infty} (V_O - v_x) \times dy = \frac{\tau_o}{\rho} \tag{3.26}$$

Now let us remember the definition of the displacement and momentum thickness:

$$\delta_1 = \int_0^\delta \left( 1 - \frac{v_x}{V_O} \right) \times dy = \int_0^{+\infty} \left( 1 - \frac{v_x}{V_O} \right) \times dy \tag{3.2}$$

$$\delta_2 = \int_0^\delta \frac{v_x}{V_O} \times \left( 1 - \frac{v_x}{V_O} \right) \times dy = \int_0^{+\infty} \frac{v_x}{V_O} \times \left( 1 - \frac{v_x}{V_O} \right) \times dy \tag{3.3}$$

The integrals in Equation (3.26) are proportional to the momentum thickness and displacement thickness respectively. Using these definitions, an abbreviated form of the momentum integral equation is:

$$\frac{\partial}{\partial x}(V_O^2 \times \delta_2) + V_O \times \delta_1 \times \frac{\partial V_O}{\partial x} = \frac{\tau_O}{\rho} \tag{3.27}$$

Equation (3.27) is called the *von Karman momentum integral equation*. It is a *fundamental equation* in boundary layer theory and it is **valid for any boundary layer** developing along a flat plate.

---

**Remarks**

1.  The von Karman momentum Equation (3.26) (and Eq. (3.27)) is a very important equation. It is valid for both laminar and turbulent boundary layer flows. Remember: on a flat plate, the developing boundary layer is laminar for $Re_x < 3 \times 10^5$ where $Re_x$ is the Reynolds number defined in terms of the distance x from the upstream edge of the plate.

2.  The von Karman momentum integral equation is valid for an incompressible boundary layer. It was first developed by Theodore von Karman (1921).

3.  Theodore von Karman (or von Kármán) (1881–1963) was a Hungarian fluid dynamicist and aerodynamicist who worked in Germany (1906 to 1929) and later in USA. He was a Ph.D. student of Ludwig Prandtl in Germany. He gave his name to the vortex shedding behind a cylinder (Karman vortex street) (see Part I, Chapter I-4).

4.  In several textbooks, the momentum integral equation is written as:

$$\frac{\partial}{\partial x}\left( \int_{y=0}^{\delta} v_x \times (V_O - v_x) \times dy \right) + \frac{\partial V_O}{\partial x} \times \int_{y=0}^{\delta} (V_O - v_x) \times dy = \frac{\tau_0}{\rho}$$

where $\delta$ is the boundary layer thickness. In the ideal fluid flow region, the velocity satisfies: $V_x = V_O$ and $\partial V_x / \partial y = 0$ for $y > \delta$. Hence the above expression is equal to Equation (3.26) because $(V_O - V_x) = 0$ for $\delta < y < +\infty$.

5.  The abbreviated form of the momentum integral equation (Eq. (3.27)) may be expressed in several identical forms:

$$\frac{\partial}{\partial x}(V_O^2 \times \delta_2) + V_O \times \delta_1 \times \frac{\partial V_O}{\partial x} = \frac{\tau_O}{\rho}$$

$$V_O^2 \times \frac{\partial \delta_2}{\partial x} + V_O \times \frac{\partial V_O}{\partial x} \times (2 \times \delta_2 + \delta_1) = \frac{\tau_O}{\rho}$$

$$\frac{\partial \delta_2}{\partial x} + \frac{1}{V_O} \times \frac{\partial V_O}{\partial x} \times (2 \times \delta_2 + \delta_1) = \frac{\tau_O}{\rho \times V_O^2}$$

where $\delta_1$ and $\delta_2$ are respectively the displacement thickness (Eq. (3.2)) and the momentum thickness (Eq. (3.3)).

6.  The present development is based upon the elegant presentation of Liggett (1994, pp. 200–201).

---

The von Karman momentum integral equation may be used to solve a boundary layer problem by assuming a particular velocity distribution. This approach is valid for both laminar and turbulent boundary layers. It was first used by Pohlhausen (1921). In modern times, the momentum integral equation is rarely used for laminar flows, but it is commonly applied to turbulent boundary layer flows.

*Table 3.2* Effects of the velocity distribution assumptions on the laminar boundary layer characteristics above a flat plate: comparison between the von Karman momentum integral equation and the Blasius equation.

| Boundary layer parameter | Approximate solutions | | Theoretical solution |
|---|---|---|---|
| Velocity distribution: | $\dfrac{V_x}{V_O} = 2 \times \dfrac{y}{\delta} - \left(\dfrac{y}{\delta}\right)^2$ | $\dfrac{V_x}{V_O} = \dfrac{y}{\delta} \times \left(2 + 2 \times \left(\dfrac{y}{\delta}\right)^2 + \left(\dfrac{y}{\delta}\right)^3\right)$ | Blasius equation |
| $\delta =$ | $5.48 \times \dfrac{x}{\sqrt{Re_x}}$ | $5.84 \times \dfrac{x}{\sqrt{Re_x}}$ | $4.91 \times \dfrac{x}{\sqrt{Re_x}}$ |
| $\delta_1 =$ | $1.83 \times \dfrac{x}{\sqrt{Re_x}}$ | $1.75 \times \dfrac{x}{\sqrt{Re_x}}$ | $1.72 \times \dfrac{x}{\sqrt{Re_x}}$ |
| $\delta_2 =$ | $0.73 \times \dfrac{x}{\sqrt{Re_x}}$ | $0.685 \times \dfrac{x}{\sqrt{Re_x}}$ | $0.664 \times \dfrac{x}{\sqrt{Re_x}}$ |
| $\dfrac{\tau_O}{\frac{1}{2} \times \rho \times V_O^2} =$ | $\dfrac{0.730}{\sqrt{Re_x}}$ | $\dfrac{0.685}{\sqrt{Re_x}}$ | $\dfrac{0.664}{\sqrt{Re_x}}$ |
| $\dfrac{\int_{x=0}^{L} \tau_O \times dx}{\frac{1}{2} \times \rho \times V_O^2 \times L} =$ | $\dfrac{1.46}{\sqrt{Re_L}}$ | $\dfrac{1.370}{\sqrt{Re_L}}$ | $\dfrac{1.328}{\sqrt{Re_L}}$ |

## 3.2   Application to a flat plate

Using the solution of the Blasius equation, the momentum integral equation may be applied to a laminar boundary layer. Let us assume that the velocity profile above a flat plate may be expressed as:

$$\frac{V_x}{V_O} = a_0 + a_1 \times \frac{y}{\delta} + a_2 \times \left(\frac{y}{\delta}\right)^2 \tag{3.28}$$

where $a_0$, $a_1$ and $a_2$ are undetermined coefficients, The coefficients are determined from the boundary conditions: $V_x(y = 0) = 0$, $V_x(y = +\infty) = V_O$ and $(\partial V_x/\partial y) = 0$ for $y = +\infty$. The velocity distribution is found to be:

$$\frac{V_x}{V_O} = 2 \times \frac{y}{\delta} - \left(\frac{y}{\delta}\right)^2 \tag{3.29}$$

The von Karman momentum equation for a flat plate becomes:

$$V_O^2 \times \frac{\partial}{\partial x}(\delta_2) = \frac{\tau_O}{\rho} \tag{3.30}$$

After the appropriate substitutions, the laminar boundary layer characteristics may be derived:

$$\delta = 5.48 \times \frac{x}{\sqrt{Re_x}} \tag{3.31}$$

$$\delta_1 = 1.83 \times \frac{x}{\sqrt{Re_x}} \tag{3.32}$$

$$\delta_2 = 0.73 \times \frac{x}{\sqrt{Re_x}} \tag{3.33}$$

The overall shear force on a plate of length L is:

$$\frac{\int_{x=0}^{L} \tau_O \times dx}{\frac{1}{2} \times \rho \times V_O^2 \times L} = \frac{1.46}{\sqrt{Re_L}} \tag{3.34}$$

The results are summarised in Table 3.2 and they are compared with the solution of the Blasius equation.

The above approach may be extended to a velocity profile which satisfies a polynomial of third degree (see Exercises). The results for a polynomial of fourth degree are listed in Table 3.2 in which they are compared with the parabolic velocity distribution results and the exact solution of the Blasius equation. The comparison shows that the momentum integral results with a polynomial of fourth degree are close to the exact solution and even the results with a quadratic polynomial are within 10% of the theoretical solution which is reasonable in many engineering applications.

## Note

A polynomial of second degree is called a quadratic polynomial, or parabolic function.

## 3.3  Discussion

This technique may be applied to other boundary layer flows including those with some longitudinal pressure gradient $\partial P / \partial x \neq 0$. Pohlhausen (1921) solved the momentum integral equation for a velocity distribution which satisfies a polynomial of fourth degree:

$$\frac{v_x}{V_O} = a_0 + a_1 \times \frac{y}{\delta} + a_2 \times \left(\frac{y}{\delta}\right)^2 + a_3 \times \left(\frac{y}{\delta}\right)^3 + a_4 \times \left(\frac{y}{\delta}\right)^4 \tag{3.35}$$

where $a_0$ to $a_4$ are undetermined constants.

At the boundary ($y = 0$), the boundary conditions are $V_x(y = 0) = 0$ and

$$\left(\frac{\partial^2 v_x}{\partial y^2}\right)_{y=0} = -\frac{\rho}{\mu} \times V_O \times \frac{\partial V_O}{\partial x} \tag{3.36}$$

The above condition comes from the boundary layer equation (see below). Outside of the boundary layer ($y \geq \delta$), the velocity profile satisfies further: $V_x = V_x(y = \delta) = V_O$, and $(\partial V_x / \partial y) = (\partial^2 V_x / \partial y^2) = 0$ for $y = \delta$.

## Note

Equation (3.36) is a simple rewriting of the boundary layer equation (Eq. (3.6)). At the boundary $(y=0)$, Equation (3.6) becomes:

$$\left( -\frac{1}{\rho} \times \frac{\partial P}{\partial x} - g \times \frac{\partial z_O}{\partial x} + \frac{\mu}{\rho} \times \frac{\partial^2 v_x}{\partial y^2} \right)_{y=0} = 0$$

Using the differential form of the Bernoulli equation along the streamline in the ideal fluid flow region:

$$V_O \times \frac{\partial V_O}{\partial x} + g \times \frac{\partial z_O}{\partial x} + \frac{1}{\rho} \times \frac{\partial P}{\partial x} = 0$$

and replacing into Equation (3.6), it yields:

$$V_O \times \frac{\partial V_O}{\partial x} + \left( \frac{\mu}{\rho} \times \frac{\partial^2 v_x}{\partial y^2} \right)_{y=0} = 0$$

Let us introduce the dimensionless parameter $\lambda$:

$$\lambda = \frac{\rho}{\mu} \times \delta^2 \times \frac{\partial V_O}{\partial x} \tag{3.37}$$

Physically the parameter $\lambda$ may be interpreted as a ratio of a longitudinal pressure gradient to some viscous force (Liggett 1994). Physically a positive value of $\lambda$ indicates a favourable pressure gradient $(\partial P/\partial x < 0)$ and a negative value denotes an adverse pressure gradient $(\partial P/\partial x > 0)$. For a flat plate in absence of pressure gradient, $\lambda = 0$.

The parameters of Equation (3.35) are found to be:

$$a_0 = 0 \tag{3.38a}$$

$$a_1 = 2 + \frac{\lambda}{6} \tag{3.38b}$$

$$a_2 = -\frac{\lambda}{2} \tag{3.38c}$$

$$a_3 = -\left( 2 - \frac{\lambda}{2} \right) \tag{3.38d}$$

$$a_4 = 1 - \frac{\lambda}{6} \tag{3.38e}$$

If $\lambda$ is a constant, the velocity profiles are self-similar. In a more general case, the parameter $\lambda$ may vary with x and the parameters $a_0$ to $a_4$ vary with distance. That is, the shape of the velocity profile changes along the boundary layer.

The boundary layer characteristics may be expressed in terms of the dimensionless parameter $\lambda$:

$$\frac{\delta_1}{\delta} = \frac{3}{10} - \frac{\lambda}{120} \tag{3.39}$$

$$\frac{\delta_2}{\delta} = \frac{37}{315} - \frac{\lambda}{945} + \frac{\lambda^2}{9072} \tag{3.40}$$

The dimensionless boundary shear stress equals:

$$\frac{\tau_O \times \delta}{\mu \times V_O} = 2 + \frac{\lambda}{6} \tag{3.41}$$

Let us consider the flow around a blunt body (Fig. 3.6). At the stagnation point, the parameter $\lambda$ equals 7.052. As the boundary layer grows around the body, the values of $\lambda$ decrease with increasing distance from stagnation. For a shape as sketched in Figure 3.6, $\lambda$ may become negative. More generally, a separation point will occur for $\lambda = -12$.

This method is called the Pohlhausen method, or Karman-Pohlhausen method. It was found to be accurate for accelerated flows, but not for decelerated flows. The calculations for $\lambda < 0$ may be suspect (Liggett 1994).

---

**Remarks**

1. The parameter $\lambda$ characterises the ratio of a longitudinal pressure gradient to some viscous force (Liggett 1994).
2. Values of $\lambda$ higher than 7.052 would represent a corner flow.
3. Physically, the values of the parameter $\lambda$ range from 7.052 at stagnation to $-12$ at separation for a blunt body, as the one illustrated in Figure 3.6.

---

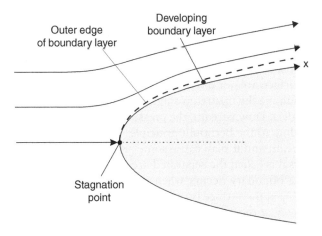

*Figure 3.6* Flow around a blunt body.

*Figure 3.7* X-31 Enhanced Fighter Maneuverability Demonstrator turning tightly over the desert floor in 1994 (Courtesy of NASA Dryden Aircraft Photo Collection, Photo Number EC94-42478-12).

## 4   SEPARATION AND STALL

Many engineering flows involve some boundary layers. In some cases, the flow may separate from the boundary. This phenomenon causes some drastic effects including some wake effects behind buildings, automobile drag and aircraft stall (Fig. 3.7). Figure 3.7 illustrates a highly-manoeuvrable experimental aircraft which could manoeuvre tightly at very high angles of attack within the post-stall regime. The movie P1040830.mov shows some flow separation at the edge of a jetty in a large estuary during a strong ebb flow (App. F).

Let us consider the flow around the curved surface sketched in Figure 3.8. Note that the velocity profiles around the surface are not drawn to scale: the vertical scale is enlarged. The pressure on the surface is maximum at the upstream stagnation point and it is minimum at the location where the body is the widest. Downstream, the pressure in the flow increases since the mean flow velocity decreases according to the Bernoulli principle. In other words, the pressure gradient satisfied $\partial P/\partial x > 0$ downstream and it may cause some reverse (upstream) flow near the boundary. The region of flow reversal is called the separated area or recirculation zone.

Flow separation at a boundary occurs when the boundary shear stress $\tau_O$ and the vertical gradient of velocity $\partial V_x/\partial y$ at the surface ($y = 0$) are zero. Downstream of the point of separation, the velocity gradient $\partial V_x/\partial y$ at the surface ($y = 0$) is negative as sketched in Figure 3.8.

The adverse pressure gradient in the flow ($\partial P/\partial x > 0$) is the main cause of separation. In any flow situation in which the pressure increases in the downstream direction, separation might occur.

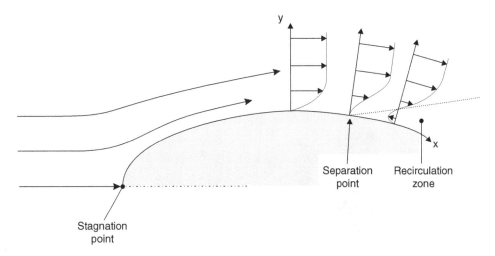

*Figure 3.8* Sketch of flow separation along a curved surface and the resulting velocity profiles.

Boundary layer calculations may predict, more and less accurately, the location of separation, but the flow downstream of the separation point cannot be estimated accurately. This is because the boundary layer calculations are based upon some ideal fluid flow solution that becomes invalid in separated flows.

Usually the exact location of separation cannot be predicted precisely. But, in flows around bodies with sharp corners, the separation takes place at the corners. Practical applications include atmospheric flow around buildings and wind flows around many vehicles. In some cases, the theory of free-streamlines and the theorem of Schwarz-Christoffel may be applied to yield ideal fluid flow solutions (Chapter I-7).

### Discussion

A classical example of separation and wake is the unsteady separation phenomenon of a steady uniform flow past a circular cylinder: i.e., the von Karman street of vortices discussed in Chapter I-4. The flow past a circular cylinder is one of the most complicated flow situations with respect to the simplicity of the geometry. For a range of Reynolds numbers $Re = 2 \times R \times V_o/\nu$ larger than 40 to 80, the separation points oscillate rearward and forward alternately. The oscillations cause some vortex shedding alternating from one side to the other of the cylinder, resulting in a pattern of vortices rotating in opposite directions behind the cylinder (Fig. 3.9). Figure 3.9 shows a von Karman street of vortices in the atmospheric boundary layer. It highlights the street of large eddies downstream of the 1,720 m high volcanic island of Rishiri-to in the northern Sea of Japan off Hokkaido with the main wind flow direction from bottom to top.

### Note

- The vortex shedding behind a cylinder was first observed by Theodore von Karman at the Technical University of Aachen. At the time (1910s), von Karman conducted a wind tunnel investigation on a streamlined spar and a cylindrical spar on biplane configurations.
- The dimensionless frequency of oscillation, or Strouhal number, is about 0.2 for a wide range of Reynolds numbers.

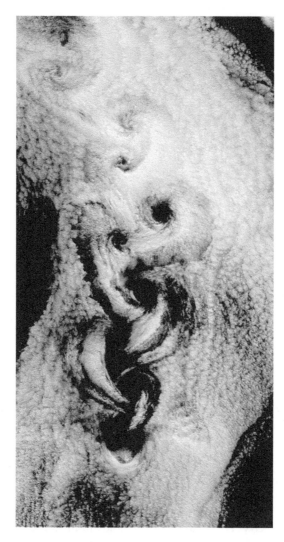

*Figure 3.9* Von Karman street of clouds downstream the volcanic island of Rishiri-to in the northern Sea of Japan off the northwest coast of Hokkaido (Japan) in 2001 (Image Courtesy of NASA STS-100 Shuttle Mission Imagery, Photograph STS100-710-182) – Wind flow direction from bottom to top.

## 5 SHEAR FLOWS

The boundary layer theory may be used for a number of boundary-free shear flows. A free shear flow is one in which two layers of fluid travel at different velocities with some transfer of momentum from the high-velocity region to the low-speed flow. Laminar flow motion applies rarely to these flow situations because these are unstable and break up into turbulence. But the basic principles and techniques used to calculate laminar flows do apply to turbulent flows also.

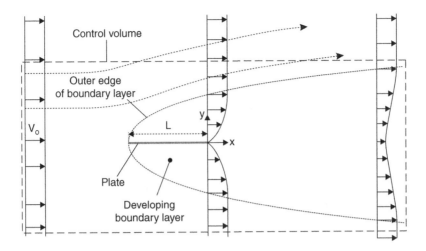

*Figure 3.10* Sketch of a wake behind a flat splitter plate.

## 5.1 Wake behind a splitter plate

Let us consider the boundary layer flows around a flat splitter plate shown in Figure 3.10. Note that the vertical scale was exaggerated in Figure 3.10. Behind the plate, there is a region of disturbed flow, called the wake flow. In that region, there is some transfer of momentum from the ideal fluid flow region to the core of the wake until the velocity profile becomes uniform again far away downstream of the plate.

For a steady flow motion, the momentum principle states that the rate of change of momentum flux equals the sum of the forces applied to the control volume sketched in Figure 3.10. The only force applied to the control volume is the drag force of the plate. Using the result of the Blasius equation, the total drag force per unit width on both sides of the plate is:

$$\text{Drag force} = 2 \times \int_{x=0}^{L} \tau_0 \times dx = 1.328 \times \sqrt{\rho \times \mu} \times \sqrt{V_O^3 \times L} \tag{3.42}$$

**Note**

A boundary layer develops at both the upper and lower side of the splitter plate (Fig. 3.10). The drag force (Eq. (3.42)) is equal to twice the shear force on a plate of length L (Eq. (3.18)).

The application of the momentum principle to the control volume sketched in Figure 3.10 gives:

$$\int_{-\infty}^{+\infty} \rho \times v_x \times (V_O - v_x) \times dy = \text{Drag force} \tag{3.43a}$$

where $(V_O - V_x)$ is called the velocity defect. In Equation (3.43a), the left handside term is taken at any position downstream of the plate $(x > 0)$. Using Equation (3.42), it becomes:

$$\int_{-\infty}^{+\infty} \rho \times v_x \times (V_O - v_x) \times dy = 1.328 \times \sqrt{\rho \times \mu} \times \sqrt{V_O^3 \times L} \tag{3.43b}$$

Assuming that $(V_O - V_x) \ll V_O$, the boundary layer equation (Eq. (3.6)) may be approximated to:

$$V_O \times \frac{\partial v_x}{\partial x} = \frac{\mu}{\rho} \times \frac{\partial^2 v_x}{\partial y^2} \tag{3.44}$$

where the boundary conditions are $V_x = V_O$ for $y = \pm\infty$ and $\partial V_x / \partial y = 0$ at $y = 0$.

The solution of Equation (3.44) has the form:

$$V_O - v_x = A \times \frac{V_O}{\sqrt{\frac{x}{L}}} \times \exp\left(-\frac{\rho \times V_O \times x}{\mu} \times \left(\frac{y}{2 \times x}\right)^2\right) \tag{3.45}$$

where A is an undetermined constant which is derived from Equation (3.43). Substituting Equation (3.45) into Equation (3.43) produces $A = 0.375$. Hence the velocity profile in the wake is:

$$v_x = V_O \times \left(1 - \frac{0.375}{\sqrt{\frac{x}{L}}} \times \exp\left(-\frac{\rho \times V_O \times x}{\mu} \times \left(\frac{y}{2 \times x}\right)^2\right)\right) \tag{3.46}$$

In dimensionless terms, it yields:

$$\frac{v_x}{V_O} = 1 - \frac{0.375}{\sqrt{\frac{x}{L}}} \times \exp\left(-Re_L \times \frac{x}{L} \times \left(\frac{1}{2} \times \frac{\frac{y}{L}}{\frac{x}{L}}\right)^2\right) \tag{3.46b}$$

where $Re_L = \rho \times V_O \times L/\mu$. Some typical results are shown in Figure 3.11.

Practically the above solution for laminar flow is rarely applicable since the flow is unstable and the wake is likely to be turbulent.

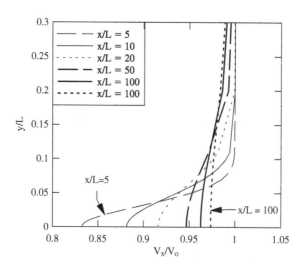

*Figure 3.11* Velocity distributions in the laminar wake behind a flat splitter plate.

**Remarks**

1. The term $(V_O - V_x)$ is often called the velocity defect.
2. The above development assumes that $(V_O - V_x) \ll V_O$ and it is only valid for $x > 3 \times L$ where x is the longitudinal distance measured from the trailing edge of the plate and L is the plate length (Fig. 3.10).
3. Equation (3.44) is a classical differential equation similar to a diffusion equation. The diffusion equation may be solved analytically for a number of basic boundary conditions. Mathematical solutions of the diffusion equation (and heat equation) were addressed in some classical references (Carslaw and Jaeger 1959, Crank 1956, Chanson 2004b). Since Equation (3.44) is linear, the theory of superposition may be used to build up solutions with more complex problems and boundary conditions.
4. Nishioka and Miyagi (1978) performed some physical experiments in steady laminar flows with Reynolds numbers between 20 and 3000. Their data matched closely Equation (3.46).
5. The substitution of Equation (3.45) into Equation (3.43) gives:

$$A = \frac{1}{2} \times \frac{1.328}{\sqrt{\pi}} = 0.375$$

6. In Equation (3.46), remember that x is the distance from the downstream edge of the plate (Fig. 3.10).

## 5.2 Two-dimensional jet

Let us consider a two-dimensional jet issued from a rectangular slot of thickness D and width B (Fig. 3.12). The following development applies to a laminar jet, but the methodology is the same for both laminar and turbulent jets.

Assuming no pressure gradient in the longitudinal direction, the momentum of the jet is conserved:

$$J = \left( \int_{y=-\infty}^{+\infty} \rho \times v_x^2 \times dy \right) \times B = \text{constant} \tag{3.47}$$

where J is the momentum flux in the x-direction. Equation (3.47) is valid at any longitudinal position $x \geq 0$, where x is the longitudinal distance from the jet nozzle. For an uniform velocity distribution at the jet nozzle, the momentum flux equals:

$$J = \rho \times V_O^2 \times D \times B \tag{3.48}$$

where $V_O$ is the jet nozzle velocity.

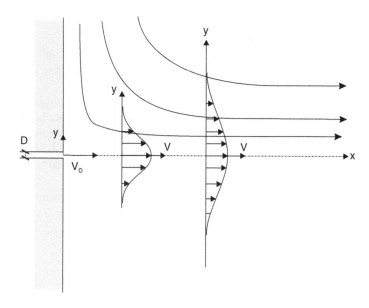

Figure 3.12   Sketch of a two-dimensional jet flow.

The solution of the boundary layer equation yields the velocity profile:

$$\frac{v_x}{V_O} = \frac{1}{6} \times \left(\frac{9}{2}\right)^{2/3} \times \frac{\left(\rho \times \frac{V_O \times D}{\mu}\right)^{1/3}}{\left(\frac{x}{D}\right)^{1/3}} \times \left(1 - \tanh^2\left(\frac{1}{2} \times \sqrt[3]{\frac{9}{2}} \times \sqrt[3]{\frac{V_O^2 \times D \times x^2}{\left(\frac{\mu}{\rho}\right)^2}} \times \frac{y}{x}\right)\right)$$

(3.49)

where tanh is the hyperbolic tangent function. The velocity distribution is symmetrical and the maximum velocity is observed on the jet centreline.

Using the above result, the discharge of entrained fluid in the jet flow per unit width may be estimated:

$$\frac{Q_{ent}}{B} = \int\limits_{y=-\infty}^{+\infty} v_x \times dy = 2 \times \sqrt[3]{\frac{9}{2}} \times \sqrt[3]{V_O^2 \times D \times \frac{\mu}{\rho}} \times \sqrt[3]{x}$$

(3.50)

The jet entrains some surrounding fluid as momentum is exchanged from the high-velocity region to the surrounding fluid at rest. The volume discharge $Q_{ent}$ and the height of the jet flow increase with increasing longitudinal distance: e.g., $Q_{ent} \propto \sqrt[3]{x}$.

(A) $V_x/V_o$ as a function of y/D – Data: Andrade (1939), D = 0.3 mm, $V_o$ = 0.045 m/s, $\rho \times V_o \times D/\mu$ = 12.8, water flow

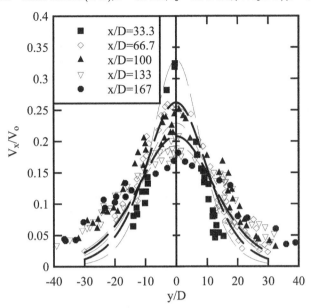

(B) $V_x/V_{max}$ as a function of $y/(\mu/\rho \times x)^{1/3}$ – Data: Sato and Sakao (1964), D = 0.2 mm, $V_o$ = 3.13 m/s, $\rho \times V_o \times D/\mu$ = 43, air flow

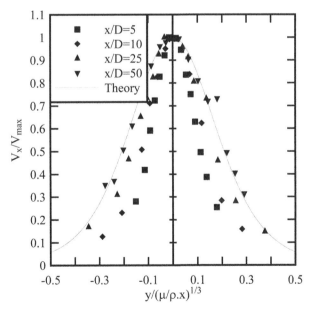

*Figure 3.13*   Dimensionless velocity distributions in two-dimensional laminar jets.

**Remarks**

1. The two-dimensional jet flow is laminar for $\rho \times V_O \times D/\mu < 30$ to $50$ (Schlichting 1979, Sato and Sakao 1964).
2. Experimental measurements showed a good agreement with Equation (3.49). Figure 3.13 presents some experimental data for $\rho \times V_o \times D/\mu = 12.8$ and 43. For the experiment shown in Figure 3.13B, the flow conditions were in the upper range of the laminar jet Reynolds number, and some light periodic velocity fluctuations were observed for $2 < x/D < 20$ although they were not observed anywhere else in the jet flow region.
3. The hyperbolic tangent function is defined as:

$$\tanh(x) = \frac{e^x - e^{-x}}{e^x + e^{-x}}$$

4. The function $(1 - \tanh^2(x))$ is unity for $x = 0$ and tends to zero for $x \to \pm\infty$.
5. The jet velocity is maximum on the centreline $(y = 0)$ and its magnitude is:

$$\frac{V_{max}}{V_o} = \frac{1}{6} \times \left(\frac{9}{2}\right)^{2/3} \times \frac{\left(\rho \times \dfrac{V_o \times D}{\mu}\right)^{1/3}}{\left(\dfrac{x}{D}\right)^{1/3}}$$

The maximum jet velocity decreases with increasing longitudinal distance: e.g., $V_{max} \propto 1/(x)^{1/3}$.

# Exercises

## Exercise No. 1

In a developing boundary layer, the velocity distribution follows:

| y (mm) | $V_x$ (m/s) | y (mm) | $V_x$ (m/s) |
|--------|-------------|--------|-------------|
| 0.5    | 0.271       | 21     | 10.79       |
| 1      | 0.518       | 35     | 11.3        |
| 2      | 1.113       | 41     | 11.52       |
| 3.5    | 1.93        | 58     | 11.55       |
| 6      | 3.137       | 60.5   | 11.51       |
| 8      | 4.157       | 66     | 11.5        |
| 10.2   | 5.32        | 70     | 11.6        |
| 14     | 7.28        | 90     | 11.59       |
| 18     | 9.22        | 100    | 11.50       |

Calculate the boundary layer thickness, the displacement thickness, the momentum thickness and the energy thickness.

### Solution

$V_O = 11.5 \, \text{m/s}$

$\delta_{99} = 37.32 \, \text{mm}$         boundary layer thickness

$\delta_1 = 11.64 \, \text{mm}$

$\delta_2 = 4.31 \, \text{mm}$

$\delta_3 = 6.72 \, \text{mm}$

## Exercise No. 2

Considering a laminar boundary layer.

(a) Using a power series, demonstrate the asymptotic solutions of the velocity distribution:

$$\frac{v_x}{V_O} = 0.332 \times \frac{y}{x} \times \sqrt{Re_x} \quad \text{for} \quad \frac{y}{x} \times \sqrt{Re_x} \ll 1$$

$$\frac{v_x}{V_O} = 1 \quad\quad\quad\quad\quad \text{for} \quad \frac{y}{x} \times \sqrt{Re_x} < 6$$

(b) For the developing boundary layer, calculate the boundary shear stress at a distance x from the plate leading edge:

$$\tau_O = \mu \times \left( \frac{\partial v_x}{\partial y} \right)_{y=0}$$

(c) Based upon the above result, calculate the shear force acting on a plate of length L and width B.

## Exercise No. 3

Let us consider a plane developing boundary layer with zero pressure gradient and a free-stream velocity $V_O = 0.1$ m/s.

(a)   Plot the velocity profile at $x = 0.12$ m. *First you must assess if the flow motion is laminar or turbulent.*
(b)   Calculate the boundary layer thickness, the displacement thickness, and the momentum thickness at $x = 0.12$ m.
(c)   Calculate and plot the boundary shear stress for $0 \leq x \leq 0.2$ m.
(d)   Compute the overall friction force on the 0.2 m long 0.3 m wide plate.

*The fluid is blood (density: $1050$ kg/m³, viscosity: 4 E-3 Pa.s).*

### Solution

(b) $\delta = 0.010$ m, $\delta_1 = 3.7$ mm, $\delta_2 = 1.4$ mm
(d) $F_{shear} = 0.0058$ N

## Exercise No. 4

Velocity measurements were conducted at 3 locations along a developing laminar boundary layer.

| y mm x (m) = | $V_x$ cm/s 0.05 | $V_x$ cm/s 0.10 | $V_x$ cm/s 0.20 |
|---|---|---|---|
| 0.5 | 0.52 | 0.45 | 0.44 |
| 1 | 1.00 | 0.58 | 0.47 |
| 2 | 1.62 | 1.02 | 0.83 |
| 3.5 | 2.80 | 1.87 | 1.31 |
| 6 | 4.51 | 3.21 | 2.17 |
| 8 | 5.88 | 4.21 | 3.10 |
| 10.2 | 7.53 | 5.13 | 3.67 |
| 14 | 10.39 | 7.24 | 5.26 |
| 18 | 11.47 | 9.09 | 6.69 |
| 21 | 11.55 | 10.71 | 7.49 |
| 35 | 11.62 | 11.46 | 10.55 |
| 41 | 11.54 | 11.69 | 11.45 |
| 58 | 11.60 | 11.52 | 11.55 |
| 60.5 | 11.67 | 11.69 | 11.67 |
| 66 | 11.58 | 11.70 | 11.67 |
| 70 | 11.66 | 11.57 | 11.66 |
| 90 | 11.58 | 11.54 | 11.69 |
| 100 | 11.69 | 11.53 | 11.67 |

(a)   Plot the velocity distributions.
(b)   Using the momentum integral equation, calculate the boundary shear stress at $x = 0.1$ m.
(c)   Based upon the momentum integral equation, integrate numerically the boundary shear stress to estimate the friction force per unit width on the 0.2 m long plate.
(d)   Compare your results with the theoretical calculations (Blasius equation) and with the approximate solution assuming a quadratic velocity distribution. *Check that the flow laminar before conducting the comparison.*

*The fluid is a bentonite suspension (density: $1,115$ kg/m³, viscosity: 0.19 Pa.s, mass concentration: 17%).*

## Solution

(a) The free-stream velocity is about 11.6 cm/s.

(b)

|  | Momentum integral equation | Blasius solution |
|---|---|---|
| $\tau_O$ (x = 0.1 m) (Pa) = | 0.41 | 0.6 |
| $F_{shear}/B$ (N/m) = | 0.083 (?) | 0.17 |

Note: The application of the momentum integral equation between x = 0 and x = 0.05 m is not possible; this region is affected by large shear stress because of the flow singularity.

# Exercise No. 5

Assuming that the velocity profile in a laminar boundary layer satisfies a polynomial of fourth degree, apply the momentum integral equation.

(a)    Derive the expression of the velocity profile. *Write carefully the boundary conditions.*
(b)    Derive mathematically the expression of the boundary layer thickness, bed shear stress and total shear force.

# Exercise No. 6

Assuming that the velocity profile in a laminar boundary layer satisfies a polynomial of third degree, apply the momentum integral equation.

(a)    Derive the expression of the velocity profile. *Write carefully the boundary conditions.*
(b)    Derive mathematically the expression of the boundary layer thickness, displacement thickness and momentum thickness.
(c)    Derive the expressions of the bed shear stress and total shear force.
(d)    Compare your results with the Blasius solution.

## Solution

(a)    Let us assume that the velocity profile above a flat plate may be expressed as:

$$\frac{v_x}{V_O} = a_0 + a_1 \times \frac{y}{\delta} + a_2 \times \left(\frac{y}{\delta}\right)^2 + a_3 \times \left(\frac{y}{\delta}\right)^3$$

where $a_0$, $a_1$, $a_2$ and $a_3$ are undetermined coefficients, The coefficients are determined from the boundary conditions: $v_x(y=0)=0$, $V_x(y=\delta)=V_O$, $(\partial v_x/\partial y)=0$ for $y=\delta$ and $(\partial^2 v_x/\partial y^2)=0$ for $y=\delta$. The velocity distribution is found to be:

$$\frac{v_x}{V_O} = 3 \times \frac{y}{\delta} - 3 \times \left(\frac{y}{\delta}\right)^2 + \left(\frac{y}{\delta}\right)^3$$

(b)    The von Karman momentum integral equation for a flat plate becomes:

$$V_O^2 \times \frac{\partial}{\partial x}(\delta_2) = \frac{\tau_O}{\rho}$$

For a velocity profile satisfying a polynomial of third degree, the momentum thickness equals:

$$\delta_2 = \int_0^{+\infty} \frac{v_x}{V_O}\left(1 - \frac{v_x}{V_O}\right) \times dy = \frac{3}{28} \times \delta$$

The bed shear stress is defined as:

$$\tau_O - \mu \times \left(\frac{\partial v_x}{\partial y}\right)_{y=0} = \frac{3 \times \mu \times V_O}{\delta}$$

The momentum integral equation yields:

$$\frac{3}{28} \times V_O^2 \times \frac{\partial \delta}{\partial x} = \frac{\mu}{\rho} \times \frac{3 \times V_O}{\delta}$$

The integration gives the expression of the boundary layer growth:

$$\delta = \sqrt{14} \times \frac{x}{\sqrt{Re_x}}$$

The displacement thickness and momentum thickness equal:

$$\delta_1 = \frac{\delta}{4} = \sqrt{\frac{7}{2}} \times \frac{x}{\sqrt{Re_x}}$$

$$\delta_2 = \frac{3}{28} \times \delta = \sqrt{\frac{9}{56}} \times \frac{x}{\sqrt{Re_x}}$$

(c)     The boundary shear stress is deduced from the momentum integral equation:

$$\frac{\tau_O}{\frac{1}{2} \times \rho \times V_O^2} = 2 \times \frac{\partial}{\partial x}(\delta_2) = \frac{9}{\sqrt{56}} \times \frac{1}{\sqrt{Re_x}}$$

The dimensionless boundary shear force per unit width equals:

$$\frac{\int_{x=0}^{L} \tau_O \times dx}{\frac{1}{2} \times \rho \times V_O^2 \times L} = \frac{9}{\sqrt{14}} \times \frac{1}{\sqrt{Re_L}}$$

(d)     The results are compared with the Blasius analytical solution below:

| Boundary layer parameter | Approximate solution | Theoretical solution |
|---|---|---|
| Velocity distribution: | $\frac{v_x}{V_O} = 3 \times \frac{y}{\delta} - 3 \times \left(\frac{y}{\delta}\right)^2 + \left(\frac{y}{\delta}\right)^3$ | Blasius equation |
| $\delta =$ | $3.74 \times \frac{x}{\sqrt{Re_x}}$ | $4.91 \times \frac{x}{\sqrt{Re_x}}$ |
| $\delta_1 =$ | $1.87 \times \frac{x}{\sqrt{Re_x}}$ | $1.72 \times \frac{x}{\sqrt{Re_x}}$ |
| $\delta_2 =$ | $0.40 \times \frac{x}{\sqrt{Re_x}}$ | $0.664 \times \frac{x}{\sqrt{Re_x}}$ |
| $\frac{\tau_O}{\frac{1}{2} \times \rho \times V_O^2}$ | $\frac{0.40}{\sqrt{Re_x}}$ | $\frac{0.664}{\sqrt{Re_x}}$ |
| $\frac{\int_{x=0}^{L} \tau_O \times dx}{\frac{1}{2} \times \rho \times V_O^2 \times L}$ | $\frac{0.80}{\sqrt{Re_L}}$ | $\frac{1.328}{\sqrt{Re_L}}$ |

## Exercise No. 7

Let us consider a laminar wake behind a 0.5 m long plate.

(a)   Calculate the total drag force (per unit width) on the plate.
(b)   Estimate at what distance, downstream of the plate, the velocity profile will recover (within 2% of the free-stream velocity)?

*The free-stream velocity is 0.35 m/s and the fluid is a viscous SAE40 oil (density: 871 kg/m³, viscosity: 0.6 Pa.s).*

### Solution

(a)   Drag per unit width $= 4.4$ N/m
(b)   $x/L = 350$ ($x = 175$ m !!!)

## Exercise No. 8

Some fluid is injected in a vast container where the surrounding fluid is at rest. The nozzle height is 0.15 mm and the injected velocity is 1 cm/s.

(a)   Calculate the maximum jet velocity at distances of 1.5 mm, 2 cm and 18 cm from the nozzle.
(b)   At a distance of 18 cm from the nozzle, estimate the volume discharge of entrained fluid (per unit width).

*Assume a two-dimensional jet.*
      *The fluid is blood (density: 1,050 kg/m³, viscosity: $4 \times 10^{-3}$ Pa.s).*

### Solution

|  | $x = 1.5$ mm | $x = 2$ cm | $x = 18$ cm |
|---|---|---|---|
| Maximum $v_x/V_O =$ | 15.5% | 6.5% | 3.1% |
| $Q/(V_O \times D) =$ | – | – | 48 |

## Exercise No. 9

Let us consider a laminar flow down an inclined plane (Fig. E3.1) at uniform equilibrium.

(a)   Derive the shear stress distribution in the direction normal to the plane.
(b)   Express the pressure distribution in the direction normal to the plane.
(c)   Deduce the laminar flow velocity.

### Solution

(a)   The problem is solved by applying the momentum principle to the control volume sketched above. At uniform equilibrium down the slope, the velocity distribution and water depth d are independent of the distance x along the slope. The application of the momentum principle along the x-direction implies that the laminar shear stress τ along the lower interface amust equal exactly the control volume weight force component along the inclined plane:

$$\tau \times L = \rho \times g \times L \times (d - y) \times \sin \theta$$

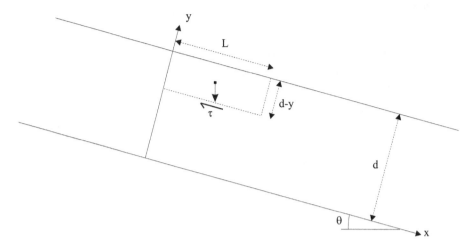

*Figure E3.1*   Laminar flow down an inclined plane.

where L is the control volume length. It yields:

$$\tau = \rho \times g \times (d - y) \times \sin\theta$$

(b)   The pressure distribution is derived from the application of the momentum principle along the y-direction implying that the pressure force acting along the lower interface must equal the weight must equal exactly the control volume weight force component along the y-direction:

$$P \times L = \rho \times g \times L \times (d - y) \times \cos\theta$$

This gives the classical result in an open channel flow with hydrostatic pressure:

$$P = \rho \times g \times (d - y) \times \cos\theta$$

(c)   In a laminar force, the shear stress is proportional to the shear rate:

$$\tau = \mu \times \frac{\partial V_x}{\partial y}$$

From the result derived in (a), the velocity distribution is:

$$V_x = \frac{\rho \times g \times \sin\theta}{\mu} \times y \times \left(d - \frac{y}{2}\right)$$

Since the boundary shear stress equals the fluid shear stress at $y = 0$, and introducing the shear velocity, the velocity distribution may be rewritten in dimensionless terms as:

$$\frac{V_x}{V_*} = \frac{\rho \times \sqrt{g \times \sin\theta \times d^3}}{\mu} \times \frac{y}{d} \times \left(1 - \frac{y}{2 \times d}\right)$$

## Exercise No. 10

The frequency of oscillation between a circular cylinder follows approximately the following relationship:

$$St = \frac{\omega_{shedding} \times D}{V_o} = 0.2 \times \left(1 - \frac{20}{Re}\right) \quad \text{for } 100 < Re < 10^5$$

where D is the cylinder diameter, $V_o$ is the approach flow velocity and Re is the Reynolds number. Estimate the vortex shedding frequency of a 10 mm wire in a 35 m/s wind flow.

### Solution

$Re = 21.400$; $St = 0.1998$; $\omega_{shedding} = 700\,Hz$

## Exercise No. 11

We consider a developing laminar boundary layer along a 1 m long 0.5 m wide flat plate. The upstream velocity profile is uniform. The velocity profile in the laminar boundary layer may be approximated by the following expression:

$$V_x = a + b \times \sin\left(c \times \frac{y}{\delta}\right) \quad \text{for } 0 \le c \times y/\delta \le \pi/2$$

(1) Based upon the momentum integral principle, derive the expression of the velocity profile.
(2) Give the expression of the boundary layer thickness, displacement thickness, momentum thickness, bed shear stress and total shear force as function of the distance x from the plate leading edge.
(3) Compare the above results with the Blasius solution.
(4) Application to a bentonite suspension
    (4.1) Plot the longitudinal distribution of boundary shear stress along the plate. Include the Blasius solution for comparison.
    (4.2) Calculate the friction force on the plate.

*The bentonite suspension properties are: $\rho = 1{,}115\,kg/m^3$, $\mu = 0.19\,Pa.s$, mass concentration: 17%. The free-stream velocity is: $V_o = 0.10\,m/s$.*

---

**Mathematical aids**

$$\int \cos x(1 - \cos x)dx = -\frac{1}{2}x + \sin x - \frac{1}{4}\sin(2x)$$

$$\int \cos x(1 - \sin x)dx = \sin x + \frac{1}{2}\cos^2(x)$$

$$\int \sin x(1 - \cos x)dx = \frac{1}{2}(\cos x - 2)\cos x$$

$$\int \sin x(1 - \sin x)dx = -\frac{1}{2}x - \cos x + \frac{1}{4}\sin(2x)$$

### Solution

The velocity distribution may be rewritten as:

$$V_x = V_o \times \sin\left(\frac{\pi}{2} \times \frac{y}{\delta}\right)$$

The application of the momentum integral principle yields:

$$\delta = 4.795 \times \frac{x}{\sqrt{Re_x}}$$

$$\delta_2 = 0.6551 \times \frac{x}{\sqrt{Re_x}}$$

$$\frac{\tau_o}{\frac{1}{2} \times \rho \times V_o^2} = \frac{0.6551}{\sqrt{Re_x}}$$

$$\frac{\int_0^L \tau_o}{\frac{1}{2} \times \rho \times V_o^2 \times L} = \frac{1.31}{\sqrt{Re_L}}$$

# Turbulent boundary layers

## SUMMARY

In this chapter, some basic turbulent flows are reviewed including developing boundary layers and jets. Their flow properties are described and discussed.

## I PRESENTATION

For fluids of low viscosity, the effects of viscosity are appreciable only in a narrow region surroundings the fluid boundaries. For incompressible flows where the boundary layer remains thin, non-viscous fluid results may be applied to real fluids to a satisfactory degree of approximation. Converging or accelerating flow situations generally have thin boundary layers, but decelerating flows may have flow separation and development of large turbulent wake that is difficult to predict with non-viscous fluid equations. In many turbulent flows, the effects of boundary friction are significant and they are discussed herein.

In Nature, three types of turbulent flows are encountered commonly: (1) jets and wakes, (2) developing boundary layers, and (3) fully-developed open channel flows. In a shear flow, a momentum flux is transferred from the region of high velocity to that of low-velocity. The shear stress characterises the fluid resistance to the transfer of momentum. In many environmental applications, the flow conditions are turbulent and the turbulent boundary layer flow properties are developed herein. A classical example of turbulent boundary layer is the atmospheric boundary layer. The velocity is zero at the ground and it increases with increasing vertical elevations. Figure 4.1 shows a wind farm drawing electricity from the wind energy. This wind farm has 6 turbines with an installed capacity of 2.3 MW each, producing annually about 17 Millions of kWh. The movie Plouarz1.mov shows a wind farm in operation in western France (App. F). Each turbine can produce 750 kW and the main characteristics are: 47 m diameter, 3 blades, rotation speed: 28.5 rpm, mast height: 38.4 m.

A key feature of turbulent flows is the highly fluctuating nature of the velocity and pressure. Figure 4.2 presents some measurements of instantaneous longitudinal velocity in a developing boundary layer above a rough plate. The data record was collected at 2 mm above the plate using a hot-wire probe sampled at 400 Hz. Figure 4.2B highlights the wide range of turbulent time scales involved.

---

### Remarks

In turbulent shear flows, the terms "shear layer" and "mixing layer" are used for the same meaning. That is, a region of high shear associated with a velocity gradient in the direction normal to the flow.

A basic reference in boundary layer flows is the text by Schlichting (1960, 1979) (also Schlichting and Gersten 2000). George (2006) reviewed the state-of-the-art in turbulent boundary layers. He presented a honest challenging opinion on "*how the world of turbulent boundary layers really works*".

---

*Figure 4.1* Wind farm near Plestan, Western France on 26 June 2008 – Each turbine can produce 2.3 MW, the rotor diameter is 90 m – A total of 6 turbines are installed producing on average 17 GW.h per year.

(A) Definition sketch

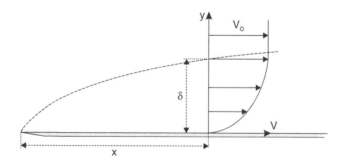

*Figure 4.2* Turbulent velocity fluctuations in a developing boundary layer above a rough plate.

(B) Experimental data: $V_o = 11.2$ m/s, $x = 0.10$ m, $y = 0.002$ m, $\overline{v_x} = 9.1$ m/s, Tu $= 21.3\%$, sampling rate: 400 Hz

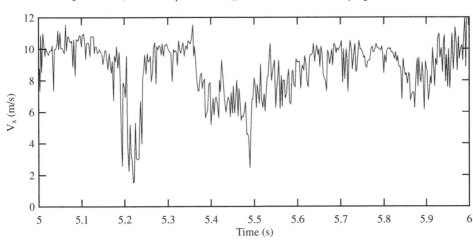

*Figure 4.2*   Continued.

## 2   TURBULENT BOUNDARY LAYERS

The boundary layer is the flow region next to a solid boundary where the flow field is affected by the presence of the boundary, and in particular the boundary shear stress. The concept is valid for both laminar and turbulent flows. The basic definitions were introduced in the previous chapter (Chap. II-3) including the free-stream velocity, boundary layer thickness, displacement thickness and momentum thickness.

### 2.1   Developing boundary layers

Considering an uniform flow past a two-dimensional smooth plate (Fig. 4.3), the presence of the plate induces the development of a boundary layer region with an ideal fluid flow region above. At the upstream edge, there is a short pseudo-laminar region, followed by a transition zone and then the turbulent region (Fig. 4.3). In the pseudo-laminar region, the flow behaves like a viscous flow but it is not a laminar flow. In many turbulent boundary layer applications, however, the pseudo-laminar region is neglected because it is a short region.

### Notes

– In the pseudo-laminar region, the boundary shear stress is approximately given by:

$$\tau_O \approx \mu \times \left(\frac{\partial v}{\partial y}\right)_{y=0}$$

– In a developing boundary layer above a rough surface, the pseudo-laminar region and transition zone do not exist.

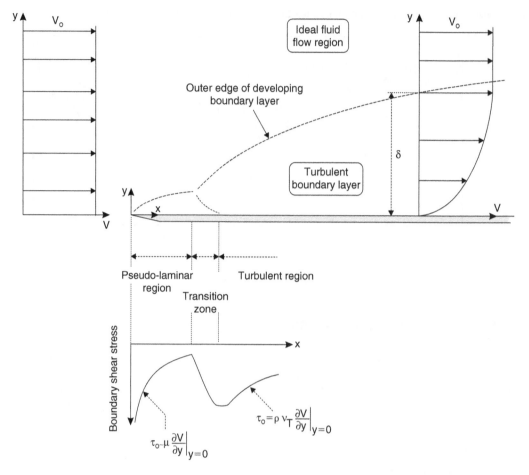

*Figure 4.3* Sketch of a developing turbulent boundary layer along a flat plate including the boundary shear stress distribution.

Some basic characteristics of developing boundary layers along a two-dimensional flat plate are (Gerhart et al. 1992):

- the streamlines of the boundary layer flow are approximately parallel to the surface and the velocity component normal to the plate is significantly smaller than the longitudinal velocity component,
- there is some shear stress at the wall ($y = 0$),
- the velocity is zero at the plate: i.e., the boundary layer flow satisfies the no-slip condition, and the velocity is equal to the free-stream velocity $V_O$ outside of the boundary layer,
- the boundary layer thickness increases with increasing distance from the plate upstream leading edge, and
- the boundary layer is thin: i.e., the ratio of the boundary layer thickness to the distance from the plate leading edge is very small ($\delta/x \ll 1$).

## Notes

-   The outer edge of the developing boundary layer is NOT a streamline. There is some mass flux from the free-stream into the boundary layer.
-   The velocity increases with the increasing distance y normal to the plate, from $v_x = 0$ at the wall $(y = 0)$ to $v_x = V_O$ outside of the boundary layer thickness $(y > \delta)$.

## 2.2  Momentum integral method

The momentum integral equation was developed in the previous chapter (Chap. II-3). It is valid in both laminar and turbulent boundary layers. Its application to turbulent boundary layers is presented below.

Considering a developing two-dimensional boundary layer with zero longitudinal pressure gradient, the application of the momentum integral method yields an ordinary differential equation involving the basic boundary layer characteristics (Fig. 4.4). The result is called the **von Karman momentum integral equation**:

$$\frac{\partial}{\partial x}\left(\int_{y=0}^{+\infty} v_x \times (V_O - v_x) \times dy\right) + \frac{\partial V_O}{\partial x} \times \int_{y=0}^{+\infty}(V_O - v_x) \times dy = \frac{\tau_O}{\rho} \tag{4.1}$$

where $\delta$ is the boundary layer thickness, $V_O$ is the free-stream velocity and $\tau_O$ is the boundary shear stress. Let us introduce the displacement thickness $\delta_1$ and momentum thickness $\delta_2$ defined respectively as:

$$\delta_1 = \int_0^\delta \left(1 - \frac{v_x}{V_O}\right) \times dy \tag{4.2}$$

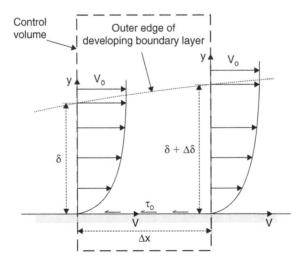

Figure 4.4  Definition sketch of the control volume for the momentum integral method.

$$\delta_2 = \int\limits_0^\delta \frac{v_x}{V_O} \times \left(1 - \frac{v_x}{V_O}\right) \times dy \tag{4.3}$$

An *abbreviated form of the momentum integral equation* is:

$$V_O^2 \times \frac{\partial \delta_2}{\partial x} + V_O \times \frac{\partial V_O}{\partial x} \times (2 \times \delta_2 + \delta_1) = \frac{\tau_O}{\rho} \tag{4.4}$$

This equation is sometimes rewritten in dimensionless form, in which the dimensionless boundary shear stress $\tau_O/(\frac{1}{8} \times \rho \times V_O^2)$ is equivalent to a Darcy-Weisbach friction factor defined in terms of the boundary shear stress $\tau_O$ and free-stream velocity $V_O$.

## Notes

– The momentum integral equation is valid for both laminar and turbulent boundary layers. It was introduced in the previous chapter on Boundary layer theory (Chap. II-3). Equations (4.1) to (4.4) are valid for both laminar and turbulent flows.
– The abbreviated form of the momentum integral equation may be expressed by several identical expressions:

$$\frac{\partial}{\partial x}(V_O^2 \times \delta_2) + V_O \times \delta_1 \times \frac{\partial V_O}{\partial x} = \frac{\tau_O}{\rho}$$

$$V_O^2 \times \frac{\partial \delta_2}{\partial x} + V_O \times \frac{\partial V_O}{\partial x} \times (2 \times \delta_2 + \delta_1) = \frac{\tau_O}{\rho}$$

$$\frac{\partial \delta_2}{\partial x} + \frac{1}{V_O} \times \frac{\partial V_O}{\partial x} \times (2 \times \delta_2 + \delta_1) = \frac{\tau_O}{\rho \times V_O^2}$$

– Let us remember that the definition of the displacement and momentum thickness satisfies:

$$\delta_1 = \int\limits_0^\delta \left(1 - \frac{v_x}{V_O}\right) \times dy = \int\limits_0^{+\infty} \left(1 - \frac{v_x}{V_O}\right) \times dy$$

$$\delta_2 = \int\limits_0^\delta \frac{v_x}{V_O} \times \left(1 - \frac{v_x}{V_O}\right) \times dy = \int\limits_0^{+\infty} \frac{v_x}{V_O} \times \left(1 - \frac{v_x}{V_O}\right) \times dy$$

since $v_x = V_O$ outside of the boundary layer ($y \geq \delta$).

## 2.3   Application to a developing boundary layer past a smooth flat plate

A classical application of the momentum integral equation is the growth of a turbulent boundary layer on a smooth flat horizontal plate. For a smooth turbulent flow, let us assume that the velocity distribution follows a 1/7 power law in the boundary layer:

$$\frac{v_x}{V_O} = \left(\frac{y}{\delta}\right)^{1/7} \quad 0 \le \frac{y}{\delta} \le 1 \tag{4.5}$$

For a 1/7th velocity power law, the displacement thickness and momentum thickness satisfy respectively: $\delta_1/\delta = 1/8$ and $\delta_2/\delta = 7/72$. Further the boundary shear stress may be estimated in first approximation by the Blasius formula:

$$\frac{\tau_O}{\rho \times V_O^2} = 0.0225 \times \left(\frac{\rho \times V_O \times \delta}{\mu}\right)^{-1/4} \tag{4.6}$$

---

**Discussion: the Blasius formula for smooth turbulent flows**

For smooth turbulent flows in circular pipes, Blasius (1911) showed that the Darcy-Weisbach friction factor equals:

$$f = \frac{0.3164}{\left(\frac{\rho \times V \times D_H}{\mu}\right)^{1/4}}$$

where V is the cross-sectional averaged velocity:

$$V = \frac{Q}{\frac{\pi}{4} \times D_H^2}$$

and $D_H$ is the hydraulic diameter or equivalent pipe diameter. The above equation, called the *Blasius formula*, is valid for smooth turbulent flows with $10^4 < \rho \times V \times D_H/\mu < 10^5$ and for small relative roughness (Fig. 4.5). Note that the Darcy friction factor is a dimensionless boundary shear stress:

$$f = \frac{\tau_O}{\frac{1}{8} \times \rho \times V^2}$$

It can be shown that the Blasius resistance formula implies a 1/7 velocity power law (Schlichting 1979, pp. 599–602). Hence, by continuity, the cross-sectional averaged velocity V and the maximum velocity $V_O$ at the pipe centreline are related by: $V/V_O = 0.817$ for a circular pipe.

The Blasius formula may be re-arranged to yield a relationship between the boundary shear stress $\tau_O$, the maximum velocity $V_O$ and a Reynolds number defined in terms of the pipe radius $(D_H/2)$:

$$\frac{\tau_O}{\rho \times V_O^2} = 0.0225 \times \left(\frac{\rho \times V_O \times \frac{D_H}{2}}{\mu}\right)^{-1/4}$$

The above equation may be compared with Equation (4.6).

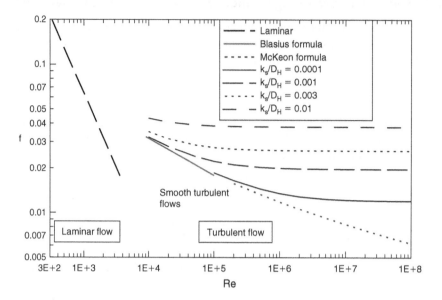

*Figure 4.5* Friction factor versus Reynolds number in circular pipe flows.

## Remarks

1. The 1/7th velocity power law (Eq. (4.6)) is an empirical relationship which fits very well the experimental data (Schlichting 1979). In Equation (4.6), $V_O$ is the free-stream velocity and $\delta$ is the boundary layer thickness.

2. For a two-dimensional boundary layer, the depth-averaged velocity V in the boundary layer and the free-stream $V_O$ in the ideal fluid flow are related by $V/V_O = 7/8$.

3. The Blasius formula was developed for smooth turbulent flows with $10^4 < \rho \times V \times D_H/\mu < 10^5$ (Blasius 1913).

4. For larger Reynolds number, the Darcy-Weisbach friction factor for smooth turbulent flow equals:

$$\frac{1}{\sqrt{f}} = 1.930 \times \log_{10}\left(\frac{\rho \times V \times D_H}{\mu} \times \sqrt{f}\right) - 0.537$$

The above relationship was based upon the "super-pipe" data within $3.1 \times 10^5 < \rho \times V \times D_H/\mu < 3.5 \times 10^7$ (Mckeon et al. 2005). It is shown in Figure 4.5.

An older expression is the Karman-Nikuradse' formula, also called Prandtl's universal law of friction for smooth pipes developed based upon data within $3.1 \times 10^3 < \rho \times V \times D_H/\mu < 3.2 \times 10^6$:

$$\frac{1}{\sqrt{f}} = 2.0 \times \log_{10}\left(\frac{\rho \times V \times D_H}{\mu} \times \sqrt{f}\right) - 0.8$$

Replacing the expressions of the displacement thickness, momentum thickness and bed shear stress (Eq. (4.6)) into Equation (4.4), the momentum integral equation becomes:

$$\frac{7}{72} \times V_O^2 \times \frac{\partial \delta}{\partial x} + \left(\frac{1}{8} + \frac{7}{36}\right) \times V_O \times \frac{\partial V_O}{\partial x} \times \delta = 0.0225 \times \frac{V_O^2}{\left(\frac{\rho \times V_O \times \delta}{\mu}\right)^{1/4}} \tag{4.7}$$

In the particular case of a developing boundary layer with zero pressure gradient (i.e. $\partial V_O/\partial x = 0$), the integration of Equation (4.7) gives:

$$\frac{\delta}{x} = 0.37 \times \left(\frac{\rho \times V_O \times x}{\mu}\right)^{-1/5} \tag{4.8}$$

$$\frac{\delta_1}{x} = 0.0462 \times \left(\frac{\rho \times V_O \times x}{\mu}\right)^{-1/5} \tag{4.9}$$

$$\frac{\delta_2}{x} = 0.036 \times \left(\frac{\rho \times V_O \times x}{\mu}\right)^{-1/5} \tag{4.10}$$

where x is the distance from the upstream edge of the plate.

The boundary layer thickness increases with $x^{4/5}$. For comparison, in laminar flow, the boundary layer thickness increases with $x^{1/2}$ (Chap. II-3).

A further integration gives the shear force per unit width on the plate of length L. In dimensionless form, the shear force per unit width equals:

$$\frac{\int_{x=0}^{L} \tau_O \times dx}{\frac{1}{2} \times \rho \times V_O^2 \times L} = 0.072 \times \left(\frac{\rho \times V_O \times L}{\mu}\right)^{-1/5} \tag{4.11}$$

## Notes

- Experimental results for turbulent boundary layers along smooth plates showed that the data followed closely:

$$\frac{\int_{x=0}^{L} \tau_O \times dx}{\frac{1}{2} \times \rho \times V_O^2 \times L} = 0.074 \times \left(\frac{\rho \times V_O \times L}{\mu}\right)^{-1/5}$$

  The result is close to Equation (4.11), and it was obtained for $5\,E+5 \le Re_L \le 2\,E+7$ where $Re_L = \rho \times V_O \times L/\mu$ (Schlichting 1979, pp. 638–639).
- Equation (4.11) was developed assuming that the boundary layer is turbulent from the leading edge onwards.
- The boundary shear stress may be estimated by the Blasius formula (Eq. (4.6)):
- Equation (4.6) and Equations (4.8) to (4.11) were derived and validated for a developing boundary layer along a smooth plate with zero pressure gradient. They are not valid for rough plates nor flows with positive/negative longitudinal pressure gradient. Schlichting (1979) discussed these specific flow situations, as well as the effects of suction and blowing.

## 2.4   Discussion

In developing boundary layers, the momentum integral equation may be applied to deduce the boundary shear stress $\tau_O$, and the shear velocity $V_*$, from some known velocity distribution data at several locations along the boundary.

For a steady flow, with zero longitudinal pressure gradient, the momentum integral equation yields an expression of the boundary shear stress:

$$\tau_O = \rho \times \left( V_O^2 \times \frac{\partial \delta_2}{\partial x} + V_O \times \frac{\partial V_O}{\partial x} \times (2 \times \delta_2 + \delta_1) \right) \tag{4.4}$$

This method may be applied to deduce the boundary shear stress in rough boundary channels based upon velocity distributions measurements.

---

**Discussion**

In turbulent boundary layers, the boundary shear stress may be deduced by several methods. A common technique is based upon the best fit of the velocity profile in the inner flow region ($y/\delta < 0.1$) with a logarithmic profile known as the law of the wall (see next paragraph).

Generally the application of the momentum integral equation (4.4) is considered a reliable and accurate technique to estimate the bed shear stress. When Equation (4.4) is applied between two sampling locations, it provides the spatial-averaged bed shear stress between the velocity measurement locations.

Koch and Chanson (2005, pp. 12–17) summarised several techniques to estimate the boundary shear stress, including the momentum integral approach. Their experimental data suggested that an accurate estimate was deduced from a Prandtl-Pitot-Preston tube. Based upon some dimensional considerations, the skin friction may be measured with a Pitot tube lying on the boundary (Preston 1954, Patel 1965). Practical experiences suggested that each Prandtl-Pitot-Preston tube must be calibrated independently, preferably in-situ, rather than relying on pre-existing correlations. (See also Exercise 11.)

---

**Note**

The shear velocity $V_*$ is defined as:

$$V_* = \sqrt{\frac{\tau_o}{\rho}}$$

where $\tau_o$ is the boundary shear stress. The shear velocity is a measure of the velocity gradient next to the boundary.

## 3   VELOCITY DISTRIBUTION IN TURBULENT BOUNDARY LAYERS

### 3.1   Velocity distribution in smooth turbulent boundary layers

Let us consider a turbulent boundary layer flow along a smooth boundary with zero pressure gradient. In the turbulent boundary layer, the flow can be divided into three regions: (a) an *inner wall region* next to the wall where the turbulent stress is negligible and the viscous stress is large, (b) a *wall region*, sometimes called a turbulent zone, where the turbulent stress is larger then the viscous stress, and (c) an *outer region* above. Above the outer edge of the boundary layer, a fourth region is the ideal-fluid flow region.

(a) Inner wall region
Very close to the wall, the viscous shear is predominant. At the boundary, the shear stress equals:

$$\tau_O = \mu \times \left(\frac{\partial v_x}{\partial y}\right)_{y=0} \approx \mu \times \frac{v_x}{y} \quad \text{very close to the wall} \tag{4.12a}$$

where $v_x$ is the local fluid velocity, y is the direction normal to the flow direction, and $\mu$ is the dynamic viscosity of the fluid. Introducing the shear velocity $V_* = \sqrt{\tau_O/\rho}$, the above equation gives the velocity field in the inner wall region:

$$\frac{v_x}{V_*} = \rho \times \frac{V_* \times y}{\mu} \quad \text{inner wall region: } \rho \times \frac{V_* \times y}{\mu} < 5 \tag{4.12b}$$

The velocity distribution follows a linear relationship in the inner wall region (also called "viscous" sub-layer region). This result is valid in the viscous layer of a smooth turbulent boundary layer.

(b) Wall region
In the wall region, a robust turbulence model is the mixing length model (Chapter II-2) which relates the "eddy viscosity" to the velocity gradient:

$$v_T = l_m^2 \times \frac{\partial v_x}{\partial y} \tag{4.13}$$

where $l_m$ is the mixing length which characterises the distance travelled by a particle of fluid before its momentum is changed by the new environment. Considering a developing boundary layer, the mixing length is proportional to the distance to the wall. It may be assumed:

$$l_m = K \times y \tag{4.14}$$

where $K$ is the von Karman constant ($K = 0.4$).
    At the boundary, the turbulent shear stress equals:

$$\tau_O = \rho \times v_T \times \left(\frac{\partial v_x}{\partial y}\right)_{y=0} \approx K^2 \times y^2 \times \left(\left(\frac{\partial v_x}{\partial y}\right)_{y=0}\right)^2 \quad \text{close to the wall} \tag{4.15a}$$

Introducing the shear velocity, it yields:

$$\frac{\partial v_x}{V_*} = \frac{1}{K} \times \frac{\partial y}{y} \tag{4.15b}$$

The integration of Equation (4.15) gives a logarithmic law:

$$\frac{v_x}{V_*} = \frac{1}{K} \times Ln\left(\frac{\rho \times V_* \times y}{\mu}\right) + D_O \quad \text{Wall region} \tag{4.16}$$

where $D_O$ is an integration constant. The above result is the well-known law of wall for a developing turbulent boundary layer that is valid is valid for smooth and rough turbulent boundary layer flows.

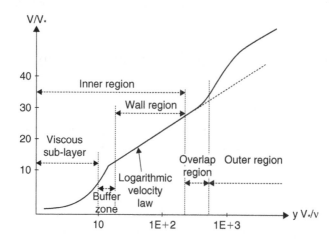

*Figure 4.6*  Sketch of the dimensionless velocity distribution in the inner region of a turbulent boundary layer.

For a smooth turbulent boundary layer, Equation (4.16) becomes:

$$\frac{v_x}{V_*} = \frac{1}{K} \times Ln\left(\frac{V_* \times y}{\nu}\right) + D_1 \quad \text{Wall region: } 30 \text{ to } 70 < \frac{V_* \times y}{\nu} \text{ and } \frac{y}{\delta} < 0.1 \text{ to } 0.15 \quad (4.17)$$

where $D_1 = 5$. The wall region is also called the turbulent zone. Equation (4.17) is called the logarithmic profile or the *law of the wall*. It is sketched in Figure 4.6 and it is compared with experimental data in Figure 4.9A. For a developing layer above a rough surface, the integration constant $D_O$ is larger than $D_1$ (section 3.2, in this chapter).

(c) Outer region
In the outer region, the distribution of velocity follows:

$$\frac{V_O - v_x}{V_*} = -\frac{1}{K} \times Ln\left(\frac{y}{\delta}\right) \qquad\qquad \text{Outer region: } \frac{y}{\delta} > 0.1 \text{ to } 0.15 \qquad (4.18)$$

where $V_O$ is the free-stream velocity and $K$ is the von Karman constant. Equation (4.18) is called the *velocity defect law* or outer law.

The logarithmic law (Eq. (4.17)) may be extended to the outer region by adding a "wake law" term to the right-handside term:

$$\frac{v_x}{V_*} = \frac{1}{K} \times Ln\left(\frac{V_* \times y}{\nu}\right) + D_1 + \frac{\Pi}{K} \times Wa\left(\frac{y}{\delta}\right)$$

$$\text{Turbulent zone and outer region: } 30 \text{ to } 70 < \frac{V_* \times y}{\nu} \qquad (4.19)$$

where $\Pi$ is the wake parameter and Wa is Coles' wake function originally estimated as (Coles 1956):

$$Wa\left(\frac{y}{\delta}\right) = 2 \times \sin^2\left(\frac{\pi}{2} \times \frac{y}{\delta}\right) \qquad\qquad (4.20)$$

The wake parameter $\Pi$ is generally flow dependent. For example, both the wake function Wa and wake parameter $\Pi$ differ substantially between open channel and pipe flows (Montes 1998, pp. 77–79).

---

**Remarks**

1. For smooth turbulent boundary layers, Schlichting (1979) gave $D_1 = 5.5$ while Liggett (1994) recommended $D_1 = 5.0$. Montes (1998) re-analysed open channel flow data yielding $D_1 = 5.0$.

2. The wall region is sometimes called the "overlap region".

3. The logarithmic law (Eq. (4.17)) is not valid in the "viscous" sub-layer ($V_* \times y/\nu < 30$ to 70). Yu and Yoon (2005) proposed a velocity formula that describes the entire inner wall region and wall region:

$$\frac{v_x}{V_*} = \left( \frac{1}{K} \times \mathrm{Ln}\left( \frac{V_* \times y}{\nu} \right) + D_1 \right) \times \left( 1 - \exp\left( -0.14 \times \frac{V_* \times y}{\nu} \right) \right) \quad \text{for } \frac{y}{\delta} < 0.1$$

4. For circular pipes, von Karman proposed the velocity defect law:

$$\frac{V_O - v_x}{V_*} = -\frac{1}{K} \times \left( \mathrm{Ln}\left( 1 - \sqrt{1 - \frac{y}{\delta}} \right) + \sqrt{1 - \frac{y}{\delta}} \right)$$

where $V_O$ is the free-stream velocity (e.g. Schlichting 1979). Prandtl derived a slightly different relationship:

$$\frac{V_O - v_x}{V_*} = -\frac{1}{K} \times \mathrm{Ln}\left( \frac{y}{\delta} \right)$$

which tended to gave a slightly better agreement with experimental results in circular pipes. Both approaches were developed for circular pipes. They are not suitable in two-dimensional flows including open channel flows.

5. The logarithmic law may be extended to the entire boundary layer thickness as:

$$\frac{v_x}{V_*} = \left( \frac{1}{K} \times \mathrm{Ln}\left( \frac{V_* \times y}{\nu} \right) + D_1 + \frac{\Pi}{K} \times \mathrm{Wa}\left( \frac{y}{\delta} \right) \right) \times \left( 1 - \exp\left( -0.14 \times \frac{V_* \times y}{\nu} \right) \right)$$

$$\text{for } \frac{y}{\delta} < 0.1$$

with $D_1 = 5$.

In practice, however, the power law is believed to be simpler and more physical expression of the velocity distribution in a developing boundary layer (see below).

6. The wake parameter $\Pi$ is flow dependent, with reported values of $+0.55$ in developing boundary layers, and between $-0.3$ to $+0.2$ in fully-developed open channel flows. In the latter case, $\Pi$ was found to increase with increasing width to depth ratio.

**Discussion**

The integration of the logarithmic law (law of wall) for a turbulent boundary layer along a smooth flat plate gives an expression of the total shear force (per unit width) on the plate. The computational results were fitted by a correlation, known as the Prandtl-Schlichting formula:

$$\frac{\int_{x=0}^{L} \tau_O \times dx}{\frac{1}{2} \times \rho \times V_O^2 \times L} = \frac{0.455}{(Ln(Re_L))^{2.58}} - \frac{1700}{Re_L}$$

where $Re_L = \rho \times V_O \times L/\mu$. Further discussions on the Prandtl-Schlichting formula include Schlichting (1979, pp. 640–642).

**Momentum exchange coefficient in a turbulent boundary layer**

In a smooth turbulent boundary layer, the mixing length $l_m$ may be estimated as:

$$\frac{l_m}{\delta} = 0.085 \times \tanh\left(\frac{K}{0.085} \times \frac{y}{\delta}\right) \quad 0 \le y \le \delta$$

where tanh is the hyperbolic tangent function, $K$ is the von Karman constant and $\delta$ is the boundary layer thickness (Schlichting and Gersten 2000, p. 557). The momentum exchange coefficient, or "eddy viscosity", $\nu_T$ derives then from:

$$\nu_T = l_m^2 \times \frac{\partial v_x}{\partial y}$$

*Power law velocity distribution in smooth turbulent boundary layers*

For a smooth turbulent flow, the velocity distribution in the entire boundary layer may be approximated by a power law:

$$\frac{v_x}{V_O} = \left(\frac{y}{\delta}\right)^{1/N} \quad 0 \le \frac{y}{\delta} \le 1 \tag{4.21}$$

Equation (4.21) is valid close to and away from the boundary without distinction between the inner and outer regions. For a smooth turbulent boundary layer, $N = 7$. Equation (4.21) is also valid for turbulent boundary layer flows along rough walls, although with a different exponent value. Barenblatt (1994, 1996) showed that the velocity power law (Eq. (4.21)) can be derived theoretically from some basic self-similarity. A self-similar process is one whose spatial distribution of properties at various times can be obtained from one another by a similarity transformation. Self-similarity is a powerful tool in turbulence flow research. Although more simple than the law of the wall, the power law velocity distribution gives results very close to the log-law (George 2006).

For a power-law velocity distribution:

$$\frac{v_x}{V_O} = \left(\frac{y}{\delta}\right)^{1/N} \tag{4.21}$$

the characteristic parameters of the boundary layers can be expressed analytically. The displacement thickness and the momentum thickness become:

$$\frac{\delta_1}{\delta} = \frac{1}{1 + N}$$

$$\frac{\delta_2}{\delta} = \frac{N}{(1 + N) \times (2 \times N)}$$

where N is the inverse of the velocity exponent. The shape factor equals:

$$\frac{\delta_1}{\delta_2} = \frac{N + 2}{N}$$

For two-dimensional turbulent boundary layers, Schlichting (1979) indicated that separation occurs for $\delta_1/\delta_2 > 1.8$ to 2.4. Such a condition implies separation for $N < 1.4$ to 2.5.

---

**Discussion**

Nikuradse (1932) performed detailed velocity measurements in fully-developed circular pipe flows. A re-analysis of his data set showed that the data were self-similar and followed closely a power law:

$$\frac{v_x}{V_*} \propto \left( \frac{\rho \times V_* \times y}{\mu} \right)^{\frac{3}{2 \times Ln(Re)}}$$

where Re is the flow Reynolds number defined in terms of the cross-sectional averaged velocity and pipe diameter $D_H$ (Barenblatt 1996). Overall the data followed closely:

$$\frac{v_x}{V_*} = \left( \frac{Ln(Re)}{\sqrt{3}} + \frac{5}{2} \right) \times \left( \frac{\rho \times V_* \times y}{\mu} \right)^{\frac{3}{2 \times Ln(Re)}}$$

for $4 \times 10^3 \leq Re \leq 3.2 \times 10^6$ (Barenblatt 1996).

Note that the 1/7th power law (N = 7) derives exactly from the Blasius resistance formula (Blasius 1913) for turbulent flows in smooth circular pipes (Schlichting 1979).

Recent developments in turbulent boundary layers suggested that *"the log and power law results seem virtually indistinguishable, at least for zero-pressure-gradient boundary layers"* (George 2006). A key difference however is the range of validity of the law: the power law is valid throughout the entire boundary layer thickness ($0 < y < \delta$).

---

### *Outer flow region in open channel flows*

Let us consider an open channel flow (Fig. 4.7). The presence of the free-surface has some impact on the velocity field as the boundary conditions at the free-surface are: $v_y(y = d) = \partial d/\partial t$ where d is the water depth. Well-known evidences of turbulence interactions with a free-surface include surface scars and boils, ship wakes and whirlpools (e.g. Fig. 4.8). The movies IMGP3647.mov, IMGP1186.avi, IMGP1289.avi and SDC11864.avi show further illustrations (App. F).

Figure 4.9 illustrates some velocity distribution data in open channel flows. In the wall region, the logarithmic law (Eq. (4.6)) provide a reasonable approximation (Fig. 4.6A). In the outer region

*Figure 4.7*  Turbulent open channel flow: the Brisbane River on 12 January 2011 during a major flood, with the flood flow direction from right to left.

(y/δ > 0.1 to 0.2), two velocity laws matched well experimental data in developing and fully-developed open channel flows (Montes 1998). Assuming a Prandtl mixing length independent of the distance to the boundary, a simplified expression proposed by Montes (1998) is:

$$\frac{V_O - v_x}{V_*} = 5.55 \times \left( \left(1 - \frac{y}{\delta}\right)^{3/2} \right) \quad 0.2 \text{ to } 0.3 \leq \frac{y}{\delta} \tag{4.22}$$

A more sophisticated profile may be based upon Zagustin's turbulence model:

$$\frac{V_O - v_x}{V_*} = \frac{2}{K} \times \tanh^{-1}\left( \left(1 - \frac{y}{\delta}\right)^{3/2} \right) \quad 0.15 \text{ to } 0.2 \leq \frac{y}{\delta} \tag{4.23}$$

Equation (4.23) and its gradient are continuous with the log-law near the wall. Both Equations (4.22) and (4.23) were compared successfully with experimental data, although the writer's experience tended to show a slightly better agreement in the overlap and wall regions with Equation (4.23) (Fig. 4.9B).

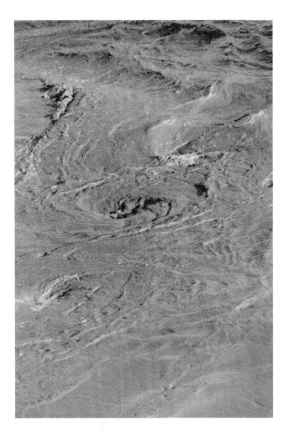

*Figure 4.8* Free-surface scars in the wake of a bridge pier during the Brisbane River flood on 12 January 2011 – Flood flow direction from foreground to background.

## Note

– The velocity distribution in fully-developed turbulent open channel flows is given approximately by Prandtl's power law:

$$\frac{v_x}{V_O} = \left(\frac{y}{d}\right)^{1/N}$$

The exponent 1/N varies typically from 1/4 down to 1/12 in depending upon the boundary friction and cross-section shape, but many open channel flow situations follow a 1/6th power law velocity distribution. A complete analysis of velocity distributions in open channels showed that, in uniform equilibrium open channel flows, the velocity distribution exponent is related to the flow resistance:

$$N = K \times \sqrt{\frac{8}{f}}$$

where f is the Darcy-Weisbach friction factor and $K$ is the von Karman constant ($K = 0.4$) (Chen 1990, Chanson 1999).

(A) Logarithmic velocity profile $v_x/V_*$ in the inner flow region as a function of $y \times V_*/\nu$ (Eq. (4.17)) – Comparison with the open channel flow data (Pitot tube and ADV) of Koch and Chanson (2005) and large wind tunnel data by Osterlund (1999)

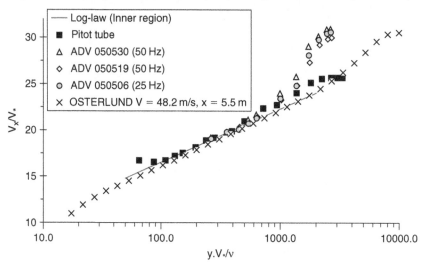

(B) Outer flow region velocity distribution $(V_O - v_x)/V_*$ as a function of $y/\delta$ – Comparison with open channel flow data (Koch and Chanson 2005, Pitot tube and ADV measurements)

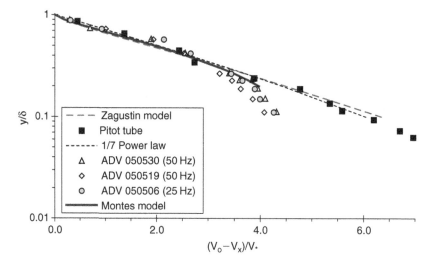

*Figure 4.9* Dimensionless velocity distributions in a developing boundary layer above a smooth boundary.

– On a dam spillway, the flow is accelerated by the gravity force component in the flow direction, while boundary friction tend to retard the flow motion. At the upstream end of the chute, a bottom boundary layer is generated by bottom friction and develops in the flow direction. When the outer edge of the boundary layer reaches the free surface, the flow becomes fully developed and free-surface aeration may take place, when the turbulent shear stress next to the surface is greater than the capillary force per unit area resisting the surface breakup.

The movie IMGP3647.mov illustrates a spillway operation (App. F). The location of the inception point of free-surface aeration is clearly seen.

## 3.2   Roughness effects

The boundary roughness has an important effect on the flow in the wall-dominated region (i.e. "viscous" sub-layer region and turbulent zone). Numerous experiments showed that, for a turbulent boundary layer along a rough plate, the "law of the wall" follows:

$$\frac{v_x}{V_*} = \frac{1}{K} \times \text{Ln}\left(\frac{V_* \times y}{\nu}\right) + D_1 + D_2 \qquad \text{Turbulent zone: } \frac{y}{\delta} < 0.1 \text{ to } 0.15 \qquad (4.24)$$

where $D_1 = 5$ and $D_2$ is a function of the type of roughness and of roughness shape, height and spacing (e.g. Schlichting 1979, Schetz 1993). For smooth turbulent boundary layer flows, $D_2$ equals zero.

In the turbulent zone, the roughness effect (i.e. $D_2 < 0$) implies a downward shift of the logarithmic velocity distribution (Fig. 4.10). Figure 4.10 illustrates some effect of the boundary roughness on the law of wall. For example, let us compare Figure 4.9A (smooth boundary) and Figure 4.11B (rough plate). For large roughness, the "viscous" sub-layer region disappears and the flow is said to be "fully-rough".

For fully-rough turbulent flows in circular pipes with uniformly distributed sand roughness, $D_2$ equals:

$$D_2 = 3 - \frac{1}{K} \times \text{Ln}\left(\frac{k_s \times V_*}{\nu}\right) \qquad \text{fully-rough turbulent flows in circular pipes} \qquad (4.25)$$

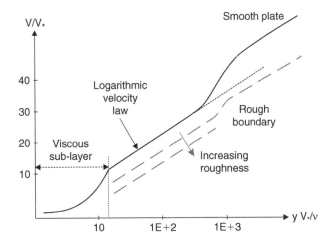

Figure 4.10   Sketch of the effect of boundary roughness on the logarithmic velocity distribution.

(A) Time-averaged velocity and turbulence intensity ($\sqrt{\overline{v_x^2}}/v_x$) profiles at $x = -0.10$ m (upstream), $+0.10$ m and $+0.98$ m

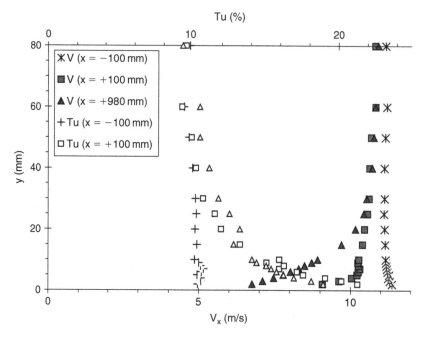

(B) Logarithmic velocity distributions at $x = +0.10$ m and $+0.98$ m – Comparison with Equation (4.26)

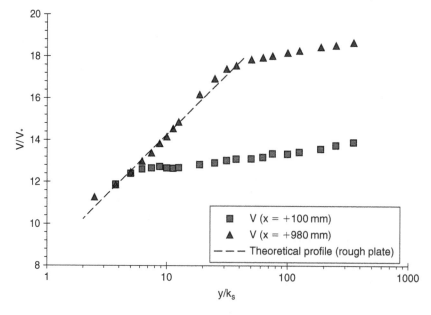

*Figure 4.11* Velocity profile in a developing boundary layer above a rough surface: $V_O = 11.2$ m/s, atmospheric boundary layer wind tunnel, rough sand paper (grade 40).

where $k_s$ is the equivalent sand roughness height. After transformation, the velocity distribution in the turbulent zone for fully-rough turbulent flow becomes:

$$\frac{v_x}{V_*} = \frac{1}{K} \times Ln\left(\frac{y}{k_s}\right) + 3 + D_1 \qquad\qquad \frac{y}{\delta} < 0.1 \text{ for fully-rough turbulent flow} \qquad (4.26)$$

Figure 4.11 illustrates some turbulent velocity measurements above a flat, rough plate. Figure 4.11A presents the time-averaged velocities and turbulent intensity distributions. Figure 4.11B compares the velocity distributions with Equation (4.26). The results show that the data followed closely Equation (4.26) for $y/k_s < 5$ to 30 depending upon the longitudinal distance x from the leading edge.

## Note

The velocity distribution in the turbulent zone for fully-rough turbulent flow becomes:

$$\frac{V_x}{V^*} = \frac{1}{K} \times Ln\left(\frac{y}{k_s}\right) + 8$$

with $D_1 = 5.0$.

## 4   SECONDARY CURRENTS

In circular straight pipes and in very-wide channels, the flow is quasi-two-dimensional. However, most channels have more complex geometries and the presence of the boundaries may induce some redistribution of the shear and affect the boundary layer characteristics.

Let us consider a developing boundary layer in a simple rectangular channel (Fig. 4.12). Next to the sidewall, a boundary layer region develops and the flow is retarded, while some complicated flow patterns develop next to the corners. In turn, some flow is generated at right angle to the longitudinal current: i.e., the secondary currents. Although the secondary flow velocities are very small (1 to 2% of mean flow velocity), the secondary currents are thought to be a main cause of discrepancies between theoretical and observed turbulent velocity measurements.

The secondary currents transport some momentum from the channel centre to the corners and sides, and the Reynolds stresses play a dominant role (Liggett 1994). Schlichting (1979, pp. 613–614) presented some schematic diagrams of secondary currents in triangular and rectangular pipes. Figure 4.13 shows some transverse distribution of boundary shear stress in a rectangular channel, with z = 0 on the channel centreline. The data for z > 0.125 m correspond to the sidewall shear stress. The findings illustrate the transverse variation of boundary shear stress in a rectangular channel. The shear distribution is fairly uniform across the bed, with a lower boundary shear stress in the vicinity of the corner. The trend is comparable with earlier studies (e.g. Knight et al. 1984).

It is well-known that secondary currents play a major role in open channel flows including both natural rivers and man-made waterways. Pertinent references include Knight et al. (1984), Nezu and Nakagawa (1993), Macintosh (1990), Apelt and Xie (1995), and Xie (1998). Montes (1998) gave a historical account. Recent observations showed further the existence of streamwise vortices with lateral length scale of between 1.5 to 3 times the depth (Tamburrino and Gulliver 2007, Trevethan et al. 2007, 2008).

---

**Remember**

In non-circular channel cross-sections, the flow resistance coefficient are defined in terms of an equivalent pipe diameter, called the hydraulic diameter. The hydraulic diameter $D_H$ is defined as:

$$D_H = 4 \times \frac{A}{P_w}$$

where A is the flow cross-section area and $P_w$ is the wetted perimeter. For a circular pipe, $D_H = D$ with D is the internal pipe diameter.

---

(A) Boundary layer experiment above a 0.5 m wide rough plate (sand paper grade 40) with wind flow from foreground to background – The 0.10 m high end walls are designed to prevent the undesirable effects of three-dimensional separation at both ends – The hot-wire probe sensor is located at x = +0.30 m and y = 0.10 m

*Figure 4.12* Examples of horizontal channels bounded by sidewalls.

(B) Winglet at the tip of a Boeing 737-800 wing – The winglets are designed to reduce the adverse impact of three-dimensional separation at wing tip

*Figure 4.12* Continued.

## 5 TURBULENT SHEAR FLOWS

Jets, wakes and shear flows are most often turbulent. Three examples of turbulent shear flows are developed below: the free shear layer, the plane jet and the circular jet.

### 5.1 Free-shear layer

A simple turbulent shear flow is the free shear layer sketched in Figure 4.14. Assuming no pressure gradient, the differential form of the momentum equation can be simplified into:

$$v_x \times \frac{\partial v_x}{\partial x} + v_y \times \frac{\partial v_x}{\partial y} = \nu_T \times \frac{\partial^2 v_x}{\partial y^2} \qquad (4.27)$$

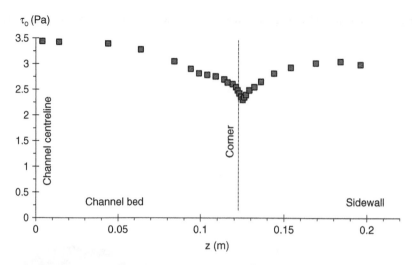

*Figure 4.13* Experimental measurements of boundary shear stress in a fully-developed open channel flow (Data: Chanson 2000) – Channel width: 0.25 m, d = 0.078 m, V = 1.29 m/s, Fr = 1.48.

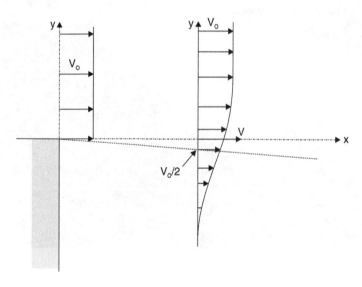

*Figure 4.14* Sketch of a free shear layer.

where $v_T$ is the turbulent momentum exchange coefficient. For a plane shear layer Goertler (1942) solved the equation of motion assuming a constant eddy viscosity $v_T$ across the shear layer at a given longitudinal distance x from the singularity:

$$v_T = \frac{1}{4 \times K^2} \times x \times V_O \tag{4.28}$$

*Table 4.1* Values of the Gaussian error function erf.

| u | erf(u) | u | erf(u) |
|---|--------|---|--------|
| 0 | 0 | 1 | 0.8427 |
| 0.1 | 0.1129 | 1.2 | 0.9103 |
| 0.2 | 0.2227 | 1.4 | 0.9523 |
| 0.3 | 0.3286 | 1.6 | 0.9763 |
| 0.4 | 0.4284 | 1.8 | 0.9891 |
| 0.5 | 0.5205 | 2 | 0.9953 |
| 0.6 | 0.6309 | 2.5 | 0.9996 |
| 0.7 | 0.6778 | 3 | 0.99998 |
| 0.8 | 0.7421 | $+\infty$ | 1 |
| 0.9 | 0.7969 | | |

where K is a constant. The constant K provides some information on the expansion rate of the momentum shear layer as the rate of expansion is proportional to 1/K. K equals between 9 and 13.5 with a generally-accepted value of 11 for monophase free shear layers (Rajaratnam 1976, Schlichting 1979, Schetz 1993).

Let us introduce the dimensionless terms $\eta$ and $\Lambda$:

$$\eta = K \times \frac{y - y_{50}}{x} \tag{4.29}$$

$$\Lambda = \frac{\psi}{x \times \frac{V_O}{2}} \tag{4.30}$$

where $y_{50}$ is the location where $v_x = V_O/2$ and $\psi$ is the stream function. $\Lambda$ is simply a dimensionless stream function. Ideally $y_{50}$ is zero, but experimental observations demonstrated that $y_{50}$ is usually negative and increases with increasing longitudinal distance x as sketched in Figure 4.14.

Basic similarity considerations show that the dimensionless stream function $\Lambda$ is a function of the dimensionless coordinate $\eta$ only, and Equation (4.27) may be rewritten as a differential function for $\Lambda(\eta)$:

$$\Lambda''' + 2 \times K^2 \times \Lambda \times \Lambda'' = 0 \tag{4.31}$$

where $\Lambda' = \partial\Lambda/\partial\eta$.

Goertler (1942) obtained the solution in the first approximation:

$$\frac{v_x}{V_O} = \frac{1}{2} \times \left(1 + \mathrm{erf}\left(K \times \frac{(y - y_{50})}{x}\right)\right) \tag{4.32}$$

where erf is the Gaussian error function defined as:

$$\mathrm{erf}(u) = \frac{2}{\sqrt{\pi}} \times \int_0^u \exp(-t^2) \times dt \tag{4.33}$$

Tabulated values of the Gaussian error function erf are listed in Table 4.1.

**Remarks**

1.  In a free-shear layer, the location $y_{50}$ where the velocity satisfies $v_x = V_O/2$ is always located towards the low-velocity region: i.e., $y_{50} < 0$ as sketched in Figure 4.14.
2.  Equation (4.26) is called the Goertler solution of a free shear layer.
3.  The Gaussian error function satisfies:

$$erf(-u) = -erf(u)$$

4.  In first approximation, the function erf(u) may be correlated by:

$$erf(u) \approx u \times (1.375511 - 0.61044 \times u + 0.088439 \times u^2) \qquad 0 \leq u < 2$$
$$erf(u) \approx tanh(1.198787 * u) \qquad -\infty < u < +\infty$$

with a normalised correlation coefficient of 0.99952 and 0.9992 respectively. In many applications, the above correlations are not accurate enough, and Table 4.1 should be used instead.

5.  While the constant K is between 9 and 13.5 in monophase free shear layers, Brattberg and Chanson (1998) showed that the value of K is affected by the air bubble entrainment rate in the developing region of plunging jet flows. They observed K to be about 5.7 for plunging jet velocities ranging from 3 to 8 m/s. It is believed that the air bubble entrainment, and the interactions between the entrained bubbles and the turbulence, are responsible for the changes in expansion rate of the momentum shear layer.

6.  An example for free-shear layer in geophysical flows is the high-velocity flow in a shallow-water strait. For example, the Naruto strait in Japan. Another example is shown in the movie P1040830.mov (App. F).

## 5.2  Two-dimensional jet

Let us consider the two-dimensional plane jet sketched in Figure 4.15. Immediately downstream of the nozzle, the jet flow is characterised by a developing flow region followed by a fully-developed flow region (Fig. 4.15). In the developing region, there is an undisturbed jet core with a velocity $v_x = V_O$, surrounded by developing shear layers in which momentum is transferred to the surrounding fluid. The transfer of momentum from the undisturbed jet core to the outer fluid is always associated with very-high levels of turbulence. The length of the developing flow region is about 6 to $12 \times D$ for two-dimensional jets, where D is the jet thickness. Further downstream, the flow becomes fully-developed and the maximum velocity on the centreline decreases with increasing distance. In the developing flow of two-dimensional jets, the velocity in the undisturbed jet core is $v_x = V_O$ and the velocity profile in the developing shear layer is given by Equation (4.26).

The developing shear layers are characterised by very intense turbulence, well in excess of levels observed in boundary layers and wakes. The developing flow region is relatively short, but the extent of the fully-developed flow region, and the influence of the jet on the surroundings, may be felt far away.

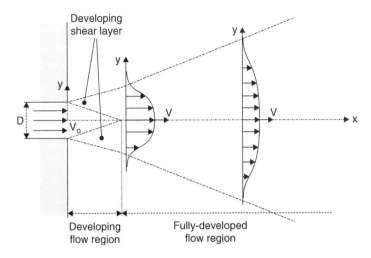

*Figure 4.15* Sketch of a two-dimensional jet.

In the fully-developed flow region, the velocity profiles follow:

$$\frac{v_x}{V_O} = \frac{2.67}{\sqrt{\frac{x}{D}}} \times \left(1 - \tanh^2\left(7.7 \times \frac{y}{x}\right)\right) \quad x/D > 12 \tag{4.34}$$

---

**Discussion**

In the fully-developed flow region, similarity considerations showed that the velocity distribution satisfies:

$$\frac{v_x}{v_x(y=0)} = \left(\cosh^2\left(\frac{y}{\sqrt{2} \times D}\right)\right)^{-1} \quad x/D > 5 \text{ to } 12$$

where $v_x(y=0)$ is the jet centreline velocity at a distance x from the nozzle and cosh is the hyperbolic cosine function (Liggett 1994, pp. 246–248). The result was verified for $y/D < 1.5$.

---

## 5.3 Circular jet

In circular jets, the length of the developing flow region is about 5 to $10 \times D$ for axi-symmetric jets discharging in a fluid at rest, where D is the jet diameter.

In the fully-developed flow region, the velocity profile follows

$$\frac{v_x}{V_O} = \frac{5.745}{\frac{x}{D}} \times \frac{1}{1 + 0.125 \times \left(18.5 \times \frac{y}{x}\right)^2} \quad x/D > 10 \tag{4.35}$$

*Table 4.2* Longitudinal variations in jet width and centreline velocity.

| Parameter | Plane Jet | | Circular Jet | |
|---|---|---|---|---|
| | Laminar jet | Turbulent jet | *Laminar jet* | Turbulent jet |
| Jet width: | $\propto x^{2/3}$ | $\propto x$ | $\propto x$ | $\propto x$ |
| Jet centreline velocity: | $\propto \dfrac{1}{\sqrt[3]{x}}$ | $\propto \dfrac{1}{\sqrt{x}}$ | $\propto \dfrac{1}{x}$ | $\propto \dfrac{1}{x}$ |

References: Schlichting (1979), Liggett (1994).

### Discussion

A comparison between laminar and turbulent jets shows some key differences in terms of the rate of expansion of the jet flow (i.e. width of the jet) and of the jet centreline velocity. The results are summarised in Table 4.2.

### Note

The movie Smoke01.mov shows a smoke plume discharging vertically in the atmosphere (App. F). The movies Smoke03.mov illustrates the effects of a strong crossflow on smoke dispersion. The movie P1060234.mov shows some water jets discharging into the atmosphere (App. F).

# Exercises

## Exercise No. 1

A turbulent boundary layer develops over the Moreton Bay as a result of 35 m/s wind storm.

(a)   Assuming a zero pressure gradient, predict the atmospheric boundary layer thickness, the displacement thickness and the momentum thickness at 1 and 5 km from the inception of the wind storm.

(b)   Plot on graph paper the vertical distribution of the longitudinal velocity.

(c)   Calculate the friction force per unit width on the 5 km long water-boundary layer interface.

*Assuming that the water free-surface is equivalent to a smooth boundary.*

### Solution

(a)   Turbulent boundary layer calculations since $Re_x > 1E+9$

$x$ (m) $= 1000$   $5000$
$\delta$ (m) $= 4.89$   $17.73$
$\delta_1$ (m) $= 0.61$   $2.21$
$\delta_2$ (m) $= 0.48$   $1.73$

(c)   $F_{shear}/B = 2500\,N/m$

## Exercise No. 2

A gust storm develops in a narrow, funnel shaped valley. The free-stream wind speed is 15 m/s at the start of that valley and it reaches 35 m/s at 10 km inside the valley.

(a)   Calculate the atmospheric boundary layer growth in the first 10 km of the valley.

(b)   Plot the longitudinal profile of the boundary layer thickness.

*Assume the boundary layer growth to initiate of the start of valley and the free-stream velocity to increase linearly with distance.*

### Solution

Assuming a linear increase in free-stream velocity and a smooth boundary, the momentum integral equation is applied. The velocity distribution is assumed to be a 1/7 power law (smooth turbulent flow). A numerical integration of Equation (4.7) with a spreadsheet may be compared with the analytical solutions for a developing boundary layer on a smooth plate with $V_O = 15$ & 35 m/s. At the beginning of the valley, the numerical solution must be close to that of a developing boundary layer solution for $V_O = 15$ m/s.

## Exercise No. 3

Turbulent velocity measurements were conducted in a developing boundary layer above a rough surface. The results at $x = 0.98$ m are listed below.

| y | $\overline{V_x}$ | |
|---|---|---|
| mm | m/s | $\sqrt{\overline{v_x^2}}/\overline{V_x}$ |
| 2 | 6.76 | 19.0 |
| 3 | 7.11 | 18.1 |
| 4 | 7.48 | 17.0 |
| 5 | 7.79 | 16.2 |
| 6 | 8.03 | 15.8 |
| 7 | 8.30 | 15.4 |
| 8 | 8.49 | 15.1 |
| 9 | 8.72 | 14.4 |
| 10 | 8.91 | 14.1 |
| 15 | 9.70 | 12.8 |
| 20 | 10.16 | 13.2 |
| 25 | 10.44 | 12.5 |
| 30 | 10.53 | 11.8 |
| 40 | 10.71 | 11.1 |
| 50 | 10.76 | 10.5 |
| 60 | 10.81 | 10.5 |
| 80 | 10.90 | 9.3 |
| 100 | 11.0 | 9.2 |
| 150 | 11.07 | 8.5 |
| 200 | 11.1 | 7.9 |
| 280 | 11.02 | 7.3 |

(a)   Plot the vertical distribution of time-averaged velocity and turbulence intensity.
(b)   Calculate the boundary layer thickness, the displacement thickness and the momentum thickness.
(c)   From the best data fit, (c1) deduce the boundary shear stress and shear velocity with the law of the wall (i.e. log-law); (c2) estimated the equivalent roughness height of the plate rugosity.

*The test were performed in an environmental wind tunnel at atmospheric pressure and ambient temperature.*

### Solution

The analysis of the velocity profile yields:

$V_O = 11.05$ m/s
$\delta = 87.4$ mm
$\delta_1 = 6.85$ mm
$\delta_2 = 4.99$ mm

Since the velocity profile was measured above a rough plate, the data are compared with Equations (4.18) and (4.20). The slope of the curve $v_x/V_* = f(Ln(y \times V_*/v))$ equals $1/K$ where $K = 0.4$ for $V_* = 0.60$ m/s ($\tau_O = 0.43$ Pa). A comparison between the data and Equation (4.20) gives $k_s = 0.8$ mm.

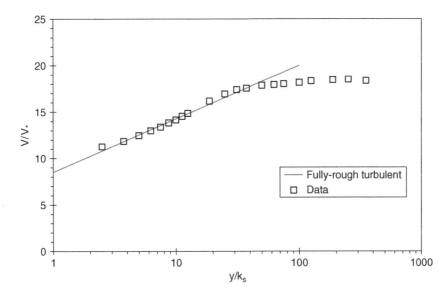

*Figure E4.1*   Comparison between experimental data and Equation (4.20).

## Exercise No. 4

In a water tunnel, turbulent velocity distributions were conducted in a developing boundary layer along a flat plate in absence of pressure gradient.

| y | $\overline{V}_x$ | $\overline{V}_x$ | $\overline{V}_x$ |
|---|---|---|---|
| mm | m/s | m/s | m/s |
| x (m) = | 0.25 | 0.50 | 0.75 |
| 0.5 | 5.57 | 3.94 | 4.14 |
| 1 | 6.23 | 5.3 | 4.82 |
| 2 | 6.60 | 6.38 | 5.89 |
| 3.5 | 7.30 | 6.61 | 6.12 |
| 6 | 7.72 | 7.02 | 6.30 |
| 8 | 7.77 | 7.20 | 6.83 |
| 10.2 | 7.70 | 7.42 | 7.07 |
| 14 | 7.64 | 7.61 | 7.13 |
| 18 | 7.69 | 7.87 | 7.38 |
| 21 | 7.68 | 7.66 | 7.87 |
| 35 | 7.84 | 7.67 | 7.69 |
| 41 | 7.69 | 7.72 | 7.71 |
| 58 | 7.76 | 7.69 | 7.66 |
| 60.5 | 7.67 | 7.72 | 7.72 |
| 66 | 7.68 | 7.70 | 7.68 |
| 70 | 7.77 | 7.73 | 7.70 |
| 90 | 7.80 | 7.65 | 7.69 |
| 100 | 7.69 | 7.68 | 7.68 |

(a)    Plot the vertical distributions of velocity.
(b)    Calculate the boundary layer thickness, displacement thickness and momentum thickness at $x = 0.25$ m, 0.5 m and 0.75 m.
(c)    Using the momentum integral equation, calculate (c1) the boundary shear stress distribution along the plate and (c2) the total force acting on the 1 m long, 0.5 m wide plate.

### Solution

| $x =$ | m | 0.25 | 0.5 | 0.75 |
|---|---|---|---|---|
| $V_O =$ | m/s | 7.72 | 7.7 | 7.71 |
| $\delta =$ | mm | 5.55 | 14.26 | 19.55 |
| $\delta_1 =$ | mm | 0.764 | 1.695 | 2.759 |
| $\delta_2 =$ | mm | 0.488 | 1.190 | 2.0340 |

The total shear force acting on the 1 m long, 0.5 m wide plate is about 0.9 N.

## Exercise No. 5

Let us consider an outfall in the sea in absence of current. The turbulent water jet is issued from a rectangular nozzle (0.1 m by 2 m). The jet velocity at the nozzle is 2.1 m/s. At the distances $x = 0.3$ m and 3.5 m, calculate the velocity distributions. Plot your results in graph paper.

   *For seawater, the fluid density, dynamic viscosity and surface tension are respectively: $\rho = 1024\ kg/m^3$, $\mu = 1.22\ E-3\ Pa.s$, and $\sigma = 0.076\ N/m$ (Pacific Ocean waters off Japan).*
   *Assume a plane jet (opening 0.1 m) with a width of 2 m.*
   *Herein x is the longitudinal distance from the nozzle.*

### Solution

At $x/D = 3$, the jet is not fully-developed. There is an ideal-fluid flow core with $V(y = 0) = V_O$ and two developing free shear layers. At $x/D = 35$, the jet flow is fully-developed and the centreline velocity equals $V(y = 0) = 0.95$ m/s.

---

**Remark**

An ocean outfall is the discharge point of a waste stream into the sea. The outfall system consists of a pipeline conveying wastewater offshore. The effluent may be released using nozzles at the end of the pipeline, or using one or more risers with nozzles located along the submerged pipe.
   There is an interesting underwater video of the submerged outfall at Bari East, Bari, Italy at {http://www.iahrmedialibrary.net/}. The video shows the 900 m long submarine pipe. The hydraulic link between the sewage plant and the submarine pipeline is made with a reinforced concrete tunnel with a circular cross section diameter of 1.4 m. Its length is 2,220 m from the shoreline. The prolongation to the sea is made with a steel conduit with a diameter of 1,200 mm, 842 m long, with a wastewater outfall 66.5 m long positioned in the end part of the pipe. The prolongation to the sea has a direction almost normal to the shoreline and in the shallow water is covered for about 2 m, in order to be protected from the wave motion field. Jets at the nozzles are visible in the video footage.

---

## Exercise No. 6

A thin plate (0.7 m wide by 2 m long) is towed through water at a velocity of 1.1 m/s. Calculate the drag force on both sides of the submerged plate assuming that (a) the boundary layer remains laminar, and (b) the boundary layer becomes turbulent at the leading edge.

## Solution

The Reynolds number defined in terms of the plate length equals: $Re_L = 2.2\ E+6$. The boundary layer flow at the end of the plate would be expected to be turbulent.

|  | Laminar boundary layer | Turbulent boundary layer |
|---|---|---|
| $\delta$ (m) at x = 2 m | 0.0066 | 0.040 |
| Total drag force (N) = | 1.5 | 6.6 |

Note: The total drag equals the force on both sides of the submerged plate.

(A) Control volume and definition sketch of the wind flow pas a turbine

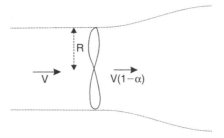

(B) Wind turbines in Plouarzel (France) on 1 March 2004 – The wind farm consists of 5 wind turbines of 750 kW – The wind farm is located 2.6 km from the coastline and each mast is 38.4 m high

*Figure E4.2* Wind turbines.

## Exercise No. 7

Considering a wind turbine (Fig. E4.2), the power taken from the wind, assuming no energy losses, is:

$$\text{Power} = 2 \times \pi \times R^2 \times \rho \times V^3 \times \alpha \times (1 - \alpha)^2$$

where R is the turbine radius, $\rho$ is the air density, V is the mean wind speed and $\alpha$ is the interference factor (Fig. E4.2A). As the wind passes through the turbine, it is decelerated down to $V \times (1 - \alpha)$ at the turbine disk. The efficiency of the wind turbine is:

$$\eta = 4 \times \alpha \times (1 - \alpha)^2$$

(a)   Calculate the maximum efficiency.
(b)   A wind turbine is located at 9.4 km from the coastline. The rotor diameter is 45 m. Assuming that the free-surface-stream velocity is 12 m/s, and that the boundary layer develops at the shoreline, calculate the optimum mast elevation to take 900 kW from the wind.

*Assume maximum wind turbine efficiency. Assume the main wind direction perpendicular to the coastline.*

### Solution

(a)   $\eta = 0.593$ for $\alpha = 1/3$
(b)   $y = 30$ m (calculations performed assuming a smooth turbulent boundary layer and a 1/7-th power law velocity distribution).

Note the approximate nature of the calculations since the velocity distribution is not uniform.

## Exercise No. 8

Considering a turbulent free-shear layer (Fig. 4.14), obtain an expression of the variation of the shear stress across the shear layer. *Assume the mixing length hypothesis.*

### Comments

The mixing length hypothesis assumes that the momentum exchange coefficient satisfies:

$$v_T = l_m^2 \times \frac{\partial v_x}{\partial y}$$

where $l_m$ is the mixing length. The turbulent shear stress equals hence:

$$\tau = \rho \times v_T \times \left(\frac{\partial v_x}{\partial y}\right) = \rho \times l_m^2 \times \left(\frac{\partial v_x}{\partial y}\right)^2$$

## Exercise No. 9

Considering a two-dimensional turbulent jet, give the expression of the discharge of entrained fluid in the jet flow. Compare the result with a two-dimensional laminar jet.

## Comment

Let us remember that the discharge of entrained fluid in the jet flow per unit width equals:

$$\frac{Q}{B} = \int\limits_{y=-\infty}^{+\infty} v_x \times dy$$

Physically, the jet entrains some surrounding fluid as momentum is exchanged from the high-velocity region to the surrounding fluid at rest. The volume discharge Q increases with increasing longitudinal distance x.

## Exercise No. 10

The nozzle of a ventilation duct is placed 12 m above the floor of a sport hall and it is directed vertically downward. The outlet discharges 45 $m^3$/minute. Calculate the diameter of the nozzle if the maximum permissible velocity at a height of 1.5 m above the floor is 1.3 m/s?

## Solution

For a circular jet, the length of the developing flow region is about 5 to 10 $\times$ D where D is the jet diameter. Let us assume a conservative estimate: 10 $\times$ D. For x > 10 $\times$ D, the jet flow is full-developed and the maximum velocity is on the jet centreline (Eq. (4.35)).

The basic equations are:

$$Q = V_O \times \frac{\pi}{4} \times D^2 \qquad\qquad \text{Continuity equation}$$

$$\frac{v_{max}}{V_O} = \frac{5.745}{\frac{x}{D}} \qquad\qquad \text{Centreline velocity, } x/D > 10 \qquad (4.35b)$$

where x = 10.5 m, Q = 45 $m^3$/minute (0.75 $m^3$/s) and $V_{max}$ = 1.3 m/s.

The results yield: D = 0.402 m. Let us verify that x > 10 $\times$ D: x/D = 26.1. Note the calculations are based upon the assumption that the floor has little effect on the jet flow at 1.5 m above it.

## Comment

In the fully-developed flow region of circular jet, the jet centreline velocity decreases as 1/x (Table 4.2). Another reasoning may be based upon the following considerations:

$$Q = V_O \times \frac{\pi}{4} \times D^2 \qquad\qquad \text{Continuity equation}$$

$$\frac{v_{max}}{V_O} = \frac{10 \times D}{x} \qquad\qquad \text{Centreline velocity, } x/D > 10 \qquad (4.35)$$

This simple approximation gives:

$$Q = V_{max} \times \frac{x}{10} \times \frac{\pi}{4} \times D$$

where x = 10.5 m, Q = 45 $m^3$/minute (0.75 $m^3$/s) and $V_{max}$ = 1.3 m/s. It yields: D = 0.7 m. Note the difference with the earlier result which was based upon Equation (4.35). Which one would you choose to be conservative?

## Exercise No. 11

A Pitot-Prandtl-Preston tube may be used to determine the shear stress at a wall in a turbulent boundary layer (Fig. E4.3). The tube is in contact with the wall and the shear stress is read from a calibration curve between the velocity head and the boundary shear stress. On the basis of the velocity distribution in the inner wall region, justify the Pitot-Prandtl-Preston tube's method.

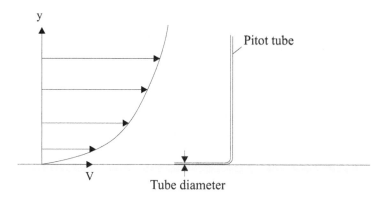

*Figure E4.3*  Sketch of a Pitot-Prandtl-Preston tube next to a solid boundary.

### Solution

When a Pitot tube is lying on the wall, it measures the velocity $v_x$ at a distance $y_O$ from the wall equal to half the Pitot tube outer diameter. Assuming that the Pitot tube is within the inner wall layer, the time-averaged longitudinal velocity satisfies:

$$\frac{v_x}{V_*} = \rho \times \frac{V_* \times y_O}{\mu} \qquad (4.12b)$$

Replacing the shear velocity $V_*$ by its expression in terms of the boundary shear stress $\tau_O$, it becomes:

$$\tau_O = \mu \times \frac{v_x}{y_O}$$

Practically, Pitot-Prandtl-Preston tubes may be also used in the wall region and in rough turbulent boundary layer with appropriate calibration.

### Comments

The Pitot tube is named after Henri Pitot (1695–1771) who invented a tubular device to measure flow velocity in the Seine river (first presentation in 1732 at the Académie des Sciences). Ludwig Prandtl (1875–1953) developed an improved Pitot tube design which provided direct measurements of the total head, piezometric head and velocity (Howe 1949, Troskolanski 1960). The accuracy of the Pitot-Prandtl tube is about 1% of the differential pressure under correct conditions of pressure recording. Based upon a dimensional analysis, Preston (1954) showed that the skin friction is measurable with a Pitot tube lying on the boundary (Fig. E4.3). He stressed that the tube diameter had to be less than 20% of the boundary layer thickness. His work was extended by Patel (1965). Macintosh (1990) and Chanson (2000) developed independently some calibration relationships for Pitot-Prandtl-Preston tubes. The experience gained at the University of Queensland (Kazemipour and Apelt 1983, Macintosh 1990, Macintosh and Isaacs 1992, Xie 1998, Chanson 2000) suggested that each tube must be calibrated independently, preferably in-situ, rather than relying on Patel's correlations.

## Exercise No. 12

A 2 m long 1.5 m wide flat plate is placed in a water tunnel. The plate acts as a splitter (Fig. E4.4) and the flow is symmetrical around both sides of the plate. Detailed velocity measurements were conducted in the developing boundary layer and the data analysis yields the following velocity profiles:

$$x = 0.8\,\text{m} \quad V = 14.98 \times (y/0.021)^{0.161} \quad y < 0.021\,\text{m}$$
$$V = 14.98\,\text{m/s} \quad y > 0.021\,\text{m}$$
$$x = 2\,\text{m} \quad V = 15.01 \times (y/0.036)^{0.154} \quad y < 0.036\,\text{m}$$
$$V = 15.01\,\text{m/s} \quad y > 0.036\,\text{m}$$

(a)  Calculate the boundary shear stress at $x = 1.4$ m.
(b)  Calculate the total drag force on the plate.

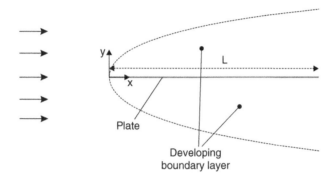

Figure E4.4  Developing boundary layers on a splitter plate – Note that the boundary layers around the plate are not drawn to scale: the vertical scale is enlarged.

## Solution

The flow is turbulent: since $Re_L = 3\,E+7$.
    The velocity distributions at $x = 0.8$ and $2$ m follow a power law:

$$\frac{V_x}{V_O} = \left(\frac{y}{\delta}\right)^{1/N} \tag{4.21}$$

Hence the displacement thickness and the momentum thickness are:

$$\frac{\delta_1}{\delta} = \frac{1}{1+N}$$

$$\frac{\delta_2}{\delta} = \frac{N}{(1+N) \times (2+N)}$$

Using the integral momentum equation:

$$\tau_O = \rho \times \left(V_O^2 \times \frac{\partial \delta_2}{\partial x} + V_O \times \frac{\partial V_O}{\partial x} \times (2 \times \delta_2 + \delta_1)\right) \tag{4.4}$$

the boundary shear stress equals 617 Pa and 279 Pa at $x = 0.4$ and $1.4$ m respectively. The total drag force on the plate equals 2.5 kN.
    Note that $x = 0$, $\delta = \delta_1 = \delta_2 = 0$.

## Exercise No. 13

Velocity measurements in a developing, turbulent boundary layer along a smooth flat plate yield the data set given in the table below.

(a)   On graph paper, plot $Ln(V_x)$ versus $Ln(y)$ at $x = 0.4\,m$.
(b)   Estimate the shear velocity and the boundary shear stress at $x = 0.4\,m$.
(c)   Using the momentum integral equation, calculate the boundary shear stress and shear force between $x = 0.4$ and $0.6\,m$. *Compare your results with the shear velocity estimate and the Blasius formula for smooth turbulent flows. Discuss your findings.*

*The fluid is air at 25 Celsius and standard pressure.*

| x mm | y mm | $V_x$ m/s | $v'_x$ m/s |
|---|---|---|---|
| 400 | 2 | 8.111 | 1.33 |
| | 4 | 8.859 | 1.362 |
| | 6 | 9.68 | 1.502 |
| | 8 | 10.283 | 1.476 |
| | 10 | 10.714 | 1.297 |
| | 15 | 10.756 | 1.366 |
| | 20 | 10.77 | 1.426 |
| | 30 | 10.8 | 1.286 |
| | 40 | 10.799 | 1.301 |
| | 50 | 10.9 | 1.03 |
| | 70 | 10.789 | 0.965 |
| | 90 | 10.98 | 1.037 |
| | 110 | 10.81 | 0.923 |
| | 130 | 10.78 | 0.919 |
| 600 | 2 | 8.591 | 1.159 |
| | 4 | 8.508 | 1.342 |
| | 6 | 9.392 | 1.249 |
| | 8 | 10.241 | 1.184 |
| | 10 | 10.327 | 1.448 |
| | 15 | 10.814 | 1.368 |
| | 20 | 11.12 | 1.341 |
| | 30 | 11.23 | 1.301 |
| | 40 | 11.157 | 1.102 |
| | 50 | 11.162 | 1.168 |
| | 70 | 11.219 | 0.943 |
| | 90 | 11.239 | 0.969 |
| | 110 | 11.186 | 1.056 |
| | 130 | 11.359 | 0.927 |

## Solution

At $x = 0.400\,m$:

$\tau_o = 0.32\,Pa$                                         Log law

$\tau_o = 0.34\,Pa$                                         Blasius formula

Between x = 0.400 m and 0.600 m:

$\tau_o = 0.43\,Pa$                                            Momentum integral equation

Overall the results are close.

## Exercise No. 14

The nozzle of a ventilation duct is placed 9.5 m above the floor of the hydraulic laboratory and it is directed vertically downward. The outlet discharges 2,300 m³/hour.

(a)  For a circular duct, calculate the diameter of the nozzle if the maximum permissible velocity at a height of 1.5 m above the floor is 0.95 m/s?

(b)  An alternative design uses a wide rectangular duct (10 m long), calculate the opening of the nozzle if the maximum permissible velocity at a height of 1.5 m above the floor is (b1) 0.95 m/s and (b2) 0.50 m/s? Discuss your results.

Assume a quasi-two-dimensional flow pattern.
  *The fluid is air at 28 Celsius and standard pressure.*

## Solution

(a)  D = 0.615 m (Circular jet)
(b)  D = 0.004 m & 0.0145 m (Two-dimensional nozzle)

A velocity of 0.95 m/s is relatively fast and may induce unpleasant working conditions in the hydraulics laboratory. The preferred design option would be a 0.0145 m thick slot.

# Appendices

# Glossary

**Abutment:** part of the valley side against which the dam is constructed. Artificial abutments are sometimes constructed to take the thrust of an arch where there is no suitable natural abutment.

**Académie des Sciences de Paris:** The Académie des Sciences, Paris, is a scientific society, part of the Institut de France formed in 1795 during the French Revolution. The academy of sciences succeeded the Académie Royale des Sciences, founded in 1666 by Jean-Baptiste Colbert.

**Acid:** a sour compound that is capable, in solution, of reacting with a base to form a salt and has a pH less than 7.

**Acidity:** having marked acid properties, more broadly having a pH of less than 7.

**Adiabatic:** thermodynamic transformation occurring without loss nor gain of heat.

**Advection:** movement of a mass of fluid which causes change in temperature or in other physical or chemical properties of fluid.

**Air:** mixture of gases comprising the atmosphere of the Earth. The principal constituents are nitrogen (78.08%) and oxygen (20.95%). The remaining gases in the atmosphere include argon, carbon dioxide, water vapour, hydrogen, ozone, methane, carbon monoxide, helium, krypton ...

**Air concentration:** concentration of undissolved air defined as the volume of air per unit volume of air and water. It is also called the void fraction.

**Alembert (d'):** Jean le Rond d'Alembert (1717–1783) was a French mathematician and philosopher. He was a friend of Leonhard Euler and Daniel Bernoulli. In 1752 he published his famous d'Alembert's paradox for an ideal-fluid flow past a cylinder (Alembert 1752).

**Alkalinity:** having marked basic properties (as a hydroxide or carbonate of an alkali metal); more broadly having a pH of more than 7.

**Analytical model:** system of mathematical equations which are the algebraic solutions of the fundamental equations.

**Angle of attack:** angle between the approaching flow velocity vector and the chordline. It is also called angle of incidence.

**Angle of incidence:** see *Angle of attack*.

**Apelt:** C.J. Apelt is an Emeritus Professor in civil engineering at the University of Queensland (Australia).

**Apron:** the area at the downstream end of a weir to protect against erosion and scouring by water.

**Archimedes:** Greek mathematician and physicist. He lived between B.C. 290–280 and B.C. 212 (or 211). He spent most of his life in Syracuse (Sicily, Italy) where he played a major role in the defence of the city against the Romans. His treaty "On Floating Bodies" is the first-known work on hydrostatics, in which he outlined the concept of buoyancy.

**Argand diagram:** graphic presentation of complex numbers, those of the form $x + i \times y$, in which x and y are real numbers and i is the square root of $-1$. The diagram was devised by the Swiss mathematician Jean Robert Argand in about 1806. Usually, the horizontal axis represents real numbers while the vertical axis represents the pure imaginary numbers. This permits the complex numbers to be plotted as points in the field defined by the two axes.

**Aristotle:** Greek philosopher and scientist (384–322 BC), student of Plato. His work "Meteorologica" is considered as the first comprehensive treatise on atmospheric and hydrological processes.

**Assyria:** land to the North of Babylon comprising, in its greatest extent, a territory between the Euphrates and the mountain slopes East of the Tigris. The Assyrian Kingdom lasted from about B.C. 2300 to B.C. 606.

**Atomic number:** The atomic number (of an atom) is defined as the number of units of positive charge in the nucleus. It determines the chemical properties of an atom.

**Atomic weight:** ratio of the average mass of a chemical element's atoms to some standard. Since 1961 the standard unit of atomic mass has been 1/12 the mass of an atom of the isotope Carbon-12.

**Avogadro number:** number of elementary entities (i.e. molecules) in one mole of a substance: $6.0221367 \, E + 23 \, \text{mole}^{-1}$. Named after the Italian physicist Amedeo Avogadro.

**Bachelor:** George Keith Bachelor (1920–2000) was an Australian Fluid dynamicist who worked on turbulence under Sir Geoffrey Ingram Taylor. In 1956 he founded the Journal of Fluid Mechanics and he was the journal editor for decades.

**Bagnold:** Ralph Alger Bagnold (1896–1990) was a British geologist and a leading expert on the physics of sediment transport by wind and water. During World War II, he founded the Long Range Desert Group and organised long-distance raids behind enemy lines across the Libyan desert.

**Bakhmeteff:** Boris Alexandrovitch Bakhmeteff (1880–1951) was a Russian hydraulician. In 1912, he developed the concept of specific energy and energy diagram for open channel flows.

**Barré de Saint-Venant:** Adhémar Jean Claude Barré de Saint-Venant (1797–1886), French engineer of the 'Corps des Ponts-et-Chaussées', developed the equation of motion of a fluid particle in terms of the shear and normal forces exerted on it (Barré de Saint-Venant 1871a, b).

**Bathymetry:** measurement of water depth at various places in water (e.g. river, ocean).

**Bazin:** Henri Emile Bazin was a French hydraulician (1829–1917) and engineer, member of the French 'Corps des Ponts-et-Chaussées' and later of the Académie des Sciences de Paris. He worked as an assistant of Henri P.G. Darcy at the beginning of his career.

**Belanger:** Jean-Baptiste Ch. Belanger (1789–1874) was a French hydraulician and professor at the Ecole Nationale Supérieure des Ponts et Chaussées (Paris). He suggested first the application of the momentum principle to hydraulic jump flow (Belanger 1849). Earlier, he presented the first 'backwater' calculation for open channel flow (Belanger 1828).

**Bélanger equation:** momentum equation applied across a hydraulic jump in a horizontal channel (named after J.B.C. Belanger).

**Belidor:** Bertrand Forêt de Belidor (1693–1761) was a teacher at the Ecole des Ponts et Chaussées. His treatise "Architecture Hydraulique" (Belidor 1737–1753) was a well-known hydraulic textbook in Europe during the 18th and 19th centuries.

**Benthic:** related to processes occurring at the bottom of the waters.

**Bernoulli:** Daniel Bernoulli (1700–1782) was a Swiss mathematician, physicist and botanist who developed the Bernoulli equation in his "Hydrodynamica, de viribus et motibus fluidorum" textbook (1st draft in 1733, 1st publication in 1738, Strasbourg).

**Bessel:** Friedrich Wilhelm Bessel (1784–1846) was a German astronomer and mathematician. In 1810 he computed the orbit of Halley's comet. As a mathematician he introduced the

Bessel functions (or circular functions) which have found wide use in physics, engineering and mathematical astronomy.

Bidone: Giorgio Bidone (1781–1839) was an Italian hydraulician. His experimental investigations on the hydraulic jump were published between 1820 and 1826.

Biesel: Francis Biesel (1920–1993) was a French hydraulic engineer and a pioneer of computational hydraulics.

Biochemical oxygen demand: The Biochemical Oxygen Demand (BOD) is the amount of oxygen used by micro-organisms in the process of breaking down organic matter in water.

Blasius: Paul Richard Heinrich Blasius (1883–1970) was a German fluid mechanician. Commonly called Heinrich Blasius, he was a student of Ludwig Prandtl and he later lectured at the University of Hamburg from 1912 to 1950.

Bod: see *Biochemical oxygen demand*.

Boltzmann: Ludwig Eduard Boltzmann (1844–1906) was an Austrian physicist.

Boltzmann constant: ratio of the universal gas constant ($8.3143 \, K \cdot J^{-1} \cdot mole^{-1}$) to the Avogadro number ($6.0221367 \, E+23 \, mole^{-1}$). It equals: $1.380662 \, E-23 \, J/K$.

Borda: Jean-Charles de Borda (1733–1799) was a French mathematician and military engineer. He achieved the rank of Capitaine de Vaisseau and participated to the U.S. War of Independence with the French Navy. He investigated the flow through orifices and developed the Borda mouthpiece.

Borda mouthpiece: a horizontal re-entrant tube in the side of a tank with a length such that the issuing jet is not affected by the presence of the walls.

Bore: a surge of tidal origin is usually termed a bore: e.g., the Mascaret in the Seine river (France), the porroroca of the Amazon river (Brazil), the Hangzhou bore on the Qiantang river (China).

Bossut: Abbé Charles Bossut (1730–1804) was a French ecclesiastic and experimental hydraulician, author of a hydrodynamic treaty (Bossut 1772).

Boundary layer: flow region next to a solid boundary where the flow field is affected by the presence of the boundary and where friction plays an essential part. A boundary layer flow is characterised by a range of velocities across the boundary layer region from zero at the boundary to the free-stream velocity at the outer edge of the boundary layer.

Boussinesq: Joseph Valentin Boussinesq (1842–1929) was a French hydrodynamicist and Professor at the Sorbonne University (Paris). His treatise "Essai sur la théorie des eaux courantes" (1877) remains an outstanding contribution in hydraulics literature.

Boussinesq coefficient: momentum correction coefficient named after J.V. Boussinesq who first proposed it (Boussinesq 1877).

Boussinesq-Favre wave: an undular surge (see Undular surge).

Bowden: Professor Kenneth F. Bowden contributed to the present understanding of dispersion in estuaries and coastal zones.

Bows: the forward part of a ship (often plural with singular meaning).

Boys: P.F.D. du Boys (1847–1924) was a French hydraulic engineer. He made a major contribution to the understanding of sediment transport and bed-load transport (Boys 1879).

Bresse: Jacques Antoine Charles Bresse (1822–1883) was a French applied mathematician and hydraulician. He was Professor at the Ecole Nationale Supérieure des Ponts et Chaussées, Paris as successor of J.B.C. Belanger. His contribution to gradually-varied flows in open channel hydraulics is considerable (Bresse 1860).

Buat: Comte Pierre Louis George du Buat (1734–1809) was a French military engineer and hydraulician. He was a friend of Abbé C. Bossut. Du Buat is considered as the pioneer of experimental hydraulics. His textbook (Buat 1779) was a major contribution to flow resistance in pipes, open channel hydraulics and sediment transport.

Bubble: small volume of gas within a liquid (e.g. air bubble in water). The term bubble is used also for a thin film of liquid inflated with gas (e.g. soap bubble) or a small air globule in a solid (e.g. gas inclusion during casting). More generally the term air bubble describes a volume of air surrounded by liquid interface(s).

Buoyancy: tendency of a body to float, to rise or to drop when submerged in a fluid at rest. The physical law of buoyancy (or Archimedes' principle) was discovered by the Greek mathematician Archimedes. It states that any body submerged in a fluid at rest is subjected to a vertical (or buoyant) force. The magnitude of the buoyant force is equal to the weight of the fluid displaced by the body.

Buoyant jet: submerged jet discharging a fluid lighter or heavier than the mainstream flow. If the jet's initial momentum is negligible, it is called a *buoyant plume*.

Camber: convexity of an airfoil curve from the leading edge to the trailing edge.

Candela: SI unit for luminous intensity, defined as the intensity in a given direction of a source emitting a monochromatic radiation of frequency $540\,E+12\,Hz$ and which has a radiant intensity in that direction of $1/683$ Watt per unit solid angle.

Carnot: Lazare N.M. Carnot (1753–1823) was a French military engineer, mathematician, general and statesman who played a key-role during the French Revolution.

Carnot: Sadi Carnot (1796–1832), eldest son of Lazare Carnot, was a French scientist who worked on steam engines and described the Carnot cycle relating to the theory of heat engines.

Cartesian co-ordinate: one of three co-ordinates that locate a point in space and measure its distance from one of three intersecting co-ordinate planes measured parallel to that one of three straight-line axes that is the intersection of the other two planes. It is named after the French mathematician René Descartes.

Casagrande: Arthur Casagrande (1902–1981) was an Austrian-born civil engineer who moved to USA in 1926 and became Professor in Geomechanics at Harvard University in 1932.

Catamaran: a boat with twin hulls and a superstructure connecting the hulls.

Cauchy: Augustin Louis de Cauchy (1789–1857) was a French engineer from the 'Corps des Ponts-et-Chaussées'. He devoted himself later to mathematics and he taught at Ecole Polytechnique, Paris, and at the Collège de France. He worked with Pierre-Simon Laplace and J. Louis Lagrange. In fluid mechanics, he contributed greatly to the analysis of wave motion.

Cavitation: formation of vapour bubbles and vapour pockets within a homogeneous liquid caused by excessive stress (Franc et al. 1995). Cavitation may occur in low-pressure regions where the liquid has been accelerated (e.g. turbines, marine propellers, baffle blocks of dissipation basin). Cavitation modifies the hydraulic characteristics of a system, and it is characterised by damaging erosion, additional noise, vibrations and energy dissipation.

Celsius: Anders Celsius (1701–1744) was a Swedish astronomer who invented the Celsius thermometer scale (or centigrade scale) in which the interval between the freezing and boiling points of water is divided into 100 degrees.

Celsius degree (or degree centigrade): temperature scale based on the freezing and boiling points of water: 0 and 100 Celsius respectively.

Chezy: Antoine Chezy (1717–1798) (or Antoine de Chezy) was a French engineer and member of the French 'Corps des Ponts-et-Chaussées'. He designed canals for the water supply of the city of Paris. In 1768 he proposed a resistance formula for open channel flows called the Chézy equation. In 1798, he became Director of the Ecole Nationale Supérieure des Ponts et Chaussées after teaching there for many years.

Chlorophyll: one of the most important classes of pigments involved in photosynthesis. Chlorophyll is found in virtually all photosynthetic organisms. It absorbs energy from ligh that is then used to convert carbon dioxide to carbohydrates. Chlorophyll occurs in several distinct forms: chlorophyll-a and chlorophyll-b are the major types found in higher plants and green algae. High concentrations of cholophyll-a occur in algual bloom.

Choke: In open channel flow, a channel contraction might obstruct the flow and induce the appearance of critical flow conditions (i.e. control section). Such a constriction is sometimes called a 'choke'.

Chord (or chord length): length of the chordline; a straight line joining two points on a curve; in aeronautics, the straight line distance joining the leading and trailing edges of an airfoil.

Chordline: straight line connecting the nose to the trailing edge of a foil.

Christoffel: Elwin Bruno Christoffel (1829–1900) was a German mathematician.

Circulation: The flow along a closed curve is called the circulation. The circulation $\Gamma$ around a closed path C is:

$$\Gamma = \int_C \vec{V} \times \vec{\delta s}$$

where $\delta s$ is an element of the closed curve C (Streeter and Wylie 1981).

Clausius: Rudolf Julius Emanuel Clausius (1822–1888) was a German physicist and thermodynamicist. In 1850 he formulated the second law of thermodynamics.

Clay: earthy material that is plastic when moist and that becomes hard when baked or fired. It is composed mainly of fine particles of a group of hydrous alumino-silicate minerals (particle sizes less than 0.05 mm usually).

Clean-air turbulence: turbulence experienced by aircraft at high-altitude above the atmospheric boundary layer. It is a form of Kelvin-Helmholtz instability occurring when a destabilising pressure gradient of the fluid become large relative to the stabilising pressure gradient.

Clepsydra: Greek name for Water clock.

Cofferdam: temporary structure enclosing all or part of the construction area so that construction can proceed in dry conditions. A diversion cofferdam diverts a stream into a pipe or channel.

Colbert: Jean-Baptiste Colbert (1619–1683) was a French statesman. Under King Louis XIV, he was the Minister of Finances, the Minister of 'Bâtiments et Manufactures' (buildings and industries) and the Minister of the Marine.

Confined flow: flow bounded between solid boundaries; in a confined aquifer, the pressure is greater than atmospheric pressure.

Conjugate depth: in open channel flow, another name for sequent depth.

Control: Considering an open channel, subcritical flows are controlled by the downstream conditions. This is called a 'downstream flow control'. Conversely supercritical flows are controlled only by the upstream flow conditions (i.e. 'upstream flow control').

Control section: in an open channel, cross-section where critical flow conditions take place. The concept of 'control' and 'control section' are used with the same meaning.

Control surface: is the boundary of a control volume.

Control volume: refers to a region in space and is used in the analysis of situations where flow occurs into and out of the space.

Convection: transport (usually) in the direction normal to the flow direction induced by hydrostatic instability: e.g. flow past a heated plate.

Coriolis: Gustave Gaspard Coriolis (1792–1843) was a French mathematician and engineer of the 'Corps des Ponts-et-Chaussées' who first described the Coriolis force (i.e. effect of motion on a rotating body).

Coriolis coefficient: kinetic energy correction coefficient named after G.G. Coriolis who introduced first the correction coefficient (Coriolis 1836).

Couette: M. Couette was a French scientist who measured experimentally the viscosity of fluids with a rotating viscosimeter (Couette 1890).

Couette flow: flow between parallel boundaries moving at different velocities, named after the Frenchman M. Couette. The most common Couette flows are the cylindrical Couette flow used to measure dynamic viscosity and the two-dimensional Couette flow between parallel plates.

Couette viscosimeter: system consisting of two co-axial cylinders of radii $r_1$ and $r_2$ rotating in opposite direction, used to measure the viscosity of the fluid placed in the space between the cylinders. In a steady state, the torque transmitted from one cylinder to another per unit length equals: $4 \times \pi \times \mu \times \omega_o \times r_1^2 \times r_2^2/(r_2^2 - r_1^2)$, where $\omega_o$ is the relative angular velocity and $\mu$ is the dynamic viscosity of the fluid.

Courant: Richard Courant (1888–1972), American mathematician born in Germany who made significant advances in the calculus of variations.

Craya: Antoine Craya was a French hydraulician and professor at the University of Grenoble.

Cunge: Born and educated in Poland, Jean A. Cunge worked in France at Sogreah in Grenoble and he lectured at the Hydraulics and Mechanical Engineering School of Grenoble, France.

Cutoff wall: In dam construction, a cutoff wall is an underground wall built to stop seepage in alluvial soils beneath the main dam wall. For example, at the Serre-Ponçon dam, the cutoff wall extends 100 m beneath the natural river bed.

Danel: Pierre Danel (1902–1966) was a French hydraulician and engineer. One of the pioneers of modern hydrodynamics, he worked from 1928 to his death for Neyrpic known prior to 1948 as 'Ateliers Neyret-Beylier-Piccard et Pictet'.

Darcy: Henri Philibert Gaspard Darcy (1805–1858) was a French civil engineer. He studied at Ecole Polytechnique between 1821 and 1823, and later at the Ecole Nationale Supérieure des Ponts et Chaussées (Brown 2002). He performed numerous experiments of flow resistance in pipes (Darcy 1858) and in open channels (Darcy and Bazin 1865), and of seepage flow in porous media (Darcy 1856). He gave his name to the Darcy-Weisbach friction factor and to the Darcy law in porous media.

Darcy law: law of groundwater flow motion which states that the seepage flow rate is proportional to the ratio of the head loss over the length of the flow path. It was discovered by H.P.G. Darcy (1856) who showed that, for a flow of liquid through a porous medium, the flow rate is directly proportional to the pressure difference.

Darcy-Weisbach friction factor: dimensionless parameter characterising the friction loss in a flow. It is named after the Frenchman H.P.G. Darcy and the German J. Weisbach.

Density-stratified flows: flow field affected by density stratification caused by temperature variations in lakes, estuaries and oceans. There is a strong feedback process: i.e., mixing is affected by density stratification, which depends in turn upon mixing.

Descartes: René Descartes (1596–1650) was a French mathematician, scientist, and philosopher. He is recognised as the father of modern philosophy. He stated: "cogito ergo sum" ('I think therefore I am').

Diffusion: the process whereby particles of liquids, gases or solids intermingle as the result of their spontaneous movement caused by thermal agitation and in dissolved substances move from a region of higher concentration to one of lower concentration. The term turbulent diffusion is used to describe the spreading of particles caused by turbulent agitation.

**Diffusion coefficient:** quantity of a substance that in diffusing from one region to another passes through each unit of cross-section per unit of time when the volume concentration is unity. The units of the diffusion coefficient are $m^2/s$.

**Diffusivity:** another name for the diffusion coefficient.

**Dimensional analysis:** organisation technique used to reduce the complexity of a study, by expressing the relevant parameters in terms of numerical magnitude and associated units, and grouping them into dimensionless numbers. The use of dimensionless numbers increases the generality of the results.

**Dirichlet:** Johann Peter Gustav Lejeune Dirichlet (1805–1859) was a German mathematician.

**Dispersion:** longitudinal scattering of particles by the combined effects of shear and diffusion.

**Dissolved oxygen content:** mass concentration of dissolved oxygen in water. It is a primary indicator of water quality: e.g., oxygenated water is considered to be of good quality.

**DOC:** see *Dissolved oxygen content.*

**Drag:** force component (on an object) in the direction of the approaching flow.

**Drag reduction:** reduction of the skin friction resistance in fluids in motion. In a broader sense, reduction in flow resistance (skin friction and form drag) in fluids in motion.

**Drainage layer:** layer of pervious material to relieve pore pressures and/or to facilitate drainage: e.g., drainage layer in an earthfill dam.

**Draught:** the depth of water a ship requires to float in.

**Drogue:** (1) sea anchor; (2) cylindrical device towed for water sampling by a boat.

**Drop:** (1) volume of liquid surrounded by gas in a free-fall motion (i.e. dropping); (2) by extension, small volume of liquid in motion in a gas; (3) a rapid change of bed elevation also called step.

**Droplet:** small drop of liquid.

**Drop structure:** single step structure characterised by a sudden decrease in bed elevation.

**Du Boys (or *Duboys*):** see P.F.D. du Boys.

**Du Buat (or *Dubuat*):** see P.L.G. du BUAT.

**Dupuit:** Arsène Jules Etienne Juvénal Dupuit (1804–1866) was a French engineer and economist. His expertise included road construction, economics, statics and hydraulics.

**Ebb:** reflux of the tide toward the sea. That is, the flow motion between a high tide and a low tide. The ebb flux is maximum at mid-tide. (The opposite is the *flood.*)

**Ecole Nationale Supérieure des Ponts et Chaussées, Paris:** French civil engineering school founded in 1747. The direct translation is: 'National School of Bridge and Road Engineering'. Among the directors there were the famous hydraulicians A. Chezy and G. de Prony. Other famous professors included B.F. de Belidor, J.B. Belanger, J.A.C. Bresse, G.G. Coriolis and L.M.H. Navier.

**Ecole Polytechnique, Paris:** Leading French engineering school founded in 1794 during the French Révolution under the leadership of Lazare Carnot and Gaspard Monge. It absorbed the state artillery school in 1802 and was transformed into a military school by Napoléon Bonaparte in 1804. Famous professors included Augustin Louis Cauchy, Jean Baptiste Joseph Fourier, Siméon-Denis Poisson, Jacques Charles François Sturm, among others.

**Eddy viscosity:** another name for the momentum exchange coefficient. It is also called 'eddy coefficient' by Schlichting (1979). (See Momentum exchange coefficient)

**Effluent:** waste water (e.g. industrial refuse, sewage) discharged into the environment, often serving as a pollutant.

**Ekman:** V. Walfrid Ekman (1874–1954) was a Swedish oceanographer best known for his studies of the dynamics of ocean currents.

**Embankment:** fill material (e.g. earth, rock) placed with sloping sides and with a length greater than its height.

**Ephemeral channel:** a river that is usually not flowing above ground except during the rainy season. Ephemeral channels are also called arroyo, wadi, wash, dry wash, oued or coulee (coulée).

**Equi-Potential lines:** are lines drawn perpendicular to streamlines so that the pattern of streamlines and equi-potentials forms a flow net. The flow potential is constant along an equi-potential line.

**Estuary:** water passage where the tide meets a river flow. An estuary may be defined as a region where salt water is diluted with fresh water.

**Euler:** Leonhard Euler (1707–1783) was a Swiss mathematician and physicist, and a close friend of Daniel Bernoulli.

**Eulerian method:** Study of a process (e.g. dispersion) from a fixed reference in space. For example, velocity measurements at a fixed point. (A different method is the *Lagangian method*.)

**Eutrophication:** process by which a body of water becomes enriched in dissolved nutrients (e.g. phosphorus, nitrogen) that stimulate the growth of aquatic plant life, often resulting in the depletion of dissolved oxygen.

**Explicit method:** calculation containing only independent variables; numerical method in which the flow properties at one point are computed as functions of known flow conditions only.

**Extrados:** upper side of a wing or exterior curve of a foil. The pressure distribution on the extrados must be smaller than that on the intrados to provide a positive lift force.

**Face:** external surface which limits a structure: e.g. air face of a dam (i.e. downstream face), water face (i.e. upstream face) of a weir.

**Favre:** H. Favre (1901–1966) was a Swiss professor at ETH-Zürich. He investigated both experimentally and analytically positive and negative surges. Undular surges are sometimes called Boussinesq-Favre waves. Later he worked on the theory of elasticity.

**Fawer jump:** undular hydraulic jump.

**Fick:** Adolf Eugen Fick was a 19th century German physiologist who developed the diffusion equation for neutral particle (Fick 1855).

**Finite differences:** approximate solutions of partial differential equations which consists essentially in replacing each partial derivative by a ratio of differences between two immediate values: e.g., $\partial V/\partial t \approx \partial V/\partial t$. The method was first introduced by Runge (1908).

**Fischer:** Hugo B. Fischer (1937–1983) was a Professor at the University of California, Berkeley. He earned his B.Sc., M.S. and Ph.D. at the California Institute of Technology. He was a professor of civil engineering at the University of California, Berkeley from 1966 until 1983. Fischer was a recognized authority in salt-water intrusion, water pollution, heat dispersion in waterways, and the mixing in rivers and oceans (e.g. Fischer et al. 1979). He died in a glider accident in May 1983.

**Flash flood:** flood of short duration with a relatively high peak flow rate.

**Flashy:** term applied to rivers and streams whose discharge can rise and fall suddenly, and is often unpredictable.

**Flettner:** Anton Flettner (1885–1961) was a German engineer. In 1924 he designed a rotor ship based on the Magnus effect. Large vertical circular cylinders were mounted on the ship. They were mechanically rotated to provide circulation and to propel the ship. During the development of his rotorship, Flettner consulted Ludwig Prandtl and the Institute of Aerodynamics in Göttingen (Ergebnisse Aerodynamische Versuchanstalt, Göttingen) (Flettner 1925). More recently a similar system was developed for the ship 'Alcyone' of Jacques-Yves Cousteau.

**Flow net:** is the combination of the streamlines and equi-potential lines such as 1- the flow between two adjacent streamlines is the same and 2- the potential increase between two adjacent equi-potential lines is the same.

**Forchheimer:** Philipp Forchheimer (1852–1933) was an Austrian hydraulician who contributed significantly to the study of groundwater hydrology.

**Fortier:** André Fortier was a French scientist and engineer. He became later Professor at the Sorbonne, Paris.

**Fourier:** Jean Baptiste Joseph Fourier (1768–1830) was a French mathematician and physicist known for his development of the Fourier series. In 1794 he was offered a professorship of mathematics at the Ecole Normale in Paris and was later appointed at the Ecole Polytechnique. In 1798 he joined the expedition to Egypt lead by (then) General Napoléon Bonaparte. His research in mathematical physics culminated with the classical study "Théorie Analytique de la Chaleur" (Fourier 1822) in which he enunciated his theory of heat conduction.

**Free-surface:** interface between a liquid and a gas. More generally a free-surface is the interface between the fluid (at rest or in motion) and the atmosphere. In two-phase gas-liquid flow, the term 'free-surface' includes also the air-water interface of gas bubbles and liquid drops.

**French revolution (Révolution Française):** revolutionary period that shook France between 1787 and 1799. It reached a turning point in 1789 and led to the destitution of the monarchy in 1791. The constitution of the First Republic was drafted in 1790 and adopted in 1791.

**Frontinus:** Sextus Julius Frontinus (A.D. 35-103 or 104) was a Roman engineer and soldier. After A.D. 97, he was 'curator aquarum' in charge of the water supply system of Rome. He dealt with discharge measurements in pipes and canals. In his analysis he correctly related the proportionality between discharge and cross-section area. His book "De Aquaeductu Urbis Romae" ('Concerning the Aqueducts of the City of Rome') described the operation and maintenance of Rome water supply system.

**Froude:** William Froude (1810–1879) was a English naval architect and hydrodynamicist who invented the dynamometer and used it for the testing of model ships in towing tanks. He was assisted by his son Robert Edmund Froude who, after the death of his father, continued some of his work. In 1868, he used Reech's law of similarity to study the resistance of model ships.

**Froude number:** The Froude number is proportional to the square root of the ratio of the inertial forces over the weight of fluid. The Froude number is used generally for scaling free surface flows, open channels and hydraulic structures. Although the dimensionless number was named after William Froude, several French researchers used it before. Dupuit (1848) and Bresse (1860) highlighted the significance of the number to differentiate the open channel flow regimes. Bazin (1865a) confirmed experimentally the findings. Ferdinand Reech introduced the dimensionless number for testing ships and propellers in 1852. The number is called the Reech-Froude number in France.

**Galileo:** Galileo Galilei (1564–1642) was an Italian mathematician, astronomer, and physicist. He demonstrated that the Earth revolves around the Sun and is not the centre of the universe, as had been believed.

**Gas transfer:** process by which gas is transferred into or out of solution: i.e., dissolution or desorption respectively.

**Gauckler:** Philippe Gaspard Gauckler (1826–1905) was a French engineer and member of the French 'Corps des Ponts-et-Chaussées'. He re-analysed the experimental data of Darcy and Bazin (1865), and in 1867 he presented a flow resistance formula for open channel flows (Gauckler-Manning formula) sometimes called improperly the Manning equation (Gauckler 1867). His son became Directeur des Antiquités et des Beaux-Arts (Director of Anquities and Fine Arts) for the French Republic in Tunisia where he directed an extensive survey of Roman hydraulic works in Tunisia (Gauckler 1897–1902, 1897, 1902).

Gay-Lussac: Joseph-Louis Gay-Lussac (1778–1850) was a French chemist and physicist.

Ghaznavid: (or Ghaznevid) one of the Moslem dynasties (10th to 12th centuries) ruling South-Western Asia. Its capital city was at Ghazni (Afghanistan).

Gulf Stream: warm ocean current flowing in the North Atlantic northeastward. The Gulf Stream is part of a general clockwise-rotating system of currents in the North Atlantic.

Hartree: Douglas R. Hartree (1897–1958) was an English physicist. He was a professor of Mathematical Physics at Cambridge. His approximation to the Schrödinger equation is the basis for the modern physical understanding of the wave mechanics of atoms. The scheme is sometimes called the Hartree-Fok method after the Russian physicist V. Fock who generalized Hartree's scheme.

Helmholtz: Hermann Ludwig Ferdinand von Helmholtz (1821–1894) was a German scientist who made basic contributions to physiology, optics, electrodynamics and meteorology.

Hennin: Georg Wilhelm Hennin (1680–1750) was a young Dutchman hired by the tsar Peter the Great to design and build several dams in Russia (Danilveskii 1940, Schnitter 1994). He went to Russia in 1698 and stayed until his death in April 1750.

Hero of Alexandria: Greek mathematician (1st century A.D.) working in Alexandria, Egypt. He wrote at least 13 books on mathematics, mechanics and physics. He designed and experimented the first steam engine. His treatise "Pneumatica" described Hero's fountain, siphons, steam-powered engines, a water organ, and hydraulic and mechanical water devices. It influenced directly the waterworks design during the Italian Renaissance. In his book "Dioptra", Hero stated rightly the concept of continuity for incompressible flow: the discharge being equal to the area of the cross-section of the flow times the speed of the flow.

Hokusai Katsushita: Japanese painter and wood engraver (1760–1849). His "Thirty-Six Views of Mount Fuji" (1826–1833) are world-known.

Huang Chun-Pi: one of the greatest masters of Chinese painting in modern China (1898–1991). Several of his paintings included mountain rivers and waterfalls: e.g., "Red trees and waterfalls", "The house by the water-falls", "Listening to the sound of flowing waters", "Water-falls".

Humboldt: Alexander von Humboldt (1769–1859) was a German explorer who was a major figure in the classical period of physical geography and biogeography.

Humboldt current: flows off the west coast of South America. It was named after Alexander von Humboldt who took measurements in 1802 that showed the coldness of the flow. It is also called Peru current.

Hydraulic diameter: is defined as the equivalent pipe diameter: i.e., four times the cross-section area divided by the wetted perimeter. The concept was first expressed by the Frenchman P.L.G. du Buat (Buat 1779).

Hydraulic jump: transition from a rapid (supercritical flow) to a slow flow motion (subcritical flow). Although the hydraulic jump was described by Leonardo da Vinci, the first experimental investigations were published by Giorgio Bidone in 1820. The present theory of the jump was developed by Belanger (1828) and it has been verified experimentally numerous researchers (e.g. Bakhmeteff and Matzke 1936).

Hydrofoil: a device that, when attached to a ship, lifts (totally or partially) the hull out of the water at speed.

Ideal fluid: frictionless and incompressible fluid. An ideal fluid has zero viscosity: i.e., it cannot sustain shear stress at any point.

**Idle discharge:** old expression for spill or waste water flow.

**Implicit method:** calculation in which the dependent variable and the one or more independent variables are not separated on opposite sides of the equation; numerical method in which the flow properties at one point are computed as functions of both independent and dependent flow conditions.

**Incidence:** the arrival of something (e.g. a projectile) at a surface.

**Inflow:** (1) upstream flow; (2) incoming flow.

**Inlet:** (1) upstream opening of a culvert, pipe or channel; (2) a tidal inlet is a narrow water passage between peninsulas or islands.

**Intake:** any structure in a reservoir through which water can be drawn into a waterway or pipe. By extension, upstream end of a channel.

**Interface:** surface forming a common boundary of two phases (e.g. gas-liquid interface) or two fluids.

**International system of units:** see Système international d'unités.

**Intrados:** lower side of a wing or interior curve of a foil.

**Invert:** (1) lowest portion of the internal cross-section of a conduit; (2) channel bed of a spillway; (3) bottom of a culvert barrel.

**Inviscid flow:** is a non-viscous flow. Since the viscosity is zero, the fluid cannot sustains shear stress and the fluid flow motion is also called frictionless.

**Ippen:** Arthur Thomas Ippen (1907–1974) was Professor in hydrodynamics and hydraulic engineering at M.I.T. (USA). Born in London of German parents, educated in Germany (Technische Hochschule in Aachen), he moved to USA in 1932, where he obtained the M.S. and Ph.D. degrees at the California Institute of Technology. There he worked on high-speed free-surface flows with Theodore von Karman. In 1945 he was appointed at M.I.T. until his retirement in 1973.

**Irrotational flow:** is defined as a zero vorticity flow. Fluid particles within a region have no rotation. If a frictionless fluid has no rotation at rest, any later motion of the fluid will be irrotational. In irrotational flow each element of the moving fluid undergoes no net rotation, with respect to chosen coordinate axes, from one instant to another.

**Isotach:** (virtual) line of constant velocity (magnitude).

**Isotherm:** (virtual) line connecting points having the same temperature at a given time.

**Jevons:** W.S. Jevons (1835–1882) was an English chemist and economist. His work on salt finger intrusions (Jevons 1858) was a significant contribution to the understanding of double-diffusive convection. He performed his experiments in Sydney, Australia, 23 years prior to Rayleigh's experiments (Rayleigh 1883).

**Joukowski:** Nikolai Egorovich Joukowski (1847–1921) was a Russian mathematician who did some research in aerodynamics. His name is also spelled Zhukovsky or Zhukoskii. In 1906 he published two papers in which he gave a mathematical expression for the lift on an airfoil. Today the theorem is known as the Kutta-Joukowski theorem, since Kutta pointed out that the equation also appears in his 1902 dissertation. In mathematics, the conformal mapping of the complex plane $\{z \rightarrow z + 1/z\}$ is called the Joukowski transformation.

**Joukowski airfoil:** family of airfoil developed by the Russian aerodynamicist Joukowski and for which he solved analytically the flow field.

**Karman:** Theodore von Karman (or von Kármán) (1881–1963) was a Hungarian fluid dynamicist and aerodynamicist who worked in Germany (1906 to 1929) and later in USA. He was a student of Ludwig Prandtl in Germany. He gave his name to the vortex shedding behind a cylinder (Karman vortex street).

Karman constant (or von Karman constant): 'universal' constant of proportionality between the Prandtl mixing length and the distance from the boundary. Experimental results indicate that $K = 0.40$.

Kelvin (Lord): William Thomson (1824–1907), Baron Kelvin of Largs, was a British physicist. He contributed to the development of the second law of thermodynamics, the absolute temperature scale (measured in Kelvin), the dynamical theory of heat, fundamental work in hydrodynamics ...

Kelvin-Helmholtz instability: instability at the interface of two ideal-fluids in relative motion. The instability can be caused by a destabilising pressure gradient of the fluid (e.g. clean-air turbulence) or free-surface shear (e.g. fluttering fountain). It is named after H.L.F. Helmoltz who solved first the problem (Helmholtz 1868) and Lord Kelvin (Kelvin 1871).

Kennedy: Professor John Fisher Kennedy (1933–1991) was a hydraulic professor at the University of Iowa. He succeeded Hunter Rouse as head of the Iowa Institute of Hydraulic Research.

Keulegan: Garbis Hovannes Keulegan (1890–1989) was an Armenian mathematician who worked as hydraulician for the US Bureau of Standards since its creation in 1932.

Kirchhoff: Gustav Robert Kirchhoff (1824–1877) was a German physicist and mathematician who made notable contributions spectrum analysis, electecricity, the study of light and astronomy.

Knot: One knot equals one nautical mile per hour.

Kuroshio: is a strong surface oceanic current of flowing northeasterly in North Pacific, between the Philippines and the east coast of Japan. It travels at rates ranging between 0.05 and 0.3 m/s and it is also called Japan current. Known to European geographers as early as 1650, it is called Kuroshio (Black Current) because it appears a deeper blue than surronding seas by Captain James COOK.

Kutta: Martin Wilhelm Kutta (1867–1944) was a German engineer and mathematician, known for his contribution to the numerical solution of differential equations: i.e., the Runge-Kutta method.

Kutta-Joukowski law: The Kutta-Joukowski law states that magnitude of the lift force on a two-dimensionalcylinder or airfoil is $\rho \times V \times \Gamma$ where $\rho$ is the fluid density, V is the fluid velocity and $\Gamma$ is the ciculation around the cylinder or airfoil.

Lagrange: Joseph-Louis Lagrange (1736–1813) was a French mathematician and astronomer. During the 1789 Revolution, he worked on the committee to reform the metric system. He was Professor of mathematics at the École Polytechnique from the start in 1795.

Lagrangian method: Study of a process in a system of coordinates moving with an individual particle. For example, the study of ocean currents with buoys. (A different method is the *Eulerian method*.)

Lamb: Sir Horace Lamb (1849–1934) was an English mathematician who made important contributions to acoustics and fluid dynamics. He was taught by George Gabriel Stokes (1819–1903) and James Clerk Maxwell (1831–1879). In 1875 Horace Lamb was appointed to the chair of mathematics at Adelaide SA, Australia where he stayed for 10 years before returning to England.

Laminar flow: is characterised by fluid particles moving along smooth paths in laminas or layers, with one layer gliding smoothly over an adjacent layer. Laminar flows are governed by Newton's law of viscosity which relates the shear stress to the rate of angular deformation: $\tau = \mu \times \partial V/\partial y$.

Langevin: Paul Langevin (1879–1946) was a French physicist, specialist in magnetism, ultrasonics, and relativity. In 1905 Einstein identified Brownian motion as due to imbalances in the forces on a particle resulting from molecular impacts from the liquid. Shortly thereafter, Paul Langevin formulated a theory in which the minute fluctuations in the position of the particle were due explicitly to a random force. His approach had great utility in describing molecular fluctuations in other systems, including non-equilibrium thermodynamics.

Laplace: Pierre-Simon Laplace (1749–1827) was a French mathematician, astronomer and physicist. He is best known for his investigations into the stability of the solar system.

Larboard: see *Port*.

LDA velocimeter: Laser Doppler Anemometer system.

Left abutment: abutment on the left-hand side of an observer when looking downstream.

Left bank (left wall): Looking downstream, the left bank or the left channel wall is on the left.

Lift: force component (on a wing or foil) perpendicular to the approaching flow.

Leonardo Da Vinci: Italian artist (painter and sculptor) who extended his interest to medicine, science, engineering and architecture (A.D. 1452–1519).

Liggett: Professor James A. Liggett is an Emeritus Professor at Cornell University, USA. He wrote several books and was formerly the Editor of the Journal of Hydraulic Engineering ASCE.

McKay: Professor Gordon M. McKay (1913–1989) was Professor in Civil Engineering at the University of Queensland.

Mach: Ernst Mach (1838–1916) was an Austrian physicist and philosopher. He established important principles of optics, mechanics and wave dynamics.

Mach number: see Sarrau-Mach number.

Magnus: H.G. Magnus (1802–1870) was a German physicist who investigated the so-called Magnus effect in 1852.

Magnus effect: A rotating cylinder, placed in a flow, is subjected to a force acting in the direction normal to the flow direction: i.e., a lift force which is proportional to the flow velocity times the rotation speed of the cylinder. This effect, called the Magnus effect, has a wide range of applications (Swanson 1961).

Manning: Robert Manning (1816–1897) was Chief Engineer of the Office of Public Works, Ireland. In 1889, he presented two formulae (Manning 1890). One was to become the so-called 'Gauckler-Manning formula' but Robert Manning did prefer to use the second formula that he gave in his paper. It must be noted that the Gauckler-Manning formula was proposed first by the Frenchman P.G. Gauckler (Gauckler 1867).

Mariotte: Abbé Edme Mariotte (1620–1684) was a French physicist and plant physiologist. He was member of the Académie des Sciences de Paris and wrote a fluid mechanics treaty published after his death (Mariotte 1686).

Mast: a tall pole or structure rising from the keel or deck of a ship.

Metric system: see Système métrique.

Mises: Richard Edler von Mises (1883–1953) was a Austrian scientist who worked on fluid mechanics, aerodynamics, aeronautics, statistics and probability theory. During World War I, he flew as test pilot and instructor in the Austro-Hungarian army. In 1921, he became the founding editor of the scientific journal Zeitschrift für Angewandte Mathematik und Mechanik.

Mixing: process by which contaminants combine into a more or less uniform whole by diffusion or dispersion.

Mixing length: The mixing length theory is a turbulence theory developed by L. Prandtl, first formulated in 1925 (Prandtl 1925). Prandtl assumed that the mixing length is the characteristic distance travelled by a particle of fluid before its momentum is changed by the new environment.

**Mole:** mass numerically equal in grams to the relative mass of a substance (i.e. 12 g for Carbon-12). The number of molecules in one mole of gas is 6.0221367 E+23 (i.e. Avogadro number).

**Momentum exchange coefficient:** In turbulent flows the apparent kinematic viscosity (or kinematic eddy viscosity) is analogous to the kinematic viscosity in laminar flows. It is called the momentum exchange coefficient, the eddy viscosity or the eddy coefficient. The momentum exchange coefficient is proportional to the shear stress divided by the strain rate. It was first introduced by the Frenchman J.V. Boussinesq (1877, 1896).

**Monge:** Gaspard Monge (1746–1818), Comte de Péluse, was a French mathematician who invented descriptive geometry and pioneered the development of analytical geometry. He was a prominent figure during the French Revolution, helping to establish the Système métrique and the École Polytechnique, and being Minister for the Navy and colonies between 1792 and 1793.

**Mud:** slimy and sticky mixture of solid material and water.

**Munk:** Walter H. Munk was an American geophysicist and oceanographer who expanded Sverdrup's work on ocean circulation.

**Navier:** Louis Marie Henri Navier (1785–1835) was a French engineer who primarily designed bridges but also extended Euler's equations of motion (Navier 1823).

**Navier-Stokes equation:** momentum equation applied to a small control volume of incompressible fluid. It is usually written in vector notation. The equation was first derived by L. Navier in 1822 and S.D. Poisson in 1829 by a different method. It was derived later in a more modern manner by A.J.C. Barré de Saint-venant in 1843 and G.G. Stokes in 1845.

**Neap tide:** tide of minimum range occurring at the first and the third quarters of the moon. (The opposite is the *spring tide*.)

**Neumann:** Carl Gottfried Neumann (1832–1925) was a German mathematician.

**Newton:** Sir Isaac Newton (1642–1727) was an English mathematician and physicist. His contributions in optics, mechanics and mathematics were fundamental.

**Nikuradse:** J. Nikuradse was a German engineer who investigated experimentally the flow field in smooth and rough pipes (Nikuradse 1932, 1933).

**Nutrient:** substance that an organism must obtain from its surroundings for growth and the sustainment of life. Nitrogen and phosphorus are important nutrients for plant growth. High levels of nitrogen and phosphorus may cause excessive growth, weed proliferation and algal bloom, leading to eutrophication.

**One-dimensional flow:** neglects the variations and changes in velocity and pressure transverse to the main flow direction. An example of one-dimensional flow can be the flow through a pipe.

**One-dimensional model:** model defined with one spatial coordinate, the variables being averaged in the other two directions.

**Organic compound:** class of chemical compounds in which one or more atoms of carbon are linked to atoms of other elements (e.g. hydrogen, oxygen, nitrogen).

**Organic matter:** substance derived from living organisms.

**Outflow:** downstream flow.

**Outlet:** (1) downstream opening of a pipe, culvert or canal; (2) artificial or natural escape channel.

**pH:** measure of acidity and alkalinity of a solution. It is a number on a scale on which a value of 7 represents neutrality, lower numbers indicate increasing acidity, and higher numbers increasing alkalinity. On the pH scale, each unit represents a tenfold change in acidity or alkalinity.

**Pascal:** Blaise Pascal (1623–1662) was a French mathematician, physicist and philosopher. He developed the modern theory of probability. Between 1646 and 1648, he formulated the concept of pressure and showed that the pressure in a fluid is transmitted through the fluid in all directions. He measured also the air pressure both in Paris and on the top of a mountain overlooking Clermont-Ferrand (France).

**Pascal:** unit of pressure named after the Frenchman B. Pascal: one Pascal equals a Newton per square-metre.

**Path of a particle:** a particle always moves tangent to a streamline. In steady flow the path of a particle is a streamline. In unsteady flow a particle follows one streamline one instant another one the next instant and so that the path of the particle may have no resemblance to any given instantaneous streamline.

**Pelton turbine (or wheel):** impulse turbine with one to six circular nozzles that deliver high-speed water jets into air which then strike the rotor blades shaped like scoop and known as bucket. A simple bucket wheel was designed by Sturm in the 17th century. The American Lester Allen Pelton patented the actual double-scoop (or double-bucket) design in 1880.

**Pervious zone:** part of the cross-section of an embankment comprising material of high permeability.

**Photosynthesis:** is the process by which green plants and certain other organisms transform light energy into chemical energy. Photosynthesis in green plants harnesses sunlight energy to convert carbon dioxide, water and minerals, into organic compounds and gaseous oxygen.

**Pitot:** Henri Pitot (1695–1771) was a French mathematician, astronomer and hydraulician. He was a member of the French Académie des Sciences from 1724. He invented the Pitot tube to measure flow velocity in the Seine river (first presentation in 1732 at the Académie des Sciences de Paris).

**Pitot tube:** device to measure flow velocity. The original Pitot tube consisted of two tubes, one with an opening facing the flow. L. Prandtl developed an improved design (e.g. Howe 1949) which provides the total head, piezometric head and velocity measurements. It is called a Prandtl-Pitot tube and more commonly a Pitot tube.

**Pitting:** formation of small pits and holes on surfaces due to erosive or corrosive action (e.g. cavitation pitting).

**Plato:** Greek philosopher (about B.C. 428–347) who influenced greatly Western philosophy.

**Poincaré:** Jules-Henri Poincaré (1854–1912), commonly known as Henri Poincaré, was a French mathematician, astronomer and philosopher of science. In 1906 he was elected President of the Académie des Sciences. In 1908 he was elected member of the Académie Française, the highest honour accorded a French writer. His first cousin was Raymond Poincaré (1860–1934), president of the French Republic during World War I.

**Poiseuille:** Jean-Louis Marie Poiseuille (1799–1869) was a French physician and physiologist who investigated the characteristics of blood flow. He carried out experiments and formulated first the expression of flow rates and friction losses in laminar fluid flow in circular pipes (Poiseuille 1839).

**Poiseuille flow:** steady laminar flow in a circular tube of constant diameter.

**Poisson:** Siméon Denis Poisson (1781–1840) was a French mathematician and scientist. He developed the theory of elasticity, a theory of electricity and a theory of magnetism.

**Port:** the left side of a ship when looking forward. It is also called *larboard*. (In French: babord.)

**Positive surge:** A positive surge results from a sudden change in flow that increases the depth. It is an abrupt wave front. The unsteady flow conditions may be solved as a quasi-steady flow situation.

**Potential flow:** Ideal-fluid flow with irrotational motion.

**Prandtl:** Ludwig Prandtl (1875–1953) was a German physicist and aerodynamicist who introduced the concept of boundary layer (Prandtl 1904) and developed the turbulent 'mixing length' theory. He was Professor at the University of Göttingen.

**Preissmann:** Alexandre Preissmann (1916–1990) was born and educated in Switzerland. From 1958, he worked on the development of hydraulic mathematical models at Sogreah in Grenoble.

**Prismatic:** A prismatic channel has an unique cross-sectional shape independent of the longitudinal distance along the flow direction. For example, a rectangular channel of constant width is prismatic.

**Prony:** Gaspard Clair François Marie Riche de Prony (1755–1839) was a French mathematician and engineer. He succeeded A. Chezy as director general of the Ecole Nationale Supérieure des Ponts et Chaussées, Paris during the French revolution.

**Rankine:** William J.M. Rankine (1820–1872) was a Scottish engineer and physicist. His contribution to thermodynamics and steam-engine was important. In fluid mechanics, he developed the theory of sources and sinks, and used it to improve ship hull contours. One ideal-fluid flow pattern, the combination of uniform flow, source and sink, is named after him: i.e., flow past a Rankine body.

**Rayleigh:** John William Strutt, Baron Rayleigh, (1842–1919) was an English scientist who made fundamental findings in acoustics and optics. His works are the basics of wave propagation theory in fluids. He received the Nobel Prize for Physics in 1904 for his work on the inert gas argon.

**Reech:** Ferdinand Reech (1805–1880) was a French naval instructor who proposed first the Reech-Froude number in 1852 for the testing of model ships and propellers.

**Rehbock:** Theodor Rehbock (1864–1950) was a German hydraulician and professor at the Technical University of Karlsruhe. His contribution to the design of hydraulic structures and physical modelling is important.

**Reynolds:** Osborne Reynolds (1842–1912) was a British physicist and mathematician who expressed first the Reynolds number (Reynolds 1883) and later the Reynolds stress (i.e. turbulent shear stress).

**Reynolds number:** dimensionless number proportional to the ratio of the inertial force over the viscous force. In pipe flows, the Reynolds number is commonly defined as:

$$Re = \rho \times \frac{V \times D}{\mu}$$

**Rheology:** science describing the deformation of fluid and matter.

**Riblet:** series of longitudinal grooves. Riblets are used to reduce skin drag (e.g. on aircraft, ship hull). The presence of longitudinal grooves along a solid boundary modifies the bottom shear stress and the turbulent bursting process. Optimum groove width and depth are about 20 to 40 times the laminar sublayer thickness (i.e. about 10 to 20 $\mu$m in air, 1 to 2 mm in water).

**Richardson:** Lewis Fry Richardson (1881–1953) was a British meteorologist who pioneered mathematical weather forecasting. It is believed that he took interest in the dispersion of smoke from shell explosion while he was an ambulance driver on the World War I battle front, leading to his classical publications (Richardson 1922,1926).

**Richardson number**: dimensionless number characterising density-stratification, commonly used to predict the occurrence of fluid turbulence and the destruction of density currents in water or air. A common definition is:

$$\text{Ri} = \frac{g}{\rho} \times \frac{\partial\rho/\partial y}{(\partial V/\partial y)^2}$$

**Richelieu**: Armand Jean du Plessis (1585–1642), Duc de Richelieu and french Cardinal, was the Prime Minister of King Louis XIII of France from 1624 to his death.

**Riemann**: Bernhard Georg Friedrich Riemann (1826–1866) was a German mathematician.

**Right abutment**: abutment on the right-hand side of an observer when looking downstream.

**Right bank (right wall)**: Looking downstream, the right bank or the right channel wall is on the right.

**Riquet**: Pierre Paul Riquet (1604–1680) was the designer and Chief Engineer of the Canal du Midi built between 1666 and 1681. The Canal provides an inland route between the Atlantic and the Mediterranean across Southern France.

**Roller**: in hydraulics, large-scale turbulent eddy: e.g., the roller of a hydraulic jump.

**Roll wave**: On steep slopes free-surface flows become unstable. The phenomenon is usually clearly visible at low flow rates. The waters flow down the chute in a series of wave fronts called roll waves.

**Rotation**: of a fluid particle about an axis is defined "as the average angular velocity of two infinitesimal line elements in the particle that are at right angles to each other and to the given axis". Rotation of a fluid particle can be caused only by a torque applied by shear forces on the sides of the particle.

**Rotation vector**: is defined as the half of the curl of the velocity vector:

$$\vec{w} = \frac{1}{2} \times \overrightarrow{\text{curl}} \; \vec{V} = \frac{1}{2} \times \left(\frac{dV_z}{dy} - \frac{dV_y}{dz}\right)\vec{i} + \frac{1}{2} \times \left(\frac{dV_x}{dz} - \frac{dV_z}{dx}\right)\vec{j} + \frac{1}{2} \times \left(\frac{dV_y}{dx} - \frac{dV_x}{dy}\right)\vec{k}$$

**Rotational flow**: has non-zero vorticity. Fluid particles within a region have rotation about any axis.

**Rouse**: Hunter Rouse (1906–1996) was an eminent hydraulician who was Professor and Director of the Iowa Institute of Hydraulic Research at the University of Iowa (USA).

**Saltation**: (1) action of leaping or jumping; (2) in sediment transport, particle motion by jumping and bouncing along the bed.

**Saint-Venant**: See Barré de Saint Venant.

**Salinity**: amount of dissolved salts in water. The definition of the salinity is based on the electrical conductivity of water relative to a specified solution of KCl and $H_2O$ (Bowie et al. 1985). In surface waters of open oceans, the salinity ranges from 33 to 37 ppt typically.

**Salt**: (1) Common salt, or sodium chloride (NaCl), is a crystalline compound that is abundant in Nature. (2) When mixed, acids and bases neutralize one another to produce salts, that is substances with a salty taste and none of the characteristic properties of either acids or bases.

**Sarrau**: Jacques Rose Ferdinand Émile Sarrau (1837–1904) was a French Professor at Ecole Polytechnique, Paris, who first introduced the Sarrau-Mach number (Sarrau 1884).

**Sarrau-Mach number**: dimensionless number proportional to the ratio of inertial forces over elastic forces. Although the number is commonly named after E. Mach who introduced it in 1887, it is often called the Sarrau number after Professor Sarrau who first highlighted the significance of the number (Sarrau 1884). The Sarrau-Mach number was once called the Cauchy number as a tribute to Cauchy's contribution to wave motion analysis.

Scalar: a quantity that has a magnitude described by a real number and no direction. A scalar means a real number rather than a vector.

Scale effect: discrepancy betwen model and prototype resulting when one or more dimenionless paremeters have different values in the model and prototype.

Schlichting: Professor Hermann Schlichting (1907–1982) was a student of Ludwig Prandtl and he became later Professor at the Technical University of Braunschweig, Germany.

Schwarz: Hermann Amandus Schwarz (1843–1921) was a German mathematician.

Scour: bed material removal caused by the eroding power of the flow.

Secondary current: is a flow generated at right angles to the primary current. It is a direct result of the Reynolds stresses and exists in any non-circular conduits (Liggett 1994, pp. 256–259). In natural rivers, they are significant at bends, and between a flood plain and the main channel.

Sediment: any material carried in suspension by the flow or as bed load which would settle to the bottom in absence of fluid motion.

Seepage: interstitial movement of water that may take place through a dam, its foundation or abutments.

Seiche: rythmic water oscillations caused by resonnance in a lake, harbour or estuary. This results in the formation standing waves In first approximation, the resonnance period ([1]) of a water body with characteristic surface length L and depth d is about: $2 \times L/\sqrt{g \times d}$.

Sennacherib (or *Akkadian Sin-Akhkheeriba*): King of Assyria (B.C. 705–681), son of Sargon II (who ruled during B.C. 722–705). He build a huge water supply for his capital city Nineveh (near the actual Mossul, Iraq) in several stages. The latest stage comprised several dams and over 75 km of canals and paved channels.

Separation: In a boundary layer, a deceleration of fluid particles leading to a reversed flow within the boundary layer is called a separation. The decelerated fluid particles are forced outwards and the boundary layer is separated from the wall. At the point of separation, the velocity gradient normal to the wall is zero:

$$\frac{\partial V_x}{\partial y}\bigg|_{y=0} = 0$$

Separation point: in a boundary layer, intersection of the solid boundary with the streamline dividing the separation zone and the deflected outer flow. The separation point is a stagnation point.

Sequent depth: In open channel flow, the solution of the momentum equation at a transition between supercritical and subcritical flow gives two flow depths (upstream and downstream flow depths). They are called sequent depths.

Shear flow: The term shear flow characterises a flow with a velocity gradient in a direction normal to the mean flow direction: e.g., in a boundary layer flow along a flat plate, the velocity is zero at the boundary and equals the free-stream velocity away from the plate. In a shear flow, momentum (i.e. per unit volume: $\rho \times V$) is transferred from the region of high velocity to that of low-velocity. The fluid tends to resist the shear associated with the transfer of momentum.

---

[1]sometimes called sloshing motion period.

Shear stress: In a shear flow, the shear stress is proportional to the rate of transfer of momentum. In laminar flows, Newton's law of viscosity states:

$$\tau = \mu \times \frac{\partial v}{\partial y}$$

where $\tau$ is the shear stress, $\mu$ is the dynamic viscosity of the flowing fluid, $v$ is the velocity and $y$ is the direction normal to the flow direction. For large shear stresses, the fluid can no longer sustain the viscous shear stress and turbulence spots develop. After apparition of turbulence spots, the turbulence expands rapidly to the entire shear flow. The apparent shear stress in turbulent flow is expressed as:

$$\tau = \rho \times (v + v_T) \times \frac{\partial v}{\partial y}$$

where $\rho$ is the fluid density, $v$ is the kinematic viscosity (i.e. $v = \mu/\rho$), and $v_T$ is a factor depending upon the fluid motion and called the eddy viscosity or momentum exchange coefficient in turbulent flow.

Shock waves: With supercritical flows, a flow disturbance (e.g. change of direction, contraction) induces the development of shock waves propagating at the free-surface across the channel (e.g. Ippen and Harleman 1956, Hager 1992). Shock waves are called also lateral shock waves, oblique hydraulic jumps, Mach waves, crosswaves, diagonal jumps.

Similitude: correspondence between the behaviour of a model and that of its prototype, with or without geometric similarity. The correspondence is usually limited by scale effects.

Slope: (1) side of a hill; (2) inclined face of a canal (e.g. trapezoidal channel); (3) inclination of the channel bottom from the horizontal.

Span: the maximum distance laterally from tip to tip of an airplane; also the extent between abutments or supports.

Spillway: opening built into a dam or the side of a reservoir to release (to spill) excess flood waters.

Spray: water droplets flying or falling through air: e.g., spray thrown up by a waterfall.

Spring tide: tide of greater-than-average range around the times of new and full moon. (The opposite of a spring tide is the *neap tide*.)

Stagnation point: is defined as the point where the velocity is zero. When a streamline intersects itself, the intersection is a stagnation point. For irrotational flow a streamline intersects itself at right-angle at a stagnation point.

Staircase: another adjective for 'stepped': e.g., a staircase cascade is a stepped cascade.

Stall: aerodynamic phenomenon causing a disruption (i.e. separation) of the flow past a wing associated with a loss of lift.

Starboard: the right side of a ship when looking forward. (In French: tribord.)

Steady flow: occurs when conditions at any point of the fluid do not change with the time:

$$\frac{\partial V}{\partial t} = 0 \quad \frac{\partial \rho}{\partial t} = 0 \quad \frac{\partial P}{\partial t} = 0 \quad \frac{\partial T}{\partial t} = 0$$

Stern: the rear end of a ship or boat.

Stokes: George Gabriel Stokes (1819–1903), British mathematician and physicist, is known for his research in hydrodynamics and a study of elasticity.

Stommel: Henry Melson Stommel (1920–1992) was an American oceanographer and meteorologist, internationally known during the 1950s for his theories on circulation patterns in the Atlantic Ocean.

Storm water: excess water running off the surface of a drainage area during and immediately following a period of rain. In urban areas, waters drained off a catchment area during or after a heavy rainfall are usually conveyed in man-made storm waterways.

Storm waterway: channel built for carrying storm waters.

Straub: L.G. Straub (1901–1963) was Professor in hydraulic engineering and Director of the St Anthony Falls Hydraulics Laboratory at the University of Minnesota (USA).

Stream function: vector function of space and time which is related to the velocity field as: $\vec{V} = -\overrightarrow{\text{curl}}\ \vec{\Psi}$. The stream function exists for steady and unsteady flow of incompressible fluid as it does satisfy the continuity equation. The stream function was introduced by the French mathematician Lagrange.

Streamline: is the line drawn so that the velocity vector is always tangential to it (i.e. no flow across a streamline). When the streamlines converge the velocity increases. The concept of streamline was first introduced by the Frenchman J.C. de Borda.

Streamline maps: should be drawn so that the flow between any two adjacent streamlines is the same.

Stream tube: is a filament of fluid bounded by streamlines.

Streeter: Victor Lyle Streeter was a Professor at the University of Michigan, USA. He made significant contributions in fluid mechanics and applied hydraulics, and authored several popular textbooks and numerous technical publications.

Strouhal: V. Strouhal (1850–1922) was a Czech physicist. In 1878, he investigated first the 'singing' of wires caused by vortex sheedding behind the wires.

Strut: a structural object designed to resist pressure in the direction of its length.

Subcritical flow: In open channel the flow is defined as subcritical if the flow depth is larger than the critical flow depth. In practice, subcritical flows are controlled by the downstream flow conditions.

Subsonic flow: compressible flow with a Sarrau-Mach number less than unity: i.e., the flow velocity is less than the sound celerity.

Supercritical flow: In open channel, when the flow depth is less than the critical flow depth, the flow is supercritical and the Froude number is larger than one. Supercritical flows are controlled from upstream.

Supersonic flow: compressible flow with a Sarrau-Mach number larger than unity: i.e., the flow velocity is larger than the sound celerity.

Surface tension: property of a liquid surface displayed by its acting as if it were a stretched elastic membrane. Surface tension depends primarily upon the attraction forces between the particles within the given liquid and also upon the gas, solid or liquid in contact with it. The action of surface tension is to increase the pressure within a water droplet or within an air bubble. For a spherical bubble of diameter $d_{ab}$, the increase of internal pressure necessary to balance the tensile force caused by surface tension equals: $\Delta P = 4 \times \sigma/d_{ab}$ is the surface tension.

Surfactant (or *surface active agent*): substance that, when added to a liquid, reduces its surface tension thereby increasing its wetting property (e.g. detergent).

Surge: A surge in an open channel is a sudden change of flow depth (i.e. abrupt increase or decrease in depth). An abrupt increase in flow depth is called a positive surge while a sudden decrease in depth is termed a negative surge. A positive surge is also called (improperly) a 'moving hydraulic jump' or a 'hydraulic bore'.

Sverdrup: Harald Ulrik Sverdrup (1888–1957) was a Norwegian meteorologist and oceanographer known for his studies of the physics, chemistry, and biology of the oceans. He explained the equatorial countercurrents and helped develop a method of predicting surf and breakers.

Sverdrup: volume discharge units in oceanic circulation: 1 Sverdrup = 1 E+6 m³/s.

Système international d'unités: international system of units adopted in 1960 based on the metre-kilogram-second (MKS) system. It is commonly called SI unit system. The basic seven units are: for length, the metre; for mass, the kilogram; for time, the second; for electric current, the ampere; for luminous intensity, the candela; for amount of substance, the mole; for thermodynamic temperature, the Kelvin. Conversion tables are given in Appendix C.

Système métrique: international decimal system of weights and measures which was adopted in 1795 during the French Révolution. Between 1791 and 1795, the Académie des Sciences de Paris prepared a logical system of units based on the metre for length and the kilogram for mass. The standard metre was defined as $1\,E-7$ times a meridional quadrant of earth. The gram was equal to the mass of $1\,cm^3$ of pure water at the temperature of its maximum density (i.e. 4 Celsius) and 1 kilogram equalled 1,000 grams. The litre was defined as the volume occupied by a cube of $1\,E+3\,cm^3$.

Taylor: Sir Geoffrey Ingram Taylor (1886–1975) was a British fluid dynamicist based in Cambridge. He established the basic developments of shear dispersion (Taylor 1953, 1954). His great-father was the British mathematician George Boole (1815–1864) who established modern symbolic logic and Boolean algebra.

Thompson: Sir Benjamin Thompson (1753–1814), also known as Count Rumford, proposed in 1797 that evaporation in the Northern Hemisphere would cause heavier, saltier water to sink and flow southward, and that a warmer northbound current would be needed to balance the southern one.

Tonne: a metric unit of weight equal to 1000 kg.

Total head: The total head is proportional to the total energy per unit mass and per gravity unit. It is expressed in metres of water.

Torricelli: Evangelista Torricelli (1608–1647) was an Italian physicist and mathematician who invented the barometer. In 1641 Torricelli worked with the elderly astronomer Galileo and was later appointed to succeed him as professor of mathematics at the Florentine Academy.

Torricelli theorem: states that the velocity V of a liquid flowing under the force of gravity out of an orifice is proportional ($\sqrt{2 \times g \times H}$) where H is the head above orifice centreline. The theorem is named after Evangelista Torricelli who discovered it in 1643.

Turbidity: opacity of water. Turbidity is a measure of the absence of clarity of the water.

Turbulence: Flow motion characterised by its unpredictable behaviour, strong mixing properties and a broad spectrum of length scales.

Turbulent flow: In turbulent flows the fluid particles move in very irregular paths, causing an exchange of momentum from one portion of the fluid to another. Turbulent flows have great mixing potential and involve a wide range of eddy length scales.

Turriano: Juanelo Turriano (1511–1585) was an Italian clockmaker, mathematician and engineer who worked for the Spanish Kings Charles V and later Philip II. It is reported that he checked the design of the Alicante dam for King Philip II.

Two-dimensional flow: all particles are assumed to flow in parallel planes along identical paths in each of these planes. There are no changes in flow normal to these planes. An example of two-dimensional flow can be an open channel flow in a wide rectangular channel.

Ukiyo-e: (or *Ukiyoe*) is a type of Japanese painting and colour woodblock prints during the period 1803–1867.

Unconfined flow: flow situation when the upper surface of the flow is at atmospheric pressure: i.e., free-surface.

Undular hydraulic jump: hydraulic jump characterised by steady stationary free-surface undulations downstream of the jump and by the absence of a formed roller. The undulations can extend far downstream of the jump with decaying wave lengths, and the undular jump occupies a significant length of the channel. It is usually observed for $1 < Fr_1 < 1.5$ to 3 (Chanson 1995). The first significant study of undular jump flow can be attributed to Fawer (1937) and undular jump flows should be called Fawer jump in homage to Fawer's work.

Undular surge: positive surge characterised by a train of secondary waves (or undulations) following the surge front. Undular surges are sometimes called Boussinesq-Favre waves in homage to the contributions of J.B. Boussinesq and H. Favre.

Uniform equilibrium flow: occurs when the velocity is identically the same at every point, in magnitude and direction, for a given instant:

$$\frac{\partial V}{\partial s} = 0$$

in which time is held constant and $\partial s$ is a displacement in any direction. That is, steady uniform flow (e.g. liquid flow through a long pipe at a constant rate) and unsteady uniform flow (e.g. liquid flow through a long pipe at a decreasing rate).

Uniform flow: occurs when the velocity is identically the same at every point, in magnitude and direction, for a given instant:

$$\frac{d V}{dx} = 0 \quad \frac{d V}{dy} = 0 \quad \frac{d V}{dz} = 0$$

Universal gas constant (also called *molar gas constant* or *perfect gas constant*): fundamental constant equal to the pressure times the volume of gas divided by the absolute temperature for one mole of perfect gas. The value of the universal gas constant is $8.31441 \, J \cdot K^{-1} \cdot mole^{-1}$.

Unsteady flow: The flow properties change with the time.

Upstream flow conditions: flow conditions measured immediately upstream of the investigated control volume.

V.O.C.: Volatile Organic Compound.

Valence: property of an element that determines the number of other atoms with which an atom of the element can combine.

Validation: comparison between model results and prototype data, to validate the model. The validation process must be conducted with prototype data that are different from that used to calibrate and to verify the model.

Vallentine: Professor Harold Rupert Vallentine (1917–2010) worked at the Department of Civil Engineering, University of New South Wales, Australia.

Vauban: Sébastien Vauban (1633–1707) was Maréchal de France. He participated to the construction of several water supply systems in France, including the extension of the feeder system of the Canal du Midi between 1686 and 1687, and parts of the water supply system of the gardens of Versailles.

Velocity potential: is defined as a scalar function of space and time such that its negative derivative with respect to any direction is the fluid velocity in that direction: $\overrightarrow{V} = -\overrightarrow{grad}\, \phi$. The velocity potential $\phi$ was introduced by the French mathematician Joseph-Louis Lagrange (Lagrange 1781). Although many applications assumes that the existence of a velocity potential implies irrotational flow of ideal-fluid, Lagrange showed that it exists for irrotational motion of real- and ideal-fluids (Chanson 2007).

Vena contracta: minimum cross-section area of the flow (e.g. jet or nappe) discharging through an orifice, sluice gate or weir.

Venturi meter: in closed pipes, smooth constriction followed by a smooth expansion. The pressure difference between the upstream location and the throat is proportional to the velocity-square. It is named after the Italian physicist Giovanni Battista Venturi (1746–1822).

Viscosity: fluid property which characterises the fluid resistance to shear: i.e. resistance to a change in shape or movement of the surroundings.

Vitruvius: Roman architect and engineer (B.C. 94–??). He built several aqueducts to supply the Roman capital with water. (Note: there are some incertitude on his full name: 'Marcus Vitruvius Pollio' or 'Lucius Vitruvius Mamurra', Garbrecht 1987a.)

Voc: Volatile Organic Compound.

Von Karman constant: see *Karman constant*.

Vortex flow: is rotational flow.

Vortex shedding: dispersion of vortical structures downstream of separation: e.g., behind a blunt body.

Vorticity: is defined as the vector whose components are those of the curl of the velocity vector (i.e. twice those of the rotation vector):

$$\overrightarrow{Vort} = \overrightarrow{curl}\, \overrightarrow{V} = \left(\frac{dV_z}{dy} - \frac{dV_y}{dz}\right) \overrightarrow{i} + \left(\frac{dV_x}{dz} - \frac{dV_z}{dx}\right) \overrightarrow{j} + \left(\frac{dV_y}{dx} - \frac{dV_x}{dy}\right) \overrightarrow{k}$$

Wake region: The separation region downstream of the streamline that separates from a boundary is called a wake or wake region.

Warrie: Australian Aboriginal name for 'rushing water'.

Water: common name applied to the liquid state of the hydrogen-oxygen combination $H_2O$. Although the molecular structure of water is simple, the physical and chemical properties of $H_2O$ are unusually complicated. Water is a colourless, tasteless, and odourless liquid at room temperature. One most important property of water is its ability to dissolve many other substances: $H_2O$ is frequently called the universal solvent. Under standard atmospheric pressure, the freezing point of water is 0 Celsius (273.16 K) and its boiling point is 100 Celsius (373.16 K).

Water clock: ancient device for measuring time by the gradual flow of water through a small orifice into a floating vessel. The Greek name is Clepsydra.

Waterfall: abrupt drop of water over a precipice characterised by a free-falling nappe of water. The highest waterfalls are the Angel fall (979-m) in Venezuela ('Churún Merú'), Tugel fall (948-m) in South Africa, Mtarazi (762-m) in Zimbabwe.

Weber: Moritz Weber (1871–1951) was a German Professor at the Polytechnic Institute of Berlin. The Weber number characterising the ratio of inertial force over surface tension force was named after him.

**Weber number:** Dimensionless number characterising the ratio of inertial forces over surface tension forces. It is relevant in problems with gas-liquid or liquid-liquid interfaces.

**Weir:** low river dam used to raise the upstream water level. Measuring weirs are built across a stream for the purpose of measuring the flow.

**Weisbach:** Julius Weisbach (1806–1871) was a German applied mathematician and hydraulician.

**Wen Cheng-Ming:** Chinese landscape painter (1470–1559). One of his famous works is the painting of "Old trees by a cold waterfall".

**Wetted perimeter:** Considering a cross-section (usually selected normal to the flow direction), the wetted perimeter is the length of wetted contact between the flowing stream and the solid boundaries. For example, in a circular pipe flowing full, the wetted perimeter equals the circle perimeter.

**Wetted surface:** In open channel, the term 'wetted surface' refers to the surface area in contact with the flowing liquid.

**Whirlpool:** a vortex of vertical axis, with a downward velocity component near its centre. A good example is the bathtub vortex. Van Dyke (1982, p. 59) presented a superb illustration. In coastal zones, whirlpools are produced by the interaction of rising and falling tides. They are often observed at the edges of straits with large tidal currents. Notable oceanic whirlpools include those of Garofalo along the coast of Calabria in southern Italy, the Maelstrøm (from Dutch for "whirling stream") located near the Lofoten Islands off the coast of Norway, the Naruto strait between Awaji and Shikoku islands in Japan (e.g. Chanson 2002).

**White waters:** non technical term used to design free-surface aerated flows. The refraction of light by the entrained air bubbles gives the 'whitish' appearance to the free-surface of the flow.

**White water sports:** include canoe, kayak and rafting racing down swift-flowing turbulent waters.

**Wind setup:** water level rise in the downwind direction caused by wind shear stress. The opposite is a wind setdown.

**Wing wall:** sidewall of an inlet or outlet.

**Wood:** I.R. Wood is an Emeritus Professor in civil engineering at the University of Canterbury (New Zealand).

**Yen:** Professor Ben Chie Yen (1935–2001) was a hydraulic professor at the University of Illinois at Urbana-Champaign, although borm and educated in Taiwan.

# Constants and fluid properties

## B.1  ACCELERATION OF GRAVITY

### Standard acceleration of gravity

The standard acceleration of gravity equals:

$$g = 9.80665 \, \text{m/s}^2 \tag{B.1}$$

This value is roughly to that at sea level and at 45-degree latitude. The gravitational acceleration varies with latitude and elevation owing to the form and rotation of the earth and may be estimated as:

$$g = 9.806056 - 0.025027 \times \cos(2 \times \text{Latitude}) - 3\,\text{E}{-}6 \times z \tag{B.2}$$

where z is the altitude with the sea level as origin, and the Latitude is in degrees (Rouse 1938).

| Altitude z (m) (1) | $g$ (m/s²) (2) | Altitude z (m) (1) | $g$ (m/s²) (2) |
|---|---|---|---|
| −1,000 | 9.810 | 4,000 | 9.794 |
| 0 (Sea level) | 9.807 | 5,000 | 9.791 |
| 1,000 | 9.804 | 6,000 | 9.788 |
| 2,000 | 9.801 | 7,000 | 9.785 |
| 3,000 | 9.797 | 10,000 | 9.776 |

### *Absolute gravity values*

The gravity varies also with the local geology and topography. Measured values of g are reported below.

| Location (1) | $g$ (m/s²) (2) | Location (1) | $g$ (m/s²) (2) | Location (1) | $g$ (m/s²) (2) |
|---|---|---|---|---|---|
| Addis Ababa, Ethiopia | 9.7743 | Helsinki, Finland | 9.81090 | Québec, Canada | 9.80726 |
| Algiers, Algeria | 9.79896 | Kuala Lumpur, Malaysia | 9.78034 | Quito, Ecuador | 9.7726 |
| Anchorage, USA | 9.81925 | La Paz, Bolivia | 9.7745 | Sapporo, Japan | 9.80476 |
| Ankara, Turkey | 9.79925 | Lisbon, Portugal | 9.8007 | Reykjavik, Iceland | 9.82265 |
| Aswan, Egypt | 9.78854 | Manila, Philippines | 9.78382 | Taipei, Taiwan | 9.7895 |
| Bangkok, Thailand | 9.7830 | Mexico city, Mexico | 9.77927 | Teheran, Iran | 9.7939 |
| Bogota, Colombia | 9.7739 | Nairobi, Kenya | 9.77526 | Thule, Greenland | 9.82914 |
| Brisbane, Australia | 9.794 | New Delhi, India | 9.79122 | Tokyo, Japan | 9.79787 |
| Buenos Aires, Argentina | 9.7949 | Paris, France | 9.80926 | Vancouver, Canada | 9.80921 |
| Christchurch, N.Z. | 9.8050 | Perth, Australia | 9.794 | Ushuaia, Argentina | 9.81465 |
| Denver, USA | 9.79598 | Port-Moresby, P.N.G. | 9.782 | | |
| Guatemala, Guatemala | 9.77967 | Pretoria, South Africa | 9.78615 | | |

Reference: Morelli (1971).

## B.2   PROPERTIES OF WATER

| Temperature Celsius (1) | Density $\rho_w$ kg/m$^3$ (2) | Dynamic viscosity $\mu_w$ Pa.s (3) | Surface tension $\sigma$ N/m (4) | Vapour pressure $P_v$ Pa (5) | Bulk modulus of elasticity $E_b$ Pa (6) |
|---|---|---|---|---|---|
| 0 | 999.9 | 1.792E−3 | 0.0762 | 0.6E+3 | 2.04E+9 |
| 5 | 1000.0 | 1.519E−3 | 0.0754 | 0.9E+3 | 2.06E+9 |
| 10 | 999.7 | 1.308E−3 | 0.0748 | 1.2E+3 | 2.11E+9 |
| 15 | 999.1 | 1.140E−3 | 0.0741 | 1.7E+3 | 2.14E+9 |
| 20 | 998.2 | 1.005E−3 | 0.0736 | 2.5E+3 | 2.20E+9 |
| 25 | 997.1 | 0.894E−3 | 0.0726 | 3.2E+3 | 2.22E+9 |
| 30 | 995.7 | 0.801E−3 | 0.0718 | 4.3E+3 | 2.23E+9 |
| 35 | 994.1 | 0.723E−3 | 0.0710 | 5.7E+3 | 2.24E+9 |
| 40 | 992.2 | 0.656E−3 | 0.0701 | 7.5E+3 | 2.27E+9 |

Reference: Streeter and Wylie (1981).

Properties of freshwater and sea water at 20 Celsius and standard atmosphere

| Fluid properties (1) | Freshwater (2) | Seawater (3) | Remarks (4) |
|---|---|---|---|
| Composition: | H$_2$O | H$_2$O, dissolved sodium and chloride ions (30 g/kg), and dissolved salts | |
| Density $\rho$ (kg/m$^3$): | 998.2 | 1,024 | At 20 Celsius. |
| Dynamic viscosity $\mu$ (Pa.s): | 1.005 E−3 | 1.22 E−3 | At 20 Celsius. |
| Surface tension between air and water $\sigma$ (N/m): | 0.0736 | 0.076 | At 20 Celsius. |
| Conductivity ($\mu$S/cm): | 87.7 | 48,800 | At 25 Celsius. |

References: Riley and Skirrow (1965), Open University Course Team (1995), Chanson et al. (2006).

## B.3   GAS PROPERTIES

### Basic equations

The *state equation* of perfect gas is:

$$P = \rho \times R \times T \tag{B.3}$$

where P is the absolute pressure (in Pascal), $\rho$ is the gas density (in kg/m$^3$), T is the absolute temperature (in Kelvin) and $R$ is the gas constant (in J/kg.K) (see below).

For a perfect gas, the *specific heat* at constant pressure $C_p$ and the specific heat at constant volume $C_v$ are related to the gas constant as:

$$C_p = \frac{\gamma}{\gamma - 1} \times R \tag{B.4a}$$

$$C_p = C_v \times R \tag{B.4b}$$

where $\gamma$ is the specific heat ratio (i.e. $\gamma = C_p/C_v$).

During an *isentropic transformation* of perfect gas, the following relationships hold:

$$\frac{P}{\rho^{\gamma}} = \text{constant} \tag{B.5a}$$

$$T \times P^{(1-\gamma)/\gamma} = \text{constant} \tag{B.5b}$$

## Physical properties

| Gas (1) | Formula (2) | Gas constant R J/kg.K (3) | Specific heat | | Specific heat ratio $\gamma$ (6) |
|---|---|---|---|---|---|
| | | | $C_p$ J/kg.K (4) | $C_v$ J/kg.K (5) | |
| **Perfect gas** | | | | | |
| Mono-atomic gas | (e.g. He) | | $\frac{5}{2} \times R$ | $\frac{3}{2} \times R$ | $\frac{5}{3}$ |
| Di-atomic gas | (e.g. $O_2$) | | $\frac{7}{2} \times R$ | $\frac{5}{2} \times R$ | $\frac{7}{5}$ |
| Poly-atomic gas | (e.g. $CH_4$) | | $4 \times R$ | $3 \times R$ | $\frac{4}{3}$ |
| **Real gas ([a])** | | | | | |
| Air | | 287 | 1.004 | 0.716 | 1.40 |
| Helium | He | 2077.4 | 5.233 | 3.153 | 1.67 |
| Nitrogen | $N_2$ | 297 | 1.038 | 0.741 | 1.40 |
| Oxygen | $O_2$ | 260 | 0.917 | 0.657 | 1.40 |
| Water vapour | $H_2O$ | 462 | 1.863 | 1.403 | 1.33 |

Notes: ([a]): at low pressures and at 299.83 K; Reference: Streeter and Wylie (1981).

## Compressibility and bulk modulus of elasticity

The compressibility of a fluid is a measure of change in volume and density when the fluid is subjected to a change of pressure. It is defined as:

$$E_{co} = \frac{1}{\rho} \times \frac{\partial \rho}{\partial P} \tag{B.6}$$

The reciprocal function of the compressibility is called the bulk modulus of elasticity:

$$E_b = \rho \times \frac{\partial P}{\partial \rho} \tag{B.7}$$

For a perfect gas, the bulk modulus of elasticity equals:

$$E_b = \gamma \times P \qquad \text{adiabatic transformation for a perfect gas} \tag{B.7b}$$

$$E_b = P \qquad \text{isothermal transformation for a perfect gas} \tag{B.7c}$$

## Celerity of sound

The celerity of sound in a medium equals:

$$C_{sound} = \sqrt{\frac{\partial P}{\partial \rho}} \qquad (B.8)$$

where P is the pressure and $\rho$ is the density. It may be rewritten in terms of the bulk modulus of elasticity $E_b$:

$$C_{sound} = \sqrt{\frac{E_b}{\rho}} \qquad (B.9)$$

Equation (B.7) applies to both liquids and gases.

For an isentropic process and a perfect gas, Equation (B.9) yields:

$$C_{sound} = \sqrt{\gamma \times R \times T} \qquad (B.10)$$

where $\gamma$ and $R$ are the specific heat ratio and gas constant respectively (see above).

The dimensionless velocity of compressible fluid is called the Sarrau-Mach number:

$$Ma = \frac{V}{C_{sound}} \qquad (B.11)$$

Classical values of the sound celerity include:

Celerity of sound in water at 20 Celsius:                       1485 m/s
Celerity of sound in dry air at sea level and 20 Celsius:    343 m/s


## B.4   ATMOSPHERIC PARAMETERS

## Air pressure

The standard atmosphere or normal pressure at sea level equals:

$$P_{std} = 1\, atm = 360\, mm\ of\ Hg = 101{,}325\, Pa \qquad (B.12)$$

where Hg is the chemical symbol of mercury. Unit conversion tables are provided in Appendix C.

The atmospheric pressure varies with the elevation above the sea level (i.e. altitude). For dry air, the atmospheric pressure at the altitude z equals

$$P_{atm} = P_{std} \times \exp\left( -\int_0^z \frac{0.0034841 \times g}{T} \times dz \right) \qquad (B.13)$$

where T is the absolute temperature in Kelvin and Equation (B.13) is expressed in SI units.

## Air temperature

In the troposphere (i.e. z < 10,000 m), the air temperature decreases with altitude, on the average, at a rate of 6.5E−3 K/m (i.e. 6.5 Kelvin per km).

Table B.1 presents the distributions of average air temperatures (Miller 1971) and the corresponding atmospheric pressures as functions of the altitude z (Eq. (B.13)).

*Table B.1*  Distributions of air temperature and air pressure as functions of the altitude for dry air and standard acceleration of gravity.

| Altitude z (m) (1) | Mean air temperature (K) (2) | Atmospheric pressure (Eq. (B.13)) (Pa) (3) | Atmospheric pressure (Eq. (B.3)) (atm) (4) |
|---|---|---|---|
| 0 | 288.2 | 1.013E+5 | 1.000 |
| 500 | 285.0 | 9.546E+4 | 0.942 |
| 1000 | 281.7 | 8.987E+4 | 0.887 |
| 1500 | 278.4 | 8.456E+4 | 0.834 |
| 2000 | 275.2 | 7.949E+4 | 0.785 |
| 2500 | 272.0 | 7.468E+4 | 0.737 |
| 3000 | 268.7 | 7.011E+4 | 0.692 |
| 3500 | 265.5 | 6.576E+4 | 0.649 |
| 4000 | 262.2 | 6.164E+4 | 0.608 |
| 4500 | 259.0 | 5.773E+4 | 0.570 |
| 5000 | 255.7 | 5.402E+4 | 0.533 |
| 5500 | 252.5 | 5.051E+4 | 0.498 |
| 6000 | 249.2 | 4.718E+4 | 0.466 |
| 6500 | 246.0 | 4.404E+4 | 0.435 |
| 7000 | 242.8 | 4.106E+4 | 0.405 |
| 7500 | 239.5 | 3.825E+4 | 0.378 |
| 8000 | 236.3 | 3.560E+4 | 0.351 |
| 8500 | 233.0 | 3.310E+4 | 0.327 |
| 9000 | 229.8 | 3.075E+4 | 0.303 |
| 9500 | 226.5 | 2.853E+4 | 0.282 |
| 10000 | 223.3 | 2.644E+4 | 0.261 |

## Viscosity of air

Viscosity and density of air at 1.0 atm:

| Temperature K (1) | $\mu_{air}$ Pa.s (3) | $\rho_{air}$ kg/m$^3$ (3) |
|---|---|---|
| 300 | 18.4E−6 | 1.177 |
| 400 | 22.7E−6 | 0.883 |
| 500 | 26.7E−6 | 0.705 |
| 600 | 29.9E−6 | 0.588 |

The viscosity of air at standard atmosphere is commonly fitted by the Sutherland formula (Sutherland 1883):

$$\mu_{air} = 17.16E{-}6 \times \left(\frac{T}{273.1}\right)^{3/2} \times \frac{383.7}{T + 110.6} \tag{B.14}$$

A simpler correlation is:

$$\frac{\mu_{air}(T)}{\mu_{air}(T_O)} = \left(\frac{T}{T_O}\right)^{0.76} \tag{B.15}$$

where $\mu_{air}$ is in Pa.s, and the temperature T and reference temperature $T_O$ are expressed in Kelvin.

# Unit conversions

## C.1 INTRODUCTION

The systems of units derived from the metric system have gradually given way to a single system, called the Système International d'Unités (SI) and the present monograph presents results expressed in SI Units.

Since a number of countries continue to use British and American units, section C.3 gives their equivalents against the SI units, while, in the next section (C.2), a table of conversion is provided for ancient (and often unused nowadays) units.

## C.2 ANCIENT UNITS AND CONVERSION TABLES

| Quantity (1) | Ancient unit (symbol) (2) | Conversion (3) | Comments (4) |
|---|---|---|---|
| Length: | 1 arshin | = 16 vershoks <br> = 711.2 E−3 m | Ancient Russian imperial unit. |
| | 1 braccio | = 0.6096 m | Equal 2 feet exactly. From the Italian 'braccia'. |
| | 1 ell | = 45 inches <br> = 1.143 m | Old English unit. |
| | 1 foot | = 0.3048 m | Ancient Russian imperial unit. Equal 1 English foot. |
| | 1 foot | ≈ 0.3 m | Roman unit. |
| | 1 palmo | = 8.25 inches <br> = 0.20955 m | Old Spanish unit. |
| | 1 sazhen | = 3 arshins <br> = 2.336 m | Ancient Russian imperial unit. <br> Equal 7 feet. |
| | 1 vara | = 33 inches <br> = 0.8382 m | Old Spanish unit. |
| | 1 vershok | = 44.45 E−3 m | Ancient Russian imperial unit. |
| | 1 versta (or verst) | = 500 sazhens <br> = 1,066.8 m | Ancient Russian imperial unit. |
| Mass/Weight: | 1 dolya | = 44.43 E−6 kg | Ancient Russian imperial unit. |
| | 1 funt | = 96 zolotniks <br> = 409.5 E−3 kg | Ancient Russian imperial unit. <br> Russian pound. |
| | 1 pud (or pood) | = 40 funts <br> = 16.38 kg | Ancient Russian imperial unit. |
| | 1 scruple | = 1.295 E−3 kg | Ancient Roman unit of weight (and also unit of coinage weight). |
| | 1 zolotnik | = 96 dolyas <br> = 4.266 E−3 kg | Ancient Russian imperial unit. |

## C.3    SI UNITS AND CONVERSION FACTORS

| Quantity (1) | Unit (symbol) (2) | Conversion (3) | Comments (4) |
|---|---|---|---|
| Length: | 1 inch (in) | $= 25.4 \, 10^{-3}$ m | Exactly. |
| | 1 foot (ft) | $= 0.3048$ m | Exactly. |
| | 1 yard (yd) | $= 0.9144$ m | Exactly. |
| | 1 mile | $= 1.609.344$ m | Exactly. |
| | 1 nautical mile | $= 1,852$ m | Exactly. |
| Area: | 1 square inch (in$^2$) | $= 6.4516 \, 10^{-4}$ m$^2$ | Exactly. |
| | 1 square foot (ft$^2$) | $= 0.09290306$ m$^2$ | Exactly. |
| Volume: | 1 cubic inch (in$^3$) | $= 16.387064 \, 10^{-6}$ m$^3$ | Exactly. |
| | 1 cubic foot (ft$^3$) | $= 28.3168 \, 10^{-3}$ m$^3$ | Exactly. |
| | 1 gallon UK (gal UK) | $= 4.54609 \, 10^{-3}$ m$^3$ | |
| | 1 gallon US (gal US) | $= 3.78541 \, 10^{-3}$ m$^3$ | |
| | 1 barrel US | $= 158.987 \, 10^{-3}$ m$^3$ | Petroleum,... |
| Velocity: | 1 foot per second (ft/s) | $= 0.3048$ m/s | Exactly. |
| | 1 mile per hour (mph) | $= 0.44704$ m/s | Exactly. |
| | 1 knot | $= 1$ naut. mile per hour | |
| Acceleration: | 1 foot per second squared (ft/s$^2$) | $= 0.3048$ m/s$^2$ | Exactly. |
| Mass: | 1 pound (lb or lbm) | $= 0.45359237$ kg | Exactly. |
| | 1 ton UK | $= 1016.05$ kg | |
| | 1 ton US | $= 907.185$ kg | |
| Density: | 1 pound per cubic foot (lb/ft$^3$) | $= 16.0185$ kg/m$^3$ | |
| Force: | 1 kilogram-force (kgf) | $= 9.80665$ N (exactly) | Exactly. |
| | 1 pound force (lbf) | $= 4.4482216152605$ N | |
| Moment of force: | 1 foot pound force (ft.lbf) | $= 1.35582$ N.m | |
| Pressure: | 1 Pascal (Pa) | $= 1$ N/m$^2$ | |
| | 1 standard atmosphere (atm) | $= 101325$ Pa | Exactly. |
| | 1 bar | $= 10^5$ Pa | Exactly. |
| | 1 Torr | $= 133.322$ Pa | |
| | 1 conventional metre of water (m of H$_2$O) | $= 9.80665 \, 10^3$ Pa | Exactly. |
| | 1 conventional metre of Mercury (m of Hg) | $= 1.33322 \, 10^5$ Pa | |
| | 1 Pound per Square Inch (PSI) | $= 6.8947572 \, 10^3$ Pa | |
| Temperature : | T (Celsius) | $= $ T (Kelvin) $- 273.16$ | 0 Celsius is 0.01 K below the temperature of the triple point of water. |
| | T (Fahrenheit) | $= $ T (Celsius) $\times \dfrac{9}{5} + 32$ | |
| | T (Rankine) | $= \dfrac{9}{5} \times$ T (Kelvin) | |
| Dynamic viscosity: | 1 Pa.s | $= 0.006720$ lbm/ft/s | |
| | 1 Pa.s | $= 10$ Poises | |
| Kinematic viscosity: | 1 square foot per second (ft$^2$/s) | $= 0.0929030$ m$^2$/s | |
| | 1 m$^2$/s | $= 10.7639$ ft$^2$/s | |
| | 1 m$^2$/s | $= 10^4$ Stokes | |

## C.3    SI UNITS AND CONVERSION FACTORS (CONTINUED)

| Quantity (1) | Unit (symbol) (2) | Conversion (3) | Comments (4) |
|---|---|---|---|
| Work energy: | 1 Joule (J) | $= 1$ N.m | |
| | 1 Joule (J) | $= 1$ W.s | |
| | 1 Watt hour (W.h) | $= 3.600\ 10^3$ J | Exactly. |
| | 1 electronvolt (eV) | $= 1.60219\ 10^{-19}$ J | |
| | 1 foot pound force (ft.lbf) | $= 1.35582$ J | |
| Power: | 1 Watt (W) | $= 1$ J/s | |
| | 1 foot pound force per second (ft.lbf/s) | $= 1.35582$ W | |
| | 1 horsepower (hp) | $= 745.700$ W | |
| Dissolved gas concentration | 1 parts per million (ppm) | $= 1$ mg/L | Exactly. |

Basic references: Degremont (1979), International Organization for Standardization (1979).

# Mathematics

## SUMMARY

## D.1   INTRODUCTION

### D.1.1   Bibliography

[D1]    CRC Standard Mathematical Tables
        W.H. Beyer
        *CRC Press Inc.*, Boca Raton, Florida, USA, 1982.
        Library of Congress call number: QA 47 .C5 1981 2

[D2]    Mathematical Handbook for Scientist and Engineers
        G.A. Korn and T.M. Korn
        *McGraw-Hill Book Comp.*, New York, USA, 1961
        Library of Congress call number: QA 27 .K74 1961 2

[D3]    Mathematical Handbook of Formulas and Tables
        M.R. Spiegel
        *McGraw-Hill Inc.*, New York, USA, 1968.
        Library of Congress call number: QA 41 .S75 1968

### D.1.2   Notation

| | |
|---|---|
| x, y, z | Cartesian coordinates; |
| r, $\theta$, z | polar coordinates; |
| $\dfrac{\partial}{\partial x}, \dfrac{\partial}{\partial y}, \dfrac{\partial}{\partial z}, \dfrac{\partial}{\partial t}$ | partial differentials; |
| $\dfrac{D}{Dt}$ | absolute derivative; |
| $\delta_{ij}$ | identity matrix element: $\delta_{ii} = 1$ and $\delta_{ij} = 0$ (for i different of j); |
| N! | N-factorial: $N! = 1 \times 2 \times 3 \times 4 \times \cdots \times (N-1) \times N$ |

### D.1.3   Constants

e     constant such as $LN(e) = 1$: $e = 2.71828182845904523536028$7;

$\pi$     $\pi = 3.14159265358979323846264$3;

$\sqrt{2}$     $\sqrt{2} = 1.414213562373095048$8;

$\sqrt{3}$     $\sqrt{3} = 1.732050807568877293$5

## D.2   VECTOR OPERATIONS

### D.2.1   Definitions

Considering a rectangular Cartesian system any point M is characterised by its coordinates (x, y, z). The cylindrical coordinates of the point M are (r, $\theta$, z) where:

$$r^2 = x^2 + y^2$$

$$\tan \theta = \frac{y}{x}$$

The Cartesian components of a vector $\overrightarrow{A}$ are: ($A_x$; $A_y$; $A_z$). The polar (or cylindrical) components of a vector $\overrightarrow{A}$ are ($A_r$; $A_\theta$; $A_z$).

### D.2.2   Vector operations

#### D.2.2.1   Scalar product of two vectors

$$\overrightarrow{A} \times \overrightarrow{B} = |A| \times |B| \times \cos(\overrightarrow{A}, \overrightarrow{B})$$

where $|A| = \sqrt{A_x^2 + A_y^2 + A_z^2}$. Two non zero vectors are perpendicular each other if and only if their scalar product is null.

#### D.2.2.2   Vector product

$$\overrightarrow{A} \wedge \overrightarrow{B} = \overrightarrow{i}\,(A_y \times B_z - A_z \times B_y) + \overrightarrow{j}\,(A_z \times B_x - A_x \times B_z) + \overrightarrow{k}\,(A_x \times B_y - A_y \times B_x)$$

## D.3   DIFFERENTIAL

### D.3.1   Absolute differential

The absolute differential $\frac{D}{Dt}$ of a scalar $\Phi(r)$ along the curve: $r = r(t)$ is, at each point (r) of the curve, the rate of change of $\Phi(r)$ with respect to the parameter t as r varies as a function of t:

$$\frac{D\Phi}{Dt} = \left(\frac{Dr}{Dt} \times \nabla\right)\Phi = \frac{\partial x}{\partial t} \times \frac{\partial \Phi}{\partial x} + \frac{\partial y}{\partial t} \times \frac{\partial \Phi}{\partial y} + \frac{\partial z}{\partial t} \times \frac{\partial \Phi}{\partial z}$$

And if $\Phi$ depends explicitly on t [$\Phi = \Phi(r, t)$] then:

$$\frac{D\Phi}{Dt} = \frac{\partial \Phi}{\partial t} + \left(\frac{Dr}{Dt} \times \nabla\right)\Phi = \frac{\partial \Phi}{\partial t} + \frac{\partial x}{\partial t} \times \frac{\partial \Phi}{\partial x} + \frac{\partial y}{\partial t} \times \frac{\partial \Phi}{\partial y} + \frac{\partial z}{\partial t} \times \frac{\partial \Phi}{\partial z}$$

and this may be rewritten as:

$$\frac{D\Phi}{Dt} = \frac{\partial\Phi}{\partial t} + \vec{V} \times \nabla\Phi = \frac{\partial\Phi}{\partial t} + V_x \times \frac{\partial\Phi}{\partial x} + V_y \times \frac{\partial\Phi}{\partial y} + V_z \times \frac{\partial\Phi}{\partial z}$$

## D.3.2 Differential operators

### D.3.2.1 Gradient

$$\overrightarrow{grad}\ \Phi(x,y,z) = \nabla\Phi(x,y,z) = \vec{i}\frac{\partial\Phi}{\partial x} + \vec{j}\frac{\partial\Phi}{\partial y} + \vec{k}\frac{\partial\Phi}{\partial z} \qquad \text{Cartesian coordinate}$$

### D.3.2.2 Divergence

$$div\ \overrightarrow{F(x,y,z)} = \nabla\overrightarrow{F(x,y,z)} = \frac{\partial F_x}{\partial x} + \frac{\partial F_y}{\partial y} + \frac{\partial F_z}{\partial z}$$

### D.3.2.3 Curl

$$\overrightarrow{curl}\ \overrightarrow{F(x,y,z)} = \nabla \wedge \overrightarrow{F(x,y,z)} = \vec{i}\left(\frac{\partial F_z}{\partial y} - \frac{\partial F_y}{\partial z}\right) + \vec{j}\left(\frac{\partial F_x}{\partial z} - \frac{\partial F_z}{\partial x}\right) + \vec{k}\left(\frac{\partial F_y}{\partial x} - \frac{\partial \vec{F}_x}{\partial y}\right)$$

### D.3.2.4 Laplacian operator

$$\Delta\Phi(x,y,z) = \nabla \times \nabla\Phi(x,y,z) = div\ \overrightarrow{grad}\ \Phi(x,y,z) = \frac{\partial^2\Phi}{\partial x^2} + \frac{\partial^2\Phi}{\partial y^2} + \frac{\partial^2\Phi}{\partial z^2} \qquad \text{Laplacian of scalar}$$

$$\Delta\overrightarrow{F(x,y,z)} = \nabla \times \nabla\overrightarrow{F(x,y,z)} = \vec{i}\,\Delta F_x + \vec{j}\,\Delta F_y + \vec{k}\,\Delta F_z \qquad \text{Laplacian of vector}$$

*Polar coordinates*

$$\overrightarrow{grad}\ \Phi(r,\theta,z) = \left(\frac{\partial\Phi}{\partial r}; \frac{1}{r} \times \frac{\partial\Phi}{\partial\theta}; \frac{\partial\Phi}{\partial z}\right)$$

$$div\ \overrightarrow{F(r,\theta,z)} = \frac{1}{r} \times \left(\frac{\partial(r \times F_r)}{\partial r} + \frac{\partial F_\theta}{\partial\theta} + r \times \frac{\partial F_z}{\partial z}\right)$$

$$\overrightarrow{curl}\ \overrightarrow{F(r,\theta,z)} = \left(\frac{1}{r} \times \frac{\partial F_z}{\partial\theta} - \frac{\partial F_\theta}{\partial z}; \frac{\partial F_r}{\partial z} - \frac{\partial F_z}{\partial r}; \frac{1}{r} \times \frac{\partial(r \times F_\theta)}{\partial r} - \frac{1}{r} \times \frac{\partial F_r}{\partial\theta}\right)$$

$$\Delta\Phi(r,\theta,z) = \frac{1}{r} \times \frac{\partial}{\partial r}\left(r \times \frac{\partial\Phi}{\partial r}\right) + \frac{1}{r^2} \times \frac{\partial^2\Phi}{\partial\theta^2} + \frac{\partial^2\Phi}{\partial z^2}$$

## D.3.3 Operator relationship

### D.3.3.1 Gradient

$$\overrightarrow{grad}(f + g) = \overrightarrow{grad}\ f + \overrightarrow{grad}\ g$$

$$\overrightarrow{grad}(f \times g) = g \times \overrightarrow{grad}\ f + f \times \overrightarrow{grad}\ g$$

where f and g are scalars.

### D.3.3.2  *Divergence*

$$\text{div}(\vec{F} + \vec{G}) = \text{div}\,\vec{F} + \text{div}\,\vec{G}$$

$$\text{div}(f \times \vec{F}) = f \times \text{div}\,\vec{F} + \vec{F} \times \overrightarrow{\text{grad}}\,f$$

where f is a scalar.

$$\text{div}(\vec{F} \wedge \vec{G}) = \vec{G} \times \overrightarrow{\text{curl}}\,\vec{F} - \vec{F} \times \overrightarrow{\text{curl}}\,\vec{G}$$

### D.3.3.3  *Curl*

$$\overrightarrow{\text{curl}}(\vec{F} + \vec{G}) = \overrightarrow{\text{curl}}\,\vec{F} + \overrightarrow{\text{curl}}\,\vec{G}$$

$$\overrightarrow{\text{curl}}(f \times \vec{F}) = f \times \overrightarrow{\text{curl}}\,\vec{F} - \vec{G} \wedge \overrightarrow{\text{grad}}\,f$$

where f is a scalar.

$$\overrightarrow{\text{curl}}(\overrightarrow{\text{grad}}\,f) = 0$$

$$\text{div}(\overrightarrow{\text{curl}}\,\vec{F}) = 0$$

### D.3.3.4  *Laplacian*

$$\Delta f = \text{div}\,\overrightarrow{\text{grad}}\,f$$

$$\Delta \vec{F} = \overrightarrow{\text{grad}}\,\text{div}\,\vec{F} - \overrightarrow{\text{curl}}(\overrightarrow{\text{curl}}\,\vec{F})$$

where f is a scalar.

$$\Delta(f + g) = \Delta f + \Delta g$$

$$\Delta(\vec{F} + \vec{G}) = \Delta \vec{F} + \Delta \vec{G}$$

$$\Delta(f \times g) = g \times \Delta f + f \times \Delta g + 2 \times \overrightarrow{\text{grad}}\,f \times \overrightarrow{\text{grad}}\,g$$

where f and g are scalars.

## D.4  TRIGONOMETRIC FUNCTIONS

### D.4.1  Definitions

The basic definitions may be stated in terms of right-angled triangle geometry:

$$\sin(x) = \frac{\text{opposite}}{\text{hypotenuse}}$$

$$\cos(x) = \frac{\text{adjacent}}{\text{hypotenuse}}$$

$$\tan(x) = \frac{\sin(x)}{\cos(x)} = \frac{\text{opposite}}{\text{adjacent}}$$

$$\cot(x) = \frac{\cos(x)}{\sin(x)} = \frac{\text{adjacent}}{\text{opposite}}$$

$$\sec(x) = \frac{1}{\cos(x)}$$

$$\csc(x) = \frac{1}{\sin(x)}$$

where sec and csc are respectively called the secant and cosecant functions.
   The power-series expansions of these functions are:

$$\sin(x) = x - \frac{x^3}{3!} + \frac{x^5}{5!} - \frac{x7}{7!} + \cdots \qquad \text{for any value of x}$$

$$\cos(x) = 1 - \frac{x^2}{2!} + \frac{x^4}{4!} - \frac{x^6}{6!} + \cdots \qquad \text{for any value of x}$$

$$\tan(x) = x + \frac{x^3}{3} + \frac{2}{15} \times x^5 + \frac{17}{315} \times x^7 + \cdots \qquad \text{for } -\pi/2 < x < \pi/2$$

$$\cot(x) = \frac{1}{x} - \frac{x}{3} - \frac{1}{45} \times x^3 - \frac{2}{945} \times x^5 + \cdots \qquad \text{for } 0 < \text{Abs}(x) < \pi$$

where: $n! = n \times (n-1) \times (n-2) \times \cdots \times 1$.

## D.4.2   Relationships

$$\tan(x) = \frac{\sin(x)}{\cos(x)}$$

$$\cot(x) = \frac{1}{\tan(x)}$$

$$\sin^2(x) + \cos^2(x) = 1$$

$$\frac{1}{\sin^2(x)} - \cot^2(x) = 1$$

$$\frac{1}{\cos^2(x)} - \tan^2(x) = 1$$

$$\sec(x) = \sqrt{1 + \tan^2(x)}$$

$$\csc(x) = \sqrt{1 + \cot^2(x)}$$

$$\sin(-x) = -\sin(x)$$

$$\cos(-x) = \cos(x)$$

$$\tan(-x) = -\tan(x)$$

$$\cot(-x) = -\cot(x)$$

$$\sec(-x) = \sec(x)$$

$$\csc(-x) = -\csc(x)$$

$$\sin(x) = \cos\left(\frac{\pi}{2} - x\right) \quad \text{for } 0 < x < \pi/2$$

$$\cos(x) = \sin\left(\frac{\pi}{2} - x\right) \quad \text{for } 0 < x < \pi/2$$

$$\tan(x) = \cot\left(\frac{\pi}{2} - x\right) \quad \text{for } 0 < x < \pi/2$$

$$\cot(x) = \tan\left(\frac{\pi}{2} - x\right) \quad \text{for } 0 < x < \pi/2$$

$$\sec\left(x + \frac{\pi}{2}\right) = -\csc(x) \quad \text{for } 0 < x < \pi/2$$

$$\csc\left(x + \frac{\pi}{2}\right) = \sec(x) \quad \text{for } 0 < x < \pi/2$$

$$\sin(x + y) = \sin(x) \times \cos(y) + \cos(x) \times \sin(y)$$

$$\sin(x - y) = \sin(x) \times \cos(y) - \cos(x) \times \sin(y)$$

$$\cos(x + y) = \cos(x) \times \cos(y) - \sin(x) \times \sin(y)$$

$$\cos(x - y) = \cos(x) \times \cos(y) + \sin(x) \times \sin(y)$$

$$\tan(x + y) = \frac{\tan(x) + \tan(y)}{1 - \tan(x) \times \tan(y)}$$

$$\tan(x - y) = \frac{\tan(x) - \tan(y)}{1 + \tan(x) \times \tan(y)}$$

$$\cot(x + y) = \frac{\cot(x) \times \cot(y) - 1}{\cot(x) + \cot(y)}$$

$$\cot(x - y) = \frac{\cot(x) \times \cot(y) + 1}{\cot(x) - \cot(y)}$$

$$\sin(x) + \sin(y) = 2 \times \sin\left(\frac{x + y}{2}\right) \times \cos\left(\frac{x - y}{2}\right)$$

$$\sin(x) - \sin(y) = 2 \times \cos\left(\frac{x + y}{2}\right) \times \sin\left(\frac{x - y}{2}\right)$$

$$\cos(x) + \cos(y) = 2 \times \cos\left(\frac{x + y}{2}\right) \times \cos\left(\frac{x - y}{2}\right)$$

$$\cos(x) - \cos(y) = -2 \times \sin\left(\frac{x + y}{2}\right) \times \sin\left(\frac{x - y}{2}\right)$$

$$\sin(x) \times \sin(y) = \frac{1}{2} \times (\cos(x - y) - \cos(x + y))$$

$$\cos(x) \times \cos(y) = \frac{1}{2} \times (\cos(x - y) + \cos(x + y))$$

$$\sin(x) \times \cos(y) = \frac{1}{2} \times (\sin(x - y) + \sin(x + y))$$

### D.4.2.1  Derivatives

$$d(\sin(x)) = \cos(x) \times dx$$

$$d(\cos(x)) = -\sin(x) \times dx$$

$$d(\tan(x)) = \frac{1}{\cos^2(x)} \times dx$$

$$d(\cot(x)) = \frac{-1}{\sin^2(x)} \times dx$$

$$d\left(\frac{1}{\cos(x)}\right) = \frac{\tan(x)}{\cos(x)} \times dx$$

$$d\left(\frac{1}{\sin(x)}\right) = \frac{-\cot(x)}{\sin(x)} \times dx$$

## D.4.3  Inverse trigonometric functions

The inverse trigonometric functions are expressed as $\sin^{-1}$, $\cos^{-1}$, $\tan^{-1}$ and $\cot^{-1}$. The power-series expansions of these functions are:

$$\sin^{-1}(x) = x + \frac{1}{2} \times \frac{x^3}{3} + \frac{1 \times 3}{2 \times 4} \times \frac{x^5}{5} + \frac{1 \times 3 \times 5}{2 \times 4 \times 6} \times \frac{x^7}{7} + \cdots \qquad \text{for } -1 < x < 1$$

$$\cos^{-1}(x) = \frac{\pi}{2} - \left(x + \frac{1}{2} \times \frac{x^3}{3} + \frac{1 \times 3}{2 \times 4} \times \frac{x^5}{5} + \frac{1 \times 3 \times 5}{2 \times 4 \times 6} \times \frac{x^7}{7} + \cdots\right) \qquad \text{for } -1 < x < 1$$

$$\tan^{-1}(x) = x - \frac{x^3}{3} + \frac{x^5}{5} - \frac{x^7}{7} + \cdots \qquad \text{for } -1 < x < 1$$

$$\cot^{-1}(x) = \frac{\pi}{2} - \left(x - \frac{x^3}{3} + \frac{x^5}{5} - \frac{x^7}{7} + \cdots\right) \qquad \text{for } -1 < x < 1$$

The following relationships can be established:

$$\sin^{-1}(-x) = -\sin^{-1}(x)$$

$$\cos^{-1}(-x) = \pi - \cos^{-1}(x)$$

$$\tan^{-1}(-x) = -\tan^{-1}(x)$$

$$\cot^{-1}(-x) = \pi - \cot^{-1}(x)$$

$$\sin^{-1}(x) + \cos^{-1}(x) = \frac{\pi}{2}$$

$$\tan^{-1}(x) + \cot^{-1}(x) = \frac{\pi}{2}$$

$$\sin^{-1}\left(\frac{1}{x}\right) + \cos^{-1}\left(\frac{1}{x}\right) = \frac{\pi}{2}$$

$$\sin^{-1}(x) = \tan^{-1}\left(\frac{1}{\sqrt{1 - x^2}}\right)$$

$$\cot^{-1}(x) = \frac{\pi}{2} - \tan^{-1}(x^2)$$

$$\sin^{-1}(x) + \sin^{-1}(y) = \sin^{-1}(x \times \sqrt{1 - y^2} + y \times \sqrt{1 - x^2})$$

$$\sin^{-1}(x) - \sin^{-1}(y) = \sin^{-1}(x \times \sqrt{1 - y^2} - y \times \sqrt{1 - x^2})$$

$$\cos^{-1}(x) + \cos^{-1}(y) = \cos^{-1}(x \times y - \sqrt{1 - x^2} \times \sqrt{1 - y^2})$$

$$\cos^{-1}(x) - \cos^{-1}(y) = \cos^{-1}(x \times y + \sqrt{1 - x^2} \times \sqrt{1 - y^2})$$

$$\tan^{-1}(x) + \tan^{-1}(y) = \tan^{-1}\left(\frac{x + y}{1 - x \times y}\right)$$

$$\tan^{-1}(x) - \tan^{-1}(y) = \tan^{-1}\left(\frac{x - y}{1 + x \times y}\right)$$

### D.4.3.1   Derivatives

$$d(\sin^{-1}(x)) = \frac{1}{\sqrt{1 - x^2}} \times dx \qquad\qquad \text{for } -\pi/2 < \sin^{-1}(x) < \pi/2$$

$$d(\cos^{-1}(x)) = \frac{-1}{\sqrt{1 - x^2}} \times dx \qquad\qquad \text{for } 0 < \cos^{-1}(x) < \pi$$

$$d(\tan^{-1}(x)) = \frac{1}{1 + x^2} \times dx \qquad\qquad \text{for } -\pi/2 < \tan^{-1}(x) < \pi/2$$

$$d(\cot^{-1}(x)) = \frac{-1}{1 + x^2} \times dx \qquad\qquad \text{for } 0 < \sin^{-1}(x) < \pi$$

## D.5   HYPERBOLIC FUNCTIONS

### D.5.1   Definitions

There are six hyperbolic functions that are comparable to the trigonometric functions. They are designated by adding the letter h to the trigonometric abbreviations: sinh, cosh, tanh, coth, sech, csch.

The basic definitions may be stated in terms of exponentials:

$$\sinh(x) = \frac{1}{\mathrm{csch}(x)} = \frac{e^x - e^{-x}}{2}$$

$$\cosh(x) = \frac{1}{\mathrm{sech}(x)} = \frac{e^x + e^{-x}}{2}$$

$$\tanh(x) = \frac{1}{\coth(x)} = \frac{e^x - e^{-x}}{e^x + e^{-x}}$$

The power-series expansion of these functions are:

$$\sinh(x) = x + \frac{x^3}{3!} + \frac{x^5}{5!} + \frac{x^7}{7!} + \cdots \qquad \text{for any value of } x$$

$$\cosh(x) = 1 + \frac{x^2}{2!} + \frac{x^4}{4!} + \frac{x^6}{6!} + \cdots \qquad \text{for any value of } x$$

$$\tanh(x) = x - \frac{x^3}{3} + \frac{2}{15} \times x^5 - \frac{17}{315} \times x^7 + \cdots \qquad \text{for } -\pi/2 < x < \pi/2$$

$$\coth(x) = \frac{1}{x} + \frac{x}{3} - \frac{1}{45} \times x^3 + \frac{2}{945} \times x^5 + \cdots \qquad \text{for } 0 < \mathrm{Abs}(x) < \pi$$

where: $n! = n \times (n-1) \times (n-2) \times \cdots \times 1$.

Typical values of the hyperbolic functions are listed below:

| x (1) | cosh(x) (2) | sinh(x) (3) | tanh(x) (4) |
|---|---|---|---|
| 0 | 1.0 | 0 | 0 |
| 0.087 | 1.004 | 0.087 | 0.087 |
| 0.175 | 1.015 | 0.175 | 0.173 |
| 0.349 | 1.062 | 0.356 | 0.336 |
| 0.524 | 1.140 | 0.548 | 0.480 |
| 0.698 | 1.254 | 0.756 | 0.603 |
| 0.873 | 1.406 | 0.988 | 0.703 |
| 1.047 | 1.600 | 1.249 | 0.781 |
| 1.222 | 1.844 | 1.549 | 0.840 |
| 1.396 | 2.144 | 1.896 | 0.885 |
| 1.571 | 2.509 | 2.301 | 0.917 |
| 1.745 | 2.951 | 2.777 | 0.941 |
| 1.920 | 3.483 | 3.337 | 0.958 |
| 2.094 | 4.122 | 3.999 | 0.970 |
| 2.269 | 4.886 | 4.783 | 0.979 |
| 2.443 | 5.800 | 5.713 | 0.985 |
| 2.618 | 6.891 | 6.818 | 0.989 |
| 2.793 | 8.192 | 8.130 | 0.993 |
| 2.967 | 9.743 | 9.692 | 0.995 |
| 3.142 | 11.592 | 11.549 | 0.996 |

## D.5.2 Relationships

$$\sinh(-x) = -\sinh(x)$$

$$\cosh(-x) = \mathrm{csch}(x)$$

$$\cosh^2(x) - \sinh^2(x) = 1$$

$$1 - \tanh^2(x) = \mathrm{sech}^2(x)$$

$$1 - \coth^2(x) = -\mathrm{csch}^2(x)$$

$$\sinh(x + y) = \sinh(x) \times \cosh(y) + \cosh(x) \times \sinh(y)$$

$$\sinh(x - y) = \sinh(x) \times \cosh(y) - \cosh(x) \times \sinh(y)$$

$$\cosh(x + y) = \cosh(x) \times \cosh(y) + \sinh(x) \times \sinh(y)$$

$$\cosh(x - y) = \cosh(x) \times \cosh(y) - \sinh(x) \times \sinh(y)$$

$$\tanh(x + y) = \frac{\tanh(x) + \tanh(y)}{1 + \tanh(x) \times \tanh(y)}$$

$$\tanh(x - y) = \frac{\tanh(x) - \tanh(y)}{1 - \tanh(x) \times \tanh(y)}$$

$$\sinh(2 \times x) = 2 \times \sinh(x) \times \cosh(x)$$

$$\cosh(2 \times x) = \cosh^2(x) + \sinh^2(x) = 2 \times \cosh^2(x) - 1 = 1 + 2 \times \sinh^2(x)$$

$$\tanh(2 \times x) = \frac{2 \times \tanh(x)}{1 + \tanh^2(x)}$$

$$\sinh\left(\frac{x}{2}\right) = \sqrt{\frac{1}{2} \times (\cosh(x) - 1)}$$

$$\cosh\left(\frac{x}{2}\right) = \sqrt{\frac{1}{2} \times (\cosh(x) + 1)}$$

$$\tanh\left(\frac{x}{2}\right) = \frac{\cosh(x) - 1}{\sinh(x)} = \frac{\sinh(x)}{\cosh(x) + 1}$$

### D.5.2.1 Derivatives

$$d(\sinh(x)) = \cosh(x) \times dx$$

$$d(\cosh(x)) = \sinh(x) \times dx$$

$$d(\tanh(x)) = \mathrm{sech}^2(x) \times dx$$

$$d(\coth(x)) = -\mathrm{csch}^2(x) \times dx$$

$$d(\mathrm{sech}(x)) = -\mathrm{sech}(x) \times \tanh(x) \times dx$$

$$d(\mathrm{csch}(x)) = -\mathrm{csch}(x) \times \coth(x) \times dx$$

### D.5.3 Inverse hyperbolic functions

The inverse hyperbolic functions are expressible in terms of logarithms.

$$\sinh^{-1}(x) = Ln(x + \sqrt{x^2 + 1}) \qquad\qquad \text{for any value of x}$$

$$\cosh^{-1}(x) = Ln(x + \sqrt{x^2 - 1}) \qquad\qquad \text{for } x > 1$$

$$\tanh^{-1}(x) = \frac{1}{2} \times Ln\left(\frac{1+x}{1-x}\right) \qquad\qquad \text{for } x^2 < 1$$

$$\coth^{-1}(x) = \frac{1}{2} \times Ln\left(\frac{x+1}{x-1}\right) \qquad\qquad \text{for } x^2 > 1$$

$$\text{sech}^{-1}(x) = Ln\left(\frac{1}{x} + \sqrt{\frac{1}{x^2} - 1}\right) \qquad\qquad \text{for } x < 1$$

$$\text{csch}^{-1}(x) = Ln\left(\frac{1}{x} + \sqrt{\frac{1}{x^2} + 1}\right) \qquad\qquad \text{for any value of x}$$

The following relationships can be established:

$$\sinh^{-1}(-x) = \sinh^{-1}(x)$$

$$\tanh^{-1}(-x) = -\tanh^{-1}(x)$$

$$\coth^{-1}(-x) = -\coth^{-1}(x)$$

$$\coth^{-1}(x) = \tanh^{-1}\left(\frac{1}{x}\right)$$

#### D.5.3.1 Derivatives

$$d(\sinh^{-1}(x)) = \frac{dx}{\sqrt{x^2 + 1}}$$

$$d(\cosh^{-1}(x)) = \frac{dx}{\sqrt{x^2 - 1}}$$

$$d(\tanh^{-1}(x)) = \frac{dx}{1 - x^2} = d(\coth^{-1}(x))$$

$$d(\text{sech}^{-1}(x)) = \frac{-dx}{x \times \sqrt{1 - x^2}}$$

$$d(\text{csch}^{-1}(x)) = \frac{-dx}{x \times \sqrt{1 + x^2}}$$

## D.6  COMPLEX NUMBERS

### D.6.1  Definition

A complex number z consists of two distinct scalar (i.e. real) parts a and b, and is written in the form:

$$z = a + i \times b$$

where: $i = \sqrt{-1}$. The first part a is called the *real part* and the second part b is called the *imaginary part* of the complex number.

The *modulus* (i.e. absolute value) of a complex number designated r and defined as:

$$r = \sqrt{x^2 + y^2}$$

The *argument* $\theta$ of the complex number is the position vector measured from the positive x-axis in an anti-clockwise direction:

$$\theta = \tan^{-1}\left(\frac{y}{x}\right)$$

An alternative mode of expressing a complex number is:

$$z = x + i \times y = r \times (\cos\theta + i \times \sin\theta) = r \times e^{i \times \theta}$$

### Notes

- The complex number i may be regarded as the complex number with a modulus of unity and argument $\pi/2$:

$$i = e^{i \times (\pi/2)}$$

- Multiplication of a complex number by i does not affect the modulus but increases the argument by $\pi/2$:

$$i \times z = i \times r \times e^{i \times \theta} = r \times e^{i \times (\theta + \pi/2)}$$

- Division of a complex number by i decreases the argument by $\pi/2$:

$$\frac{z}{i} = \frac{r \times e^{i \times \theta}}{i} = r \times e^{i \times (\theta - \pi/2)}$$

- If $i = \sqrt{-1}$, it satisfies:

$$i = -\frac{1}{i}$$

The complex number $-1/i$ is the conjugate of the complex number i.

### D.6.2  Basic properties

Various operations involving complex numbers are:

- addition:

$$z_1 + z_2 = (x_1 + x_2) + i \times (y_1 + y_2)$$

- multiplication:

$$z_1 \times z_2 = r_1 \times r_2 \times e^{i \times (\theta_1 + \theta_2)}$$

- square:

$$z^2 = r^2 \times e^{i \times 2 \times \theta} = (x^2 - y^2) + i \times x \times y$$

- power:

$$z^n = r^n \times e^{i \times n\theta}$$

- division:

$$\frac{z_1}{z_2} = \frac{r_1}{r_2} \times e^{i \times (\theta_1 - \theta_2)}$$

- multiplication by i:

$$i \times z = -y + i \times x = r \times e^{i \times (\theta + \pi/2)}$$

- division by i:

$$\frac{1}{i} \times z = y - i \times x = r \times e^{i \times (\theta - \pi/2)}$$

- logarithm:

$$Ln(z) = Ln(r) + i \times \theta = \frac{1}{2} \times Ln(x^2 + y^2) + i \times \tan^{-1}\left(\frac{y}{x}\right) \qquad \text{for } 0 \le \theta < 2 * \pi$$

### D.6.3   Conjugate

If a complex number is: $z = x + i \times y$, the **conjugate number** is defined as:

$$\bar{z} = r \times e^{i \times (\theta - \pi/2)}$$

The main properties of a conjugate are:

$$z \times \bar{z} = r^2 \qquad \text{(real)}$$
$$z + \bar{z} = 2 \times x \qquad \text{(real)}$$
$$z - \bar{z} = 2 \times i \times y \qquad \text{(imaginary)}$$

### D.6.4   Advanced properties

Advanced operations involving complex numbers are:

$$\sinh(i \times x) = i \times \sin(x) \qquad\qquad \sinh(x) = -i \times \sin(i \times x)$$
$$\cosh(i \times x) = \cos(x) \qquad\qquad \cosh(x) = \cos(i \times x)$$

$$\sin(z) = \sin(x) \times \cosh(y) + i \times \cos(x) \times \sinh(y)$$

$$\cos(z) = \cos(x) \times \cosh(y) - i \times \sin(x) \times \sinh(y)$$

$$\sinh(z) = \sinh(x) \times \cos(y) + i \times \cosh(x) \times \sin(y)$$

$$\cosh(z) = \cosh(x) \times \cos(y) + i \times \sinh(x) \times \sin(y)$$

Further properties are:

$$\sinh(x + i \times 2 \times \pi \times n) = \sinh(x)$$

$$\cosh(x + i \times 2 \times \pi \times n) = \cosh(x)$$

$$\tanh(x + i \times \pi \times n) = \tanh(x)$$

where n is an integer.

# The Software 2d Flow+

## E.1  PRESENTATION

The software 2DFlow+ is a Windows-based product developed by Dynaflow Inc. {http://www.dynaflow-inc.com/}. It provides a graphical solution of the Laplace equations for irrotational flow motion of ideal fluid. It was "specially designed for students of fluid mechanics who want to understand and experiment with potential flows".

Based upon the principle of superposition, simple flow patterns (source, vortex, uniform flow, doublet) may be placed using a mouse click. The software automatically calculates the stream function, velocity potential, velocity and pressure fields. The two-dimensional results are presented on-screen as contours, iso-contour and trajectories.

### Demonstration version

A demonstration version may be downloaded from Dynaflow Inc. website {www.dynaflow-inc.com/2DFlow/}. The demonstration version has most features of the software, but it cannot print the results.

The system requirements are: Windows 95 or 98, Windows NT 3.51 (Service Pack 5) and Windows NT 4.0 (Service Pack 3). The software works also with Windows Millenium, 2000 and XP.

## E.2  APPLICATION

Figure E.1 illustrates a typical application: i.e., the flow past a Rankine body. The settings are:

– uniform velocity V = 1 (angle $\alpha = 0$),
– a source (strength 5) at $(-3, 0)$, and
– a sink (strength 5) at $(+3, 0)$.

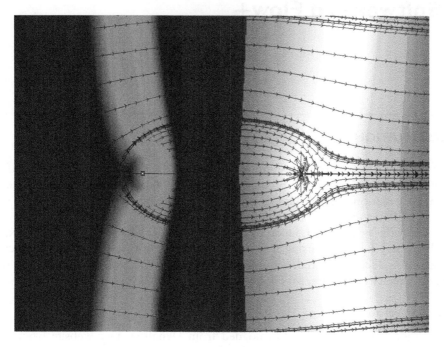

*Figure E.1*   Flow past a Rankine body.

## E.3   REFERENCES

{http://www.dynaflow-inc.com/}                          Dynaflow Inc.
{http://www.dynaflow-inc.com/DFlow/2dflow.htm}   Dynaflow 2DFlow+ software download

# Digital video movies

## F.I  PRESENTATION

The book aims to facilitate the advancement and diffusion of knowledge of the flow physics in ideal and real fluid flow applications. It contains a number of photographs and videos showing a range of flow phenomena. This appendix presents a series of movies taken by the author (unless stated). The movies illustrate some aspects of applied hydrodynamics and its applications into the real world. Each movie is described below.

All the movies are Copyrights Hubert Chanson.

### Video movie access

To view the supporting, please click on the URL below. A video player will open in your web browser.

## F.2  VIDEO MOVIES

| Title: | Butterfly flight |
|---|---|
| URL: | http://goo.gl/fRHQV |
| QR-code | |
| Duration: | 9 s |
| Description: | Butterfly near Zhoushui River on 11 Nov. 2008 in Nantou county (Taiwan). The wing area of a butterfly is very large relative to the body mass. The wing motion is basically limited to flapping, and the flapping frequency is relatively low and it can be seen in the video. |
| Relevant book sections: | Chapter 1 |
| Reference: | Tanaka, H., and Shimoyama, I. (2010). "Forward flight of swallowtail butterfly with simple flapping motion." *Bioinspiration & Biomimetics*, Vol. 5, paper 026003, 9 pages (DOI:10.1088/1748-3182/5/2/026003). |

| Title: | Para-gliding |
|---|---|
| URL: | http://goo.gl/EDbef |
| QR-code | |

| Duration: | 11 s |
|---|---|
| Description: | Para-gliding and landing in Pingtung county (Taiwan) on 16 Dec. 2008. The video shows a landing. Note how the sail angle of incidence increases shortly before touch down to increase both lift and drag. |
| Relevant book sections: | Chapters 1, I-1, I-3, I-6 |

| Title: | Minimum energy loss spillway inlet |
|---|---|
| URL: | http://goo.gl/vhsqq |
| QR-code | |
| Duration: | 10 s |
| Description: | Lake Kurwongbah and spillway (Brisbane, Australia) in operation on 29 Jan. 2013. The video illustrates the minimum energy loss (MEL) inlet design. The concept of Minimum Energy Loss (MEL) weir was developed to pass large floods with minimum energy loss, hence with minimum afflux. The flow in the approach channel is contracted through a streamlined chute and the channel width is minimum at the chute toe, just before impinging into thedownstream natural channel. The inlet and chute are streamlined to avoid significant form losses. Their design is often based upon a flow net analysis. |
| Relevant book sections: | Chapter I-3 |
| References: | McKay, G.R. (1971). "Design of Minimum Energy Culverts." *Research Report*, Dept. of Civil Eng., Univ. of Queensland, Brisbane, Australia, 29 pages and 7 plates. Chanson, H. (2003). "Minimum Energy Loss Structures in Australia: Historical Development and Experience." *Proc. 12th Nat. Eng. Heritage Conf.*, IEAust., Toowoomba Qld, Australia, N. Sheridan Ed., pp. 22–28 (ISBN 0-646-42775-X). |

| Title: | Hele-Shaw cell apparatus | |
|---|---|---|
| URL: | http://goo.gl/kgzZy (Low Resolution) | http://goo.gl/ITbaC (High Resolution) |
| QR-code | | |
| Duration: | 15 s | |
| Description: | Uniform flow (from right to left) past a 15% cambered foil perpendicular to the main flow direction at the University of Queensland on 27 March 2003 – The dye is injected upstream of the foil and the video illustrates the dye motion around the object. Note some slight dye dispersion downstream of body. Rounded edges were used to prevent any flow singularities. The fluid was water and vegetable dye was used. The foil shape was made out of rubber. | |
| Relevant book sections: | Chapter I-3 | |

| Title: | Hele-Shaw cell apparatus | |
|---|---|---|
| URL: | http://goo.gl/282DKc (Low Resolution) | http://goo.gl/SUC6J (High Resolution) |
| QR-code | | |
| Duration: | 15 s | |
| Description: | Uniform flow (from right to left) past a (thick rounded) plate perpendicular to flow direction at the University of Queensland on 27 March 2003. The video illustrates some dye dispersion downstream of body, while rounded edges were used to prevent singularities. The flow pattern may be compared with the flow normal to a flat plate with separation (Chapter I-7). | |
| Relevant book sections: | Chapters I-3, I-5, I-7 | |

| Title: | Shark swim |
|---|---|
| URL: | http://goo.gl/KPhn2q |
| QR-code | |
| Duration: | 7 s |
| Description: | Baby whale shark at National Museum of Marine Biology and Aquarium, Kenting, Pingtung county (Taiwan) on 15 Dec. 2008. The body of a whale shark is streamlined with a broad flattened head and a very large mouth. This baby whale shark weighted 700 kg at the time of the video. |
| Relevant book sections: | Chapter I-4 |

| Title: | Stingray swimming motion |
|---|---|
| URL: | http://goo.gl/oqRvg |
| QR-code | |
| Duration: | 23 s |
| Description: | Stingray at National Museum of Marine Biology and Aquarium, Kenting, Pingtung county (Taiwan) on 15 Dec. 2008. Stingray locomotion thrust is generated by undulatory waves passing down the enlarged pectoral fins. Swimming velocity is increased by increasing fin-beat frequency and wave speed. This video illustrates nicely the undulating fin motion of the stingray. |
| Relevant book sections: | Chapter I-4 |
| Reference: | Rosenberger, L.J., and Westneat, M.W. (1999). "Functional Morphology of Undulatory Pectoral Fin Locomotion in the Stingray *Taeniura Lymna* (Chondrichthyes: Dasyatidae)." *The Journal of Experimental Biology*, Vol. 202, pp. 3523–3539. |

| Title: | Penguin swimming |
|---|---|
| URL: | http://goo.gl/aVhNM |
| QR-code | |
| Duration: | 18 s |
| Description: | Penguin swimming at the National Museum of Marine Biology and Aquarium, Kenting, Pingtung county (Taiwan) on 15 Dec. 2008. Penguins use their wings to generate a swimming motion. They tend to keep their head hunched into the shoulders to maintain a streamlined body with minimum hydrodynamic resistance, when swimming under water. |
| Relevant book sections: | Chapter I-4 |

| Title: | Old windmill |
|---|---|
| URL: | http://goo.gl/VhCe7 |
| QR-code | |
| Duration: | 20 s |
| Description: | 1806 windmill at Moidrey, Bay of Mont Saint Michel (France) on 24 June 2008. The video shows the windmill operation after the roof was rotated to face the wind and the wings/blades were unfolded. The windmill is equipped with 1840 wings/blades based upon a system developed by the French engineer Berton. The Berton system consists of wooden-slatted sails which can be deployed or folded from inside the mill. The windmill was restored in 2003. |
| Relevant book sections: | Chapters I-1, I-3, I-6, II-4 |

| Title: | Sharp-crested weir overflow |
|---|---|
| URL: | http://goo.gl/mrQQar |
| QR-code | |
| Duration: | 53 s |
| Description: | Large 90° V-notch weir experiments at the University of Queensland in May 2012. The video shows an unsteady experiment with the rapid opening of the gate (in less than 0.15 s) followed by the weir overflow. The example illustrates a three-dimensional, unsteady flow with separation. |
| Relevant book sections: | Chapter I-7 |
| Reference: | Chanson, H., and Wang, H. (2013). "Unsteady Discharge Calibration of a Large V-notch Weir." *Flow Measurement and Instrumentation*, Vol. 29, pp. 19–24 & 2 videos (DOI: 10.1016/j.flowmeasinst.2012.10.010). |

| Title: | Laminar and turbulent flows in a Reynolds experiment | |
|---|---|---|
| URL: | http://goo.gl/hsN9F (Low Resolution) | http://goo.gl/8gTYH (High Resolution) |
| QR-code | | |
| Duration: | 25 s | |
| Description: | Reynolds experiment: laminar-turbulent flow transition in a circular pipe (D = 12.7 mm) with water and blue dye at the University of Queensland on 12 May 2006. With the flow direction from right to left, the video starts by showing a laminar motion in which the dye flows as a ribbon and there is no dye dispersion. The discharge is increased. When the flow becomes turbulent, the dye is rapidly mixed across the entire flow cross-section. | |
| Relevant book sections: | Chapters II-1, II-2 | |

| Title: | Developing shear layer and large vortical structures |
|---|---|
| URL: | http://goo.gl/mm206 |
| QR-code | |
| Duration: | 14 s |
| Description: | Eddies formation, development and vortex shedding at a jetty at Pointe de Grave, Gironde estuary (France) during a strong ebb flow on 5 July 2004 at 9:30 am. During the ebb flow, separation at the end of the jetty creates a shear layer development. The video illustrates the development of large vortical structures by vortex pairing as the eddies are advected in the shear layer. |
| Relevant book sections: | Chapters II-2, II-3, II-4 |

| Title: | Smoke dispersion in a cross-flow | |
|---|---|---|
| URL: | http://goo.gl/m8cMeV (Low Resolution) | http://goo.gl/7To6w (High Resolution) |
| QR-code | | |
| Duration: | 15 s | |
| Description: | Dispersion in a cross-flow, downstream of an incineration plant chimney in Tempaku-cho, Toyohashi (Japan) on 27 Nov. 2001 around 7:00 am at sunrise. On the day, the strong cold winds led to the rapid dispersion of hot smoke. The video highlights the large vortical structures in the atmospheric boundary layer. On 27 Nov. 2001, a cold front was passing over Japan associated with strong cold winds: the weather forecast in Toyohashi was 5 C (night time)/11 C (daytime), compared to 6 C/18 C on 26 Nov. 2001. | |
| Relevant book sections: | Chapters II-2, II-4 | |

| Title: | Turbulence and free-surface vortices in an inundated urban environment |
| --- | --- |
| URL: | http://goo.gl/NvCAi |
| QR-code | |
| Duration: | 60 s |
| Description: | Brisbane River flood and flooding of the University of Queensland St Lucia campus on 12 Jan. 2011 afternoon. Note fast flowing waters between the building towards the tennis courses in the background. |
| Relevant book sections: | Chapters II-2, II-4 |
| Reference | Brown, R., and Chanson, H. (2012). "Suspended Sediment Properties and Suspended Sediment Flux Estimates in an Urban Environment during a Major Flood Event." *Water Resources Research*, AGU, Vol. 48, Paper W11523, 15 pages (DOI: 10.1029/2012WR012381) |

| Title: | Smoke plume in the atmosphere | |
| --- | --- | --- |
| URL: | http://goo.gl/Mwqz9 (Low Resolution) | http://goo.gl/lpqrj (High Resolution) |
| QR-code | | |
| Duration: | 15 s | |
| Description: | Large smoke plume in the Vallée de l'Isère near Grenoble (France) on 10 Feb. 2004. The video illustrates the turbulent jet spreading in the atmosphere in absence of any cross-flow. Note the differences in flow patterns with the movie Smoke03.mov (Smoke dispersion in a cross-flow), in which the strong cross-flow and associated turbulent shear led to the rapid dispersion of the jet. | |
| Relevant book sections: | Chapter II-4 | |

| Title: | Turbulence and free-surface vortices during a major flood |
| --- | --- |
| URL: | http://goo.gl/UppTb |
| QR-code | |
| Duration: | 65 s |
| Description: | Brisbane River in flood at Indooroopilly, Brisbane (Australia) on 12 Jan. 2011 afternoon. View from Indooroopilly Bridge looking upstream. (Courtesy of André Chanson) |
| Relevant book sections: | Chapter II-4 |

| Title: | Developing boundary layer above a spillway | |
|---|---|---|
| URL: | http://goo.gl/5pwSd | |
| QR-code | | |
| Duration: | 13 s | |
| Description: | Somerset dam and spillway (Australia) in operation on 28 Jan. 2013 – Gates fully-opened (Q ~ 450 m³/s (q ~ 8 m²/s)). At the spillway crest, a turbulent boundary layer is generated by bottom friction and develops in flow direction; the flow consists of a turbulent boundary layer and an ideal-fluid flow region above; when its outer edge is close to the free-surface, the turbulence next to the air-water interface may induce some air-water exchanges; downstream of the inception point of surface aeration, the flow is fully-developed and aerated. | |
| Relevant book sections: | Chapter II-4 | |
| Reference: | Wood, I.R., Ackers, P., and Loveless, J. (1983). "General Method for Critical Point on Spillways." *Jl. of Hyd. Eng.*, ASCE, Vol. 109, No. 2, pp. 308–312. | |

| Title: | Wind farm operation | |
|---|---|---|
| URL: | http://goo.gl/1QC13 (Low Resolution) | http://goo.gl/bICMa (High Resolution) |
| QR-code | | |
| Duration: | 10 s | |
| Description: | Wind farm at Plouarzel (France) on 1 March 2004. The video shows the 4 turbines in operation. Each turbine can produce 750 kW. The main characteristics are: 47 m diameter, type Neg-Micon (3 blades), rotation speed: 28.5 rpm, mast: 38.4 m high. Weights: mast: 28.9 tons, nacelle: 20.5 tons, rotor: 7.2 tons. The installed capacity of the wind farm is 3.3 MW, and the annual production is 9.8 Millions of kWh. | |
| Relevant book sections: | Chapter I-7, II-4 | |

| Title: | Turbulent water jets discharging into the atmosphere | |
|---|---|---|
| URL: | http://goo.gl/pUCRe | |
| QR-code | | |
| Duration: | 8 s | |
| Description: | Bassin de Latone and water jets in Jardins du Chateau de Versailles (France) on 27 July 2008. First built in 1670, the Bassin was re-designed by Jules Hardouin-Mansart between 1687 and 1689. It includes a total of 74 water jets. The video shows the operation of the Bassin with the Grand Canal in the background. | |
| Relevant book sections: | Chapter II-4 | |
| Reference | {http://en.chateauversailles.fr/homepage} | |

| Title: | Unsteady boundary layer and sediment motion beneath waves | |
|---|---|---|
| URL: | http://goo.gl/fKOg3 (Low Resolution) | http://goo.gl/OSbxq (High Resolution) |
| QR-code | | |
| Duration: | 14 s | |
| Description: | Sediment motion beneath waves in a laboratory flume, Grenoble (France), on 10 Feb. 2004. The wave propagation is from left to right. The sediment material (red colour) has the following characteristics: $d_{50} = 0.6$ mm, relative density $= 1.19$. The video shows the sediment re-suspension beneath the waves. | |
| Relevant book sections: | Chapter II-4 | |

| Title: | Wave runup and run-down in a swash zone (high tide) | | Wave runup and run-down in a swash zone (low tide) | |
|---|---|---|---|---|
| URL: | http://goo.gl/xTgwF (Low Resolution) | http://goo.gl/RmCzs (High Resolution) | http://goo.gl/P852x (Low Resolution) | http://goo.gl/7eSDj (High Resolution) |
| QR-code | | | | |
| Duration: | 5 s | | | |
| Description: | Surfers in swash zone at Terasawa beach, Toyohashi (Japan), on 23 Nov. 2001 around 12:00 noon (high tide). Sunset over Terasawa beach, Toyohashi (Japan), on 23 Nov. 2001 around 4:00 pm (low tide). On 23 Nov. 2001, there were long period waves (T = 3 to 5 minutes), strong wave reflection, and edges waves, particularly during the long period wave crest. | | | |
| Relevant book sections: | | | | |

| Title: | Turbulence and free-surface vortices during a major flood | |
|---|---|---|
| URL: | http://goo.gl/Qvno9 | |
| QR-code | | |
| Duration: | 14 s | |
| Description: | Brisbane River in flood at Toowong, Brisbane (Australia) on 12 Jan. 2011 morning. View from the left bank. Note submerged bicycle path (foreground) as well as the amount and types of debris in the main river channel (background). | |
| Relevant book sections: | Chapter II-4 | |

# Assignments

# Application to the design of the Alcyone 2

## INTRODUCTION

The assignment includes 2 parts: a study of an auxiliary propulsion system for a ship, and the analysis of the hydrofoil system of the same boat. It is based upon the characteristics of a real ship, the "Alcyone", built by the Equipe Cousteau (see Chapter I–4, section 3.7) (Fig. 1 & 2). It is proposed to design a new version of the ship, called the "Alcyone 2". Appendix 1 provides information on naval hydrodynamics (drag on ship, vocabulary, units). Appendix 2 gives the characteristics of and summarises the principle of the turbosail of the first "Alcyone".

You consider the design of the new ship "Alcyone 2" that will be built on the Gold Coast for the Jacques Cousteau Foundation. The new boat will be 25 meters long, 7 meters wide and the maximum draught (or draft) will be 1.5 meters (excluding the foils). The maximum weight will be 260 tonnes. The ship is designed with two hydrofoils below the hull to reduce the drag force (Fig. 3). Each foil is a flat plate with a inverted T-shape. Auxiliary propulsion is supplied by 2 masts, each being a rotating cylinder of radius 0.75 m and 9 metres height. The first mast is located 4 metres from the bows and the second mast is 6 metres from the stern. The maximum speed is expected to be 25 knots.

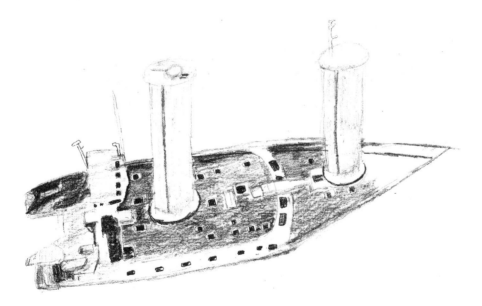

*Figure 1*  Sketch of the Alcyone at sea.

*Figure 2* The Alcyone arriving in New York, USA (Courtesy of Equipe Cousteau, France).

### Hypothesis

You will assume that the flow around the hull of the ship is a two-dimensional and irrotational flow of ideal fluid. Typical sea temperature is 15 Celsius.

You will assume that the wind flow around the masts is a two-dimensional and irrotational flow of ideal fluid. The atmospheric conditions are: $P = P_{atm} = 10^5$ Pa; $T = 20$ Celsius.

### Question 1 – Design of the hull

You will design the hull with a shape of a Rankine body (obtained for the maximum speed).

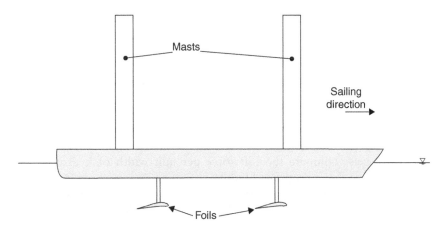

*Figure 3*  Cross-section of the Alcyone 2.

(a) Apply the theorem of superposition to obtain the stream function and the velocity potential of the resulting flow pattern. Explain the basic flow patterns used to obtain the Rankine body. Define the complete characteristics of the Rankine body. Give the numerical results of these main parameters, in SI units. On a sketch, explain the origin and the directions of the axes used to define the flow pattern. Show the direction of motion of the ship on the sketch.

(b) On graph paper, draw the complete flow net. Indicate **clearly** on the sketch the location of the free surface and the contour of the hull. Using the scale: $1 \, \text{cm} = 1 \, \text{m}$, draw the flow net with the appropriate value of $\Delta \psi$. Indicate the values of $\psi$ and $\phi$ of each streamline and equipotential line.

(c) What is the maximum velocity on the hull? Explain and justify your answer.

(d) Neglecting the Drag force due to the waves, compute the drag force on the ship hull. Explain your calculations in detail. Justify your choice(s).

Note: A formula for the drag coefficient is given in Appendix 1.

## Question 2 – Design of the hydrofoils

Two identical hydrofoils will be installed 3 metres below the bottom of the hull. Each hydrofoil is a flat plate of 0.9 m chord length ([1]). You will design the foils for the maximum speed.
    The effects of the hull on the flow past the foils will be neglected.

(a) In a text book on hydrofoils, you read that a hydrofoil is mounted with an angle of attack varying from 5 to 20 degrees. You will study the flow past a flat hydrofoil at angle of attack ([2]) **15 degrees**. On a sketch, show the hull of the ship, the position of the foils, the direction of motion of the ship and the angle of attack of each foil.

---

[1] Length of the straight line connecting the nose to the trailing edge of the foil.
[2] Angle between the approaching flow velocity vector and the chordline.

(b)    On graph paper (A4 size), draw the complete flow net for the flow past a flat plate at angle of attack 15 degrees. Assuming that the discharge between two streamlines is $\Delta q = 3\ \text{m}^2/\text{s}$, indicate what value you choose for $\Delta\psi$. Using the scale: $1\ \text{cm} = 0.1\ \text{m}$, draw the flow net with the appropriate value of $\Delta\psi$. Indicate on the graph the direction of motion of the foil (i.e. the flat plate).

(c)    Far away (from the foil), the pressure is equal to the atmospheric pressure plus 4.5 metres of water. Using the flow net construction, determine the velocity and pressure distributions on each side of the foil: i.e. the velocity and the pressure at the locations 0.1125, 0.3375, 0.5625 and 0.7875 m from the leading edge of the foil. Explain carefully in words your method.

(d)    Using the flow nets, compute the Lift force per unit width of hydrofoil. Explain your calculations and use the appropriate SI units.

(e)    The hydrofoils are introduced to lift the hull totally above the water surface at maximum speed. (i) Compute the width of each foil required to lift the ship at maximum speed and maximum weight. (ii) What is the total lift force resulting from the two hydrofoils? Neglect the drag induced by the foils.

The hydrofoils are replaced by two NACA2415 profiles with the performances shown in Figure 6.5 (Chap. I-6).

(f)    Calculate the total lift force at maximum speed for a 3° angle of incidence. What is the drag force on the foils?

(g)    Calculate the optimum angle of incidence and the foil width to lift the hull totally above the water surface at maximum speed. That is, the optimum angle of incidence is that for which the ratio of lift to drag forces is maximum.

## Question 3 – Auxiliary propulsion

You will now consider the auxiliary propulsion system consisting of two circular rotating cylinders.

(a)    Apply the theorem of superposition to describe the flow past a rotating cylinder. Give the stream function and velocity potential of the flow around a rotating cylinder as a function of the uniform flow velocity $V_O$, the cylinder radius R and the speed of rotation of the cylinder $\omega$.

(b)    Give the expression for the velocity at any point on the surface of the cylinder as a function of the uniform flow velocity $V_O$, the cylinder radius R and the cylinder's rotational speed $\omega$. What is the pressure distribution at any point on the surface of the cylinder as a function of the uniform flow velocity $V_O$, the cylinder radius R and the speed of rotation of the cylinder $\omega$? Integrate the pressure distribution around the cylinder to obtain the expression of the lift force and drag force as a function of the uniform flow velocity $V_O$, the cylinder radius R and the speed of rotation of the cylinder $\omega$. Detail and justify your calculations.

(c)    Sketch the flow pattern(s) of the flow past a rotating cylinder. Indicate the location of the stagnation point(s). If there are several cases, sketch the flow pattern for each case and indicate the flow conditions.

(d)    In operation, the speed of rotation of the cylinder will be 50 revolutions per minute. Assume that the absolute wind speed is 30 knots, blowing perpendicularly to the ship direction, and that the ship is at maximum speed. On a sketch, show the direction of the absolute wind as well as the apparent wind direction, ship motion direction and the sense of rotation of each cylinder. Considering the first cylinder (i.e. the forward mast), sketch the flow pattern and indicate the ship motion direction, the sense of rotation of the cylinder and the wind velocity. How many stagnation point would occur at the surface of the first cylinder? Explain what wind speed and direction you will use. Neglect the interference between the masts. Explain and justify your calculations.

## APPENDIX 1. NAVAL HYDRODYNAMICS

### Drag on a ship hull – Drag coefficient

Dimensional analysis shows that the drag on a ship hull can be written as:

$$\text{Drag} = \frac{1}{2} \times C_D \times A_w \times V_O^2$$

where $C_D$ is the drag coefficient, $A_w$ is the wetted-surface area, $V_O$ is the ship velocity. The drag coefficient is a function of both the Reynolds and Froude number.

To a first approximation, it is assumed that the drag coefficient can be expressed as the sum of a frictional-drag coefficient, depending on the Reynolds number only, plus a residual-drag coefficient which depends on the Froude number only (Newman 1977, pp. 27–29).

Neglecting the residual-drag coefficient, the drag coefficient equals the frictional drag coefficient. This may be estimated as the flat plate drag coefficient. Experiments showed that the frictional-drag coefficient of a flat plate is determined by Schoenherr's correlation (Newman 1977, pp. 16–18):

$$\frac{0.242}{\sqrt{C_D}} = \log_{10}(\text{Re} \times C_F)$$

where Re is the Reynolds number: $\text{Re} = V_O \times L/\nu$, L is the ship length, $\nu$ is the fluid kinematic viscosity.

### Units

knots                     1 knot      = 1 nautical mile per hour
nautical mile      1 naut. mile = 1852 m

### Glossary

*Angle of attack*: angle between the approaching flow velocity vector and the chordline.
*Bow*: the forward part of a ship (often plural with singular meaning).
*Chord* (or chord length): length of the chordline.
*Chordline*: straight line connecting the nose to the trailing edge (of a foil).
*Drag*: force component (on an object) in the direction of the approaching flow.
*Draught*: the depth of water a ship requires to float in.
*Hydrofoil*: a device that, when attached to a ship, lifts (totally or partially) the hull out of the water at speed.
*Lift*: force component (on a wing or foil) perpendicular to the approaching flow.
*Mast*: a tall pole or structure rising from the keel or deck of a ship.

*Table 1*   Characteristics of the Alcyone ship.

| Section (1) | Characteristics (2) | Remarks (3) |
|---|---|---|
| Hull | Built in 1984–85. Overall length: 31.1 m. Maximum draught: 2.34 m. Maximum width: 8.92 m. Half-load displacement: 76.8 t. Cruise speed: 10.5 knots | Designed by A. Mauric and J.C. Nahon (Fra.) |
| Turbosails™ | 2 identical fixed masts equipped with a boundary layer suction system. Height: 10.2 m. Maximum chord: 2.05 m. Width: 1.35 m. Surface area: 21 m$^2$ (each). Suction fan: 25 hp. Performances: $C_L = 5$ to 6.5 and $C_D = 1.2$ to 1.8. | Developed by L. Malavard. Cousteau-Pechiney system built in aluminium. |

*Stern*: the rear end of a ship or boat.
*Tonne*: a metric unit of weight equal to 1000 kg.

## APPENDIX 2. THE ALCYONE

The propulsion system of the Alcyone was developed by the Frenchman L. Malavard for Jacques-Yves Cousteau (Chapter I-4, section 3.7). It consists of two identical fixed masts (i.e. Turbosail™). Boundary layer suction is generated on one side of the cylinder by a suction fan installed inside the mast, while a flap controls flow separation downstream (Fig. 4.17B, Chapter I-4). The resulting effect is a large fluid circulation around the sail and a significant lift force ($C_L \sim 5$ to 6). Optimum performances are obtained when the angle between wind and ship directions ranges between 50 and 140°.

Dimensions and characteristics of the ship are listed in Table 1.

# Applications to civil design on the **Gold Coast**

## INTRODUCTION

The assignment deals with the construction of a new building on the Gold Coast (Fig. 1) and it includes 2 parts: a wind impact study and a groundwater study.

(A) Gold Coast on 20 July 2009 looking South from the Q1 building

*Figure 1* Photographs of the Gold Coast (Australia).

(B) Revolving restaurant at the 26th floor of the Crown Plaza building on 3 Dec. 2007 – The restaurant rotates a full rotation in 1 hour 10 minutes

*Figure 1* Continued.

## PART 1

You consider the "Daiko" building on the Gold coast (Surfers' Paradise). This a circular cylinder building (Height: 100 m – Diameter: 20 m). The building is in front of the beach facing East (the axis of the building is 15 m West of the beach). Most the winds are South-East wind (Fig. 2).

You will assume that the wind flow around the building is a two-dimensional and irrotational flow of ideal fluid. The atmospheric conditions are: $P = P_{atm} = 10^5$ Pa; $T = 20$ Celsius.

## Part 1A

### *Flow net*

1.1   On a A4 page of graph paper, draw a (top view) sketch of the building, indicate the North, South, East and West as well as the main wind direction. Sketch the beach (15 m East of the

(A) View in elevation

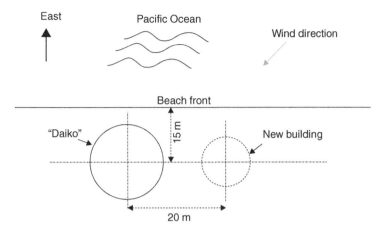

(B) Side in elevation

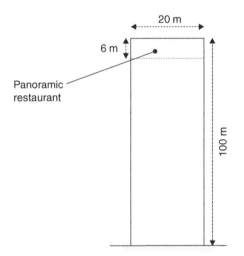

*Figure 2* Sketch of the "Daiko" building location.

centre of the building; Fig. 2), **using the South-East direction as x-axis and the North-East direction as y-axis.** Use the scale 1/200 (i.e. 1 cm = 2 m). Clearly indicate the x-axis and y-axis, their direction and use the centre of the building as the origin of your system of coordinates.

1.2   On graph paper, draw the complete flow net using ten (10) streamlines and the appropriate equipotentials for a **20 m/s South-East wind.** Assuming that the total discharge between two streamlines is $\Delta Q = 6000\,\text{m}^3/\text{s}$ over the height of the building (i.e. 100 m), indicate what value you choose for $\Delta \psi$. Using the scale: 1 cm = 2 m, draw the flow net with the appropriate value of $\Delta \psi$.

1.3   Using the above flow net, determine the wind velocity and the wind direction for people standing on the beach, just in front of the building (15 m East of the centre of the building)? For SE wind 20 m/s.

*Flow pattern*

1.4   What standard flow pattern would you use to describe the wind around the building? What are the main characteristics (Flow, strength, ...) of the resulting flow pattern? Give the numerical results of these main parameters, in SI units.

1.5   Write the stream function and the velocity potential of the resulting flow pattern. Use this result to compute the velocity components on the beach in front of the building (see question 1.3) and compare your result with the above question. What is the pressure on the beach, assuming that far away on sea, the pressure is atmospheric (i.e. $P_{atm} = 10^5$ Pa).

1.6   What is the expression of the pressure and the velocity at any point at the surface of the building?

1.7   What is the total lift and drag force on the building. Discuss your results and explain clearly your answer.

1.8   (a)   Is the above result realistic? Why?
      (b)   Compute the Reynolds number of such a flow. Is it laminar, turbulent, ...?
      (c)   From the lecture material, sketch the most probable flow pattern, for a real fluid, around and behind the building.
      (d)   Use the lecture notes to estimate of: {d1} the drag coefficient of the building, and {d2} compute the real drag force on the "Daiko" building (for a real fluid flow situation) for a 20 m/s SE wind.
      (e)   Compare your result with the question 1.7.

## Part 1B

The top floor of the building (height 6 m – diameter 20 m) is a panoramic restaurant, rotating **clockwise** at a speed of 1 rotation per minute (Fig. 2B). Considering a **20 m/s SE wind**:

1.9   Sketch the flow pattern around the top floor using 4 streamlines;

1.10  what standard flow pattern would you use to describe the wind around the building? What are the main characteristics (Flow, strength, ...) of the resulting flow pattern? Give the numerical results of these main parameters, in SI units.

1.11  what is the velocity and the pressure at any point at the surface of the restaurant? Neglect the effect on the roof of the restaurant.

1.12  is there any stagnation point on the surface of the restaurant? Why? In the affirmative, give the location(s) of the stagnation point(s).

1.13  what is the total drag force on the upper level of the building (i.e. the restaurant)?

1.14  what is the total lift force on the restaurant?

## PART 2

The Gold Coast City Council authorised the construction of a second building next to the Daiko building. The new building will be a 10 m diameter circular tower (100 m height). The axis of the new building will be 15 m West of the beach, and 20 m South of the present "Daiko" building (Fig. 2A).

## Part 2A

Assuming that the ground is 4 meter above the sea level, and that the sea level is constant, you now consider the excavation for the foundation of the new building. The foundations must reach the rock level, 6 m below the surface and the excavation pit must be drained (Fig. 3). A 6 m deep well will be installed to pump the groundwater flow and to reduce the water level in the pit to 0.5 m above the rock level.

You will assume that:

(A)   the hydraulic conductivity of the ground is $K = 1 \times 10^{-5}$ m/s,
(B)   the water level in the well is 0.5 m above the rock floor (Fig. 3),
(C)   the ground water table, before the installation of the well was equal to the sea level (i.e. 4 m below the ground surface), and
(D)   the axis of the circular well is the axis of the new building.

You are required to compute the discharge to be pumped from a 0.5 m diameter well.

2.1   Sketch, with a scale 1/200 (i.e. 1 cm = 2 m), the piezometric head between the sea and the well and on the West of the well. Sketch a sectional elevation East-West through the well, and a top view with the streamlines and equipotentials.
2.2   Select an appropriate flow pattern and explain, in words, your choice.
2.3   What method would to choose to compute the total discharge Q to be pumped? Explain your method.

You must size the submersible pump to be installed at the bottom of the well. What discharge will be pumped from the well during the excavation work? Explain your answer in words.

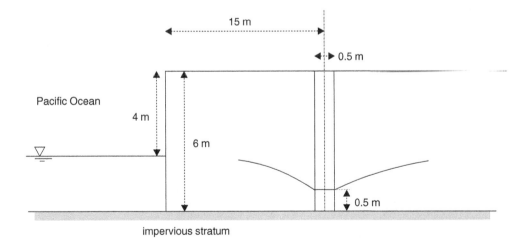

*Figure 3*  Sketch of the excavation works.

## Part 2B

You now consider the building itself and for a 20 m/s SE wind you are required to study the interactions between the Daiko building (Part 2A) and the new building (Fig. 2). You will neglect the rotating action of the restaurant at the top of the Daiko building (i.e. no rotating top floor) and you will assume that the wind flow around the buildings is a two-dimensional and irrotational flow of ideal fluid.

The atmospheric conditions are: $P = P_{atm} = 10^5$ Pa; $T = 20$ Celsius.

2.4 On graph paper, sketch the flow net (with the two building) for a **20 m/s SE wind**, using 10 streamlines equally spaced 4 m between apart (i.e. 4 m between streamlines in uniform flow). What is the discharge between two streamlines? **Use the South-East direction as x-axis and the North-East direction as y-axis**; use the scale 1/200 (i.e. 1 cm = 2 m). Clearly indicate the x-axis and y-axis, their direction and use the centre of the Daiko building as the origin of your system of coordinates.

2.5 Explain what standard flow patterns you would use to describe the flow around these two buildings.

2.6 Write the stream function and the velocity potential as a function of the wind speed $V_O$ (20 m/s) and the two building diameters $D_1$ (20 m for the Daiko building) and $D_2$ (10 m for the new building).

The Gold Coast City Council is concerned about the wind velocity between the buildings that may damage cars or knock over people:

2.7 Compute the wind velocity and the pressure at any point on the line joining the axis of the two building. Where is located the point of maximum velocity and minimum pressure and what is the maximum velocity and minimum pressure between the buildings? Indicate that point on your sketch.

Discuss your results. Do you think that this result is realistic? Why?

*Using the software 2DFlow Plus:*

2.8 What is the velocity and the pressure at any point at the surface of the Daiko building?
2.9 What is the velocity and the pressure at any point at the surface of the new building?

## SOLUTIONS

## Part 1A

1.1 $\Delta\psi = 60\,\mathrm{m^2/s}$ and $V_O = 20\,\mathrm{m/s} \Rightarrow \Delta n = 3\,\mathrm{m}$ in uniform flow away from the building.

   At a scale of 1/200, the distance between streamlines far away from the building is 1.5 cm.

1.3 $V \sim 24$ m/s (graphical solution)
1.4 Doublet: $\mu = 2000\,\mathrm{m^3/s}$
1.5 Beach: $V = 21.9$ m/s and $P = 0.995\,\mathrm{E+5}$ Pa
1.8 The flow situation is not realistic for turbulent flow (Re = 2.9 E+7 in this case). For real fluid flows, the drag on the building is about: Drag = 1.44 E+5 N.

## Part 1B

1.10   $\mu = 2000\,\mathrm{m^3/s}$ and $K = 65.8\,\mathrm{m^2/s}$
1.12   Flow pattern with 2 stagnation points at $\theta = 1.5°$ and $178.5°$
1.14   Lift $= 31.6\,\mathrm{E+3\,N}$ for all the top floor, ignoring three-dimensional effects at the roof of the building.

## Part 2A

2.2   Sink + Image Source
2.3   $Q \sim 1\,\mathrm{E-5\,m^3/s}$

## Part 2B

2.6   $\mu_1 = 2000\,\mathrm{m^3/s}$ and $\mu_2 = 500\,\mathrm{m^3/s}$
2.7   Maximum velocity: $28.9\,\mathrm{m/s} \Rightarrow P = 0.9974\,\mathrm{E+5\,Pa}$

# Wind flow past a series of circular buildings

Let us consider a new architectural landmark to be built at the Mount Cootha Lookout, Brisbane QLD (Australia). The structure consists of three circular cylinders (Height: 25 m, Diameters: 2, 3 and 5 m). The landmark will be facing North-East, while the dominant winds are Easterlies (Fig. 1).

(A) Sketch of the Mt Cootha landmark – View in elevation

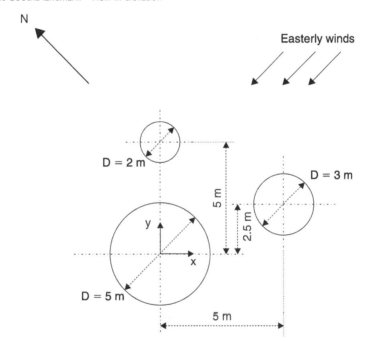

(B) Brisbane viewed from Mount Cootha on 4 July 2011

*Figure 1*  Mount Cootha, Brisbane (Australia).

You will assume that the wind flow around the structure is a two-dimensional irrotational flow of ideal fluid. The atmospheric conditions are: $P = P_{atm} = 10^5$ Pa; $T = 25$ Celsius.

1.  On graph paper, sketch the flow net with the landmark for a 25 m/s Easterly wind. Indicate clearly on the graph the discharge between two streamlines, the x-axis and y-axis, their direction, and use the centre of the 5-m diameter cylinder as the origin of your system of coordinates (with x in the South-East direction and y in the North-East direction).
2.  The Brisbane City Council is concerned about the wind velocities between the buildings that may blow down tourist pedestrians and damage cars.
    (a)  From the flow net, compute the wind velocity and the pressure at:

    $$x = 1.5\,m, \quad y = 6\,m$$
    $$x = 5\,m, \quad\;\; y = 4.5\,m$$
    $$x = 2.1\,m, \quad y = 2.1\,m$$

    *These locations would be typical of tourists standing in front of the vertical cylinders.*
    (b)  Where is located the point of maximum velocity and minimum pressure?
    (c)  What is the maximum velocity and minimum pressure between the cylinders? Indicate that location on the flow net.
    (d)  Discuss the results. Do you think that this result is realistic? Why?
3.  Explain what standard flow patterns you would use to describe the flow around these three buildings.
4.  Write the stream function and the velocity potential as a function of the wind speed $V_O$ (25 m/s) and the two cylinder diameters $D_1$ (5 m), $D_2$ (3 m) and $D_3$ (2 m).
    *Do not use numbers. Express the results as functions of the above symbols.*
5.  For a real fluid flow, what is (are) the drag force(s) on each cylinder (Height: 25 m)?

**Solution**

2.  (a1)  $x = 1.5$ m, $y = 6$ m $\quad$ $V \sim 23.2$ m/s $\quad$ $P - P_{atm} = +52$ Pa
    (a2)  $x = 5$ m, $y = 4.5$ m $\quad$ $V \sim 28.5$ m/s $\quad$ $P - P_{atm} = -112$ Pa
    (a3)  $x = 2.1$ m, $y = 2.1$ m $\quad$ $V \sim 10.5$ m/s $\quad$ $P - P_{atm} = +309$ Pa
    (b)  between the 5-m and 2-m diameter cylinders
    (c)  $V \sim 60$ m/s, $P - P_{atm} = -1785$ Pa

    Such a large maximum wind velocity may cause a potential hazard for pedestrians and cyclists.
    (d)  The flow is turbulent: $\rho \times V \times D/\mu \sim 3.2\,E{+}6$ ($D = 2$ m). Separation is likely to occur behind the cylinder. However, since the location of maximum velocity is likely to be outside of a wake region, the above results are very likely representative.
3.  Use 3 doublet patterns ($\mu = 100, 225, 625$ m$^3$/s)
5.

| D (m) = | 2 | 3 | 5 |
|---|---|---|---|
| Re = | 3.2 E+6 | 4.8 E+6 | 8 E+6 |
| $C_D$ = | 0.7 | 0.75 | 0.75 |
| Drag (N) = | 1.9 E+4 | 2.1 E+4 | 3.5 E+4 |

# Prototype freighter testing

## PRESENTATION

A large towing tank facility is used to conduct drag tests on a 1:40 scale model of a submerged freighter to be used for trans-oceanic shipping (Fig. 1). The maximum speed of the prototype is

*Figure 1* Photograph of the 1:40 scale model of a submerged freighter tested at Hiroshima University (Japan) in October 2001.

expected to be 54 knots. Since it is impossible to achieve simultaneously both Froude and Reynolds similitudes, the tests are conducted at identical Froude number. The magnitude of the corrected surface drag for the prototype will be deduced by means of the boundary layer equations.

The prototype freighter has a total length of 140 m, a length of 32 m at the waterline, and a wetted area of 5,100 m$^2$. What would be the total drag on the prototype at maximum speed, if the corresponding model drag is 95 N.

*For seawater, the fluid density, dynamic viscosity and surface tension are respectively:*
$\rho = 1024\,kg/m^3$, $\mu = 1.22 \times 10^{-3}$ *Pa.s, and* $\sigma = 0.076\,N/m$.
*The hydraulic model tests are conducted in freshwater.*

## Preliminary comments

This application involves a real-case study and its physical modelling (Fig. 1). In a geometrically-scaled model, it is impossible to achieve simultaneously both Froude and Reynolds similitudes when the same fluid(s) is(are) used in model and prototype (Henderson 1966, Chanson 2004). Herein a Froude similitude is chosen, but note carefully the relevant length scale (next paragraph).

In ship model tests, the total drag force is roughly the sum of the wave drag and the friction resistance of the hull. The wave drag is a free-surface phenomenon that satisfies a Froude similarity, and the relevant Froude number is based upon the ship waterline length. The friction resistance is a viscous phenomenon that is scaled herein using the boundary layer equations. The relevant Reynolds number $Re_L$ is based upon the overall length of the submerged hull.

Note that the model tests are conducted in freshwater while the prototype operates in seawater. Units: 1 nautical mile = 1,852 m; 1 knot = 1 nautical mile per hour.

## Solution

The maximum speed of the prototype is $V_p = 27.8$ m/s and the corresponding Froude number is:

$$Fr_p = \frac{V_p}{\sqrt{g \times (L_{waterline})_p}}$$

where the subscript p refers to the prototype conditions. Based upon a Froude similitude, the model speed satisfies:

$$\frac{V_m}{\sqrt{g \times (L_{waterline})_m}} = Fr_m = Fr_p = \frac{V_p}{\sqrt{g \times (L_{waterline})_p}}$$

where the subscript m refers to the prototype conditions. The prototype waterline length is 32 m and the 1:40 scale model length is 0.8 m. It yields a model speed: $V_m = 4.4$ m/s.

The prototype and model Reynolds numbers are respectively:

$$Re_p = \rho_p \times \frac{V_p \times (L_{total})_p}{\mu_p} = 3.3 \times 10^9$$

$$Re_m = \rho_m \times \frac{V_m \times (L_{total})_m}{\mu_m} = 1.5 \times 10^7$$

Both flows are turbulent and this is a requirement to minimise viscous scale effects.

In both model and prototype, the surface friction force is deduced from the application of the Blasius formula to smooth turbulent developing boundary layers:

$$\frac{\iint \tau_o \times dA}{\frac{1}{2} \times \rho \times V_O^2 \times A} = 0.072 \times \left(\frac{\rho \times V_O \times L}{\mu}\right)^{-1/5} \tag{4.11b}$$

The model friction force equals:

$$(F_{friction})_m = \left(\iint \tau_o \times dA\right)_m = 81\,N$$

In the prototype, the friction force equals:

$$(F_{friction})_p = \left(\iint \tau_o \times dA\right)_p = 1.8 \times 10^6\,N$$

Assuming that the total drag is the sum of the friction force and wave resistance, the force required by the model to overcome the wave drag is $95 - 81 = 15\,N$. The corresponding surface resistance in the prototype is deduced from the Froude similarity:

$$\frac{(F_{wave})_p}{(F_{wave})_p} = \frac{\rho_p}{\rho_m} \times \left(\frac{(L_{waterline})_p}{(L_{waterline})_m}\right)^3$$

It yields a surface resistance force of 9.3 E+5 N in prototype and a total drag force of $2.7 \times 10^6\,N$. The required thrust power is 76 MW in the prototype freighter.

### Discussion

The above results for a submerged freighter may be compared with the drag for a conventional freighter with a waterline length equals to the total length (140 m) for an identical wetted surface area and maximum speed. With a conventional freighter, the wave resistance is scaled to $L_{total}^3$ based upon a Froude similitude. For the above example, the wave resistance would be about $1.5 \times 10^8\,N$, that is 80 times larger than for the submerged freighter.

The difference explains partly why submerged submarines and torpedoes can reach larger speeds than conventional ships for an identical power plant and wetted area.

### Comments

Note that the Froude number is supercritical: $Fr = 1.6$. For such flow conditions, the wave resistance is always important, although it affects herein the waterline only, not the submerged body (Fig. 1).

# References

Ackerman, J.D., and Hoover, T.M. (2001). "Measurements of Local Bed Shear Stress in Streams using a Preston-Static Tube." *Limnology & Oceanography*, Vol. 46, No. 8, pp. 2080–2087.

Alembert, Jean le Rond d' (1752). "Essai d'une Nouvelle Théorie de la Résistance des Fluides." ('Essay on a New Theory on the Resistance of Fluids.') *David*, Paris, France (in French).

Allen, H.P.G. (1968). "Field Mapping with Conducting Paper." *Physics Education*, Vol. 3, No. 5, pp. 266–272.

Andrade, E.N. da C. (1939). "The Velocity Distribution in a Liquid-into-liquid Jet. Part 2: the Plane Jet." *Proc. Phys. Soc.*, Vol. 51, pp. 784–792. Discussion: Vol. 51, pp. 792–793.

Apelt, C.G., and Xie, Q. (1995). "Turbulent Flow in Irregular Channels." *Proc. 12th Australasian Fluid Mech. Conf. AFMC*, Sydney, Australia, Vol. 1, pp. 179–182.

Arrian (1976). "Arrian." *Harvard University Press*, Cambridge, USA, translated by P.A. Brunt, 2 volumes.

Bakhmeteff, B.A., and Matzke, A.E. (1936). "The Hydraulic Jump in Terms of Dynamic Similarity." *Transactions*, ASCE, Vol. 101, pp. 630–647. Discussion: Vol. 101, pp. 648–680.

Barenblatt, G.I. (1996). "Scaling, Self-Similarity, and Intermediate Asymptotics." *Cambridge University Press*, UK, 386 pages.

Barré de Saint-Venant, A.J.C. (1871a). "Théorie et Equations Générales du Mouvement Non Permanent des Eaux Courantes." *Comptes Rendus des séances de l'Académie des Sciences*, Paris, France, Séance 17 July 1871, Vol. 73, pp. 147–154 (in French).

Barré de Saint-Venant, A.J.C. (1871b). "Théorie du Mouvement Non Permanent des Eaux, avec Application aux Crues de Rivières et à l'Introduction des Marées dans leur Lit." *Comptes Rendus des séances de l'Académie des Sciences*, Paris, France, Vol. 73, No. 4, pp. 237–240 (in French).

Barrett, D.S., Triantafyllou, M.S., Yue, D.K.P., Grosenbaugh, M.A., and Wolfgang, M.J. (1999). "Drag reduction in Fish-Like Locomotion." *Jl of Fluid Mech.*, Vol. 392, pp. 183–212.

Bazin, H. (1865a). "Recherches Expérimentales sur l'Ecoulement de l'Eau dans les Canaux Découverts." ('Experimental Research on Water Flow in Open Channels.') *Mémoires présentés par divers savants à l'Académie des Sciences*, Paris, France, Vol. 19, pp. 1–494 (in French).

Belanger, J.B. (1828). "Essai sur la Solution Numérique de quelques Problèmes Relatifs au Mouvement Permanent des Eaux Courantes." ('Essay on the Numerical Solution of Some Problems relative to Steady Flow of Water.') *Carilian-Goeury*, Paris, France (in French).

Belanger, J.B. (1849). "Notes sur le Cours d'Hydraulique." ('Notes on a Course in Hydraulics.') *Mém. Ecole Nat. Ponts et Chaussées*, Paris, France, session 1849–1850, 222 pages (in French).

Belidor, B.F. de (1737–1753). "Architecture Hydraulique." ('Hydraulic Architecture.') *Charles-Antoine Jombert*, Paris, France, 4 volumes (in French).

Beyer, W.H. (1982). "CRC Standard Mathematical Tables." *CRC Press Inc.*, Boca Raton, Florida, USA.

Blasius, H. (1907). "Grenzschichten in Flüssigkeiten mit kleiner Reibing." *Ph.D. Dissertation*, University of Göttingen, Germany.

Blasius, H. (1908). "Grenzschichten in Flüssigkeiten mit kleiner Reibing." *Z. Math. Phys.*, Vol. 56, pp. 1–37 (in German) (also NACA Tech. Memo. No. 1256).

Blasius, H. (1913). "Das Ähnlichkeitsgesetz bei Reibungsvorgängen in Flüssigkeiten." *Forschg. Arb. Ing.-Wes.*, No. 134, Berlin, Germany (in German).

Bois, P.A. (2007). "Joseph Boussinesq (1842–1929): a pioneer of mechanical modelling at the end of the 19th Century." Comptes Rendus Mécanique, Vol. 335, No. 9–10, pp. 479–495.

Bossut, Abbé C. (1772). "Traité Elémentaire d'Hydrodynamique." (Elementary Treaty on Hydrodynamics.') *Imprimerie Chardon*, Paris, France, 1st ed. (in French). (2nd ed.: 1786, Paris, France; 3rd ed.: 1796, Paris, France.)

Boussinesq, J.V. (1877). "Essai sur la Théorie des Eaux Courantes." ('Essay on the Theory of Water Flow.') *Mémoires présentés par divers savants à l'Académie des Sciences*, Paris, France, Vol. 23, ser. 3, No. 1, supplément 24, pp. 1–680 (in French).

Boussinesq, J.V. (1896). "Théorie de l'Ecoulement Tourbillonnant et Tumultueux des Liquides dans les Lits Rectilignes à Grande Section (Tuyaux de Conduite et Canaux Découverts) quand cet Ecoulement s'est régularisé en un Régime Uniforme, c'est-à-dire, moyennement pareil à travers toutes les Sections Normales du Lit." ('Theory of Turbulent and Tumultuous Flow of Liquids in prismatic Channels of Large Cross-sections (Pipes and Open Channels) when the Flow is Uniform, i.e., Constant in Average at each Cross-section along the Flow Direction.') *Comptes Rendus des séances de l'Académie des Sciences*, Paris, France, Vol. 122, pp. 1290–1295 (in French).

Boys, P.F.D. du (1879). "Etude du Régime et de l'Action exercée par les Eaux sur un Lit à Fond de Graviers indéfiniment affouillable." ('Study of Flow Regime and Force exerted on a Gravel Bed of infinite Depth.') *Ann. Ponts et Chaussées*, Paris, France, série 5, vol. 19, pp. 141–195 (in French).

Bradshaw, P. (1971). "An Introduction to Turbulence and its Measurement." *Pergamon Press*, Oxford, UK, The Commonwealth and International Library of Science and technology Engineering and Liberal Studies, Thermodynamics and Fluid Mechanics Division, 218 pages.

Bradshaw, P. (1976). "Turbulence." *Springer-Verlag*, Topics in Applied Physics, Vol. 12, Berlin, Germany, 335 pages.

Brattberg, T., and Chanson, H. (1998). "Air Entrapment and Air Bubble Dispersion at Two-Dimensional Plunging Water Jets." *Chemical Engineering Science*, Vol. 53, No. 24, Dec., pp. 4113–4127. Errata: 1999, Vol. 54, No. 12, p. 1925 (ISSN 0009-2509).

Bresse, J.A. (1860). "Cours de Mécanique Appliquée Professé à l'Ecole des Ponts et Chaussées." ('Course in Applied Mechanics lectured at the Pont-et-Chaussées Engineering School.') *Mallet-Bachelier*, Paris, France (in French).

Brown, G.O. (2002). "Henry Darcy and the Making of a Law." *Water Res. Res.*, Vol. 38, No. 7, paper 11, pp. 11-1 to 11-12.

Brown, R., and Chanson, H. (2013). "Turbulence and Suspended Sediment Measurements in an Urban Environment during the Brisbane River Flood of January 2011." Journal of Hydraulic Engineering, ASCE, Vol. 139, No. 2, pp. 244–252 (DOI: 10.1061/(ASCE)HY.1943-7900.0000666).

Buat, P.L.G. du (1779). "Principes d'Hydraulique, vérifiés par un grand nombre d'expériences faites par ordre du gouvernement." ('Hydraulic Principles, verified by a large number of experiments.') *Imprimerie de Monsieur*, Paris, France, 1st ed. (in French). (2nd ed.: 1786, Paris, France, 2 volumes; 3rd ed.: 1816, Paris, France, 3 volumes)

Carslaw, H.S., and Jaeger, J.C. (1959). "Conduction of Heat in Solids." *Oxford University Press*, London, UK, 2nd ed., 510 pages.

Carvill, J. (1981). "Famous Names in Engineering." *Butterworths*, London, UK, 1981.

Chanson, H. (1995). "Flow Characteristics of Undular Hydraulic Jumps. Comparison with Near-Critical Flows." *Report CH45/95*, Dept. of Civil Engineering, University of Queensland, Australia, June, 202 pages.

Chanson, H. (1999). "The Hydraulics of Open Channel Flows: An Introduction." *Butterworth-Heinemann*, Oxford, UK, 512 pages.

Chanson, H. (2000). "Boundary Shear Stress Measurements in Undular Flows: Application to Standing Wave Bed Forms." *Water Res. Res.*, Vol. 36, No. 10, pp. 3063–3076.

Chanson, H. (2001). "Whirlpools. Experiencing Naruto Whirlpools." *Internet reference*. {http://www.uq.edu.au/~e2hchans/whirlpl.html}

Chanson, H. (2002). "Whirlpools. Experiencing Naruto Whirlpools." *IAHR Newsletter*, Vol. 40, No. 2, pp. 17 & 28–29.

Chanson, H. (2003). "A Hydraulic, Environmental and Ecological Assessment of a Sub-tropical Stream in Eastern Australia: Eprapah Creek, Victoria Point QLD on 4 April 2003." *Report No. CH52/03*, Dept. of Civil Engineering, The University of Queensland, Brisbane, Australia, June, 189 pages.

Chanson, H. (2004). "The Hydraulics of Open Channel Flow: An Introduction." *Butterworth-Heinemann*, Oxford, UK, 2nd edition, 630 pages.

Chanson, H. (2004b). "Environmental Hydraulics of Open Channel Flows." *Elsevier Butterworth-Heinemann*, Oxford, UK, 483 pages.

Chanson, H. (2004d). "Wind Farms in Brittany (Bretagne), France." *Shore & Beach*, Vol. 72, No. 3, pp. 15–16.

Chanson, H. (2005). "Mascaret, Aegir, Pororoca, Tidal Bore. Quid ? Où? Quand? Comment? Pourquoi?" *Jl La Houille Blanche*, No. 3, pp. 103–114.

Chanson, H. (2006). "Discharge through a Permeable Rubble Mound Weir. Discussion." Journal of Hydraulic Engineering, ASCE, Vol. 132, No. 4, pp. 432–433.

Chanson, H. (2007). "Le Potentiel de Vitesse pour les Ecoulements de Fluides Réels: la Contribution de Joseph-Louis Lagrange." ('Velocity Potential in Real Fluid Flows: Joseph-Louis Lagrange's Contribution.') *Jl La Houille Blanche*, No. 5, pp. 127–131 (DOI: 10.1051/ lhb:2007072).

Chanson, H. (2009). "Applied Hydrodynamics: An Introduction to Ideal and Real Fluid Flows." CRC Press, Taylor & Francis Group, Leiden, The Netherlands, 478 pages.

Chanson, H. (2011). "Tidal Bores, Aegir, Eagre, Mascaret, Pororoca: Theory and Observations." *World Scientific*, Singapore, 220 pages (ISBN 9789814335416).

Chanson, H., Aoki, S., and Maruyama, M. (2002). "Unsteady Two-Dimensional Orifice Flow: a Large-Size Experimental Investigation." *Jl of Hyd. Res.*, IAHR, Vol. 40, No. 1, pp. 63–71.

Chanson, H., Aoki, S., and Hoque, A. (2006). "Bubble Entrainment and Dispersion in Plunging Jet Flows: Freshwater versus Seawater." *Journal of Coastal Research*, Vol. 22, No. 3, May, pp. 664–677 (DOI: 10.2112/03-0112.1).

Chanson, H., Trevethan, M., and Koch, C. (2007). "Turbulence Measurements with Acoustic Doppler Velocimeters. Discussion." *Journal of Hydraulic Engineering*, ASCE, Vol. 133, No. 11, pp. 1283–1286 (DOI: 10.1061/(ASCE)0733-9429(2005)131:12(1062)).

Chen, C.L. (1990). "Unified Theory on Power Laws for Flow Resistance." *Jl of Hyd. Engrg.*, ASCE, Vol. 117, No. 3, pp. 371–389.

Colas des Francs, E. (1975). "Sealing and Drainage Facilities for the Ste Croix and Quinson Dams." *Intl Water Power & Dam Construction*, Vol. 27, Oct., pp. 356–361.

Coles, D. (1956). "The Law of Wake in the Turbulent Boundary Layer." *Jl of Fluid Mech.*, Vol. 1, pp. 191–226.

Coriolis, G.G. (1836). "Sur l'établissement de la formule qui donne la figure des remous et sur la correction qu'on doit introduire pour tenir compte des différences de vitesses dans les divers points d'une même section d'un courant." ('On the establishment of the formula giving the backwater curves and on the correction to be introduced to take into account the velocity differences at various points in a cross-section of a stream.') *Annales des Ponts et Chaussées*, 1st Semester, Series 1, Vol. 11, pp. 314–335 (in French).

Couette, M. (1890). "Etude sur les Frottements des Liquides." ('Study on the Frictions of Liquids.') *Ann. Chim. Phys.*, Paris, France, Vol 21, pp. 433–510 (in French).

Cousteau, J.Y., and Paccalet, Y. (1987). "Méditerranée: la mer blessée" *Flammarion*, 192 pages.

Crank, J. (1956). "The Mathematics of Diffusion." *Oxford University Press*, London, UK.

Danilevslkii, V.V. (1940). "History of Hydroengineering in Russia before the Nineteenth Century." *Gosudarstvennoe Energeticheskoe Izdatel'stvo*, Leningrad, USSR (in Russian) (English translation: *Israel Program for Scientific Translation*, IPST No. 1896, Jerusalem, Israel, 1968, 190 pages).

Darcy, H.P.G. (1856). "Les Fontaines Publiques de la Ville de Dijon." ('The Public Fountains of the City of Dijon.') *Victor Dalmont*, Paris, France, 647 pages (in French).

Darcy, H.P.G. (1858). "Recherches Expérimentales relatives aux Mouvements de l'Eau dans les Tuyaux." ('Experimental Research on the Motion of Water in Pipes.') *Mémoires Présentés à l'Académie des Sciences de l'Institut de France*, Vol. 14, p. 141 (in French).

Darcy, H.P.G., and Bazin, H. (1865). "Recherches Hydrauliques." ('Hydraulic Research.') *Imprimerie Impériales*, Paris, France, Parties 1ère et 2ème (in French).

Degremont (1979). "Water Treatment Handbook." *Halsted Press Book*, John Wiley & Sons, 5th edition, New York, USA.

Donnelly, C., and Chanson, H. (2005). "Environmental Impact of Undular Tidal Bores in Tropical Rivers." *Environmental Fluid Mechanics*, Vol. 5, No. 5, pp. 481–494.

Dupuit, A.J.E. (1848). "Etudes Théoriques et Pratiques sur le Mouvement des Eaux Courantes." ('Theoretical and Practical Studies on Flow of Water.') *Dunod*, Paris, France (in French).

Eckert, E.R.G., and Drake, R.M. (1987). "Analysis of Heat and Mass Transfer." *Hemisphere Publishing Corp.*, New York, USA.

Fang, T., GUO, F., and LEE, C.F.F. (2006). "A Note on the Extended Blasius Equation." *Applied. Math. Letters*, Vol. 19, pp. 613–617.

Fawer, C. (1937). "Etude de Quelques Ecoulements Permanents à Filets Courbes." ('Study of some Steady Flows with Curved Streamlines.') *Thesis*, Lausanne, Switzerland, Imprimerie La Concorde, 127 pages (in French).

Fick, A.E. (1855). "On Liquid Diffusion." *Philos. Mag.*, Vol. 4, No. 10, pp. 30–39.

Fischer, H.B., List, E.J., Koh, R.C.Y., Imberger, J., and Brooks, N.H. (1979). "Mixing in Inland and Coastal Waters." *Academic Press*, New York, USA.

Flettner, A. (1925). "The Flettner Rotor Ship. The Utilisation of Aerodynamical Knowledge for the propulsion of Ships by Wind." *Engineering*, Jan. 23, pp. 117–120.

Fourier, J.B.J. (1822). "Théorie Analytique de la Chaleur." ('Analytical Theory of Heat.') *Didot*, Paris, France (in French).

Fox, J.F., Papanicolaou, A.N., and Kjos, L. (2005). "Eddy Taxonomy Methodology around Submerged Barb Obstacle within a Fixed Rough Bed." *Jl of Eng. Mech.*, ASCE, Vol. 131, No. 10, pp. 1082–1101.

Franc, J.P., Avellan, F., Belahadji, B., Billard, J.Y., Briancon-Marjollet, L., Frechou, D., Fruman, D.H., Karimi, A., Kueny, J.L., and Michel, J.M. (1995). "La Cavitation. Mécanismes Physiques et Aspects Industriels." ('The Cavitation. Physical Mechanisms and Industrial Aspects.') *Presses Universitaires de Grenoble*, Collection Grenoble Sciences, France, 581 pages (in French).

Fransson, J.H.M., Matsubara, M., and Alfredsson, P.H. (2005). "Transition Induced by Free-stream Turbulence." *Jl of Fluid Mech.*, Vol. 527, pp. 1–25.

Garbrecht, G. (1987a). "Hydraulics and Hydraulic Research: a Historical Review." *Balkema Publ.*, Rotterdam, The Netherlands.

Gauckler, P.G. (1867). "Etudes Théoriques et Pratiques sur l'Ecoulement et le Mouvement des Eaux." ('Theoretical and Practical Studies of the Flow and Motion of Waters.') *Comptes Rendues de l'Académie des Sciences*, Paris, France, Tome 64, pp. 818–822 (in French).

Gauckler, P. (1897–1902). "Enquête sur les Installations Hydrauliques Romaines en Tunisie." ('Study of the Roman Hydraulic Works in Tunisia.') *Imprimerie Rapide (Louis Nicolas et Co.)*, Tunis, Tunisia, 2 volumes (in French).

Gauckler, P. (1897). "Enquête sur les Installations Hydrauliques Romaines en Tunisie. Tome I." ('Study of the Roman Hydraulic Works in Tunisia. Volume 1.') *Imprimerie Rapide (Louis Nicolas et Co.)*, Tunis, Tunisia, Volume 1, 298 pages (in French).

Gauckler, P. (1902). "Enquête sur les Installations Hydrauliques Romaines en Tunisie. Tome II." ('Study of the Roman Hydraulic Works in Tunisia. Volume 2.') *Imprimerie Rapide (Louis Nicolas et Co.)*, Tunis, Tunisia, Volume 2, 174 pages (in French).

George, W.K. (2006). "Recent Advancements towards the Understanding of Turbulent Boundary Layers." *AIAA Jl*, Vol. 44, No. 11, pp. 2435–2449.

Gerhart, P.M., GROSS, R.J., and Hochstein, J.I. (1992). "Fundamentals of Fluid Mechanics." *Addison-Wesley Publ.*, Reading MA, USA, 2nd edition, 983 pages.

Gjevik, G., Moe, H., and Ommundsen, A. (1997). "Sources of the Maelstrom." *Nature*, Vol. 388, 28 Aug 1997, pp 837–838.

Glauert, H. (1924). "A Generalised Type of Joukowski Aerofoil." *Report and Memoranda No. 911*, Aeronautical Research Committee, Jan., 6 pages & 3 plates.

Goertler, H. (1942). "Berechnung von Aufgaben der freien Turbulenz auf Grund eines neuen Näherungsansatzes." Z.A.M.M., 22, pp. 244–254 (in German).

Hager, W.H. (1992). "Spillways, Shockwaves and Air Entrainment – Review and Recommendations." *ICOLD Bulletin*, No. 81, Jan., 117 pages.

Hager, W.H. (2003). "Blasius: A Life in Research and Education. *Experiments in Fluids*, Vol. 34, No. 5, pp. 566–571.

Hallback, M., Groth, J., and Johansson, A.V. (1989). "A Reynolds Stress Closure for the Dissipation in Anisotropic Turbulent Flows." *Proc. 7th Symp. Turbulent Shear Flows*, Stanford University, USA, Vol. 2, pp. 17.2.1–17.2.6.

Hele-Shaw, H.J.S. (1898). "Investigation of the nature of the surface resistance of water and of stream-line motion under certain experimental conditions." *Trans. Inst. Naval Architects*, Vol. 40.

Helmholtz, H.L.F. (1868). "Über discontinuirliche Flüssigkeits-Bewegungen." *Monatsberichte der königlich preussichen Akademie der Wissenschaft zu Berlin*, pp. 215–228 (in German).

Henderson, F.M. (1966). "Open Channel Flow." *MacMillan Company*, New York, USA.

Hinze, J.O. (1975). "Turbulence." *McGraw-Hill Publ.*, 2nd Edition, New York, USA.

Hokusai, K. (1826–1833). "Thirty-Six Views of Mount Fuji." Japan.

Homsy, G.M. (2000). "Multi-Media Fluid Mechanics." *Cambridge University Press*, Cambridge, UK, CD-ROM.

Homsy, G.M. (2004). "Multi-Media Fluid Mechanics: Multingual Version." *Cambridge University Press*, Cambridge, UK, CD-ROM.

Howe, J.W. (1949). "Flow Measurement." *Proc 4th Hydraulic Conf.*, Iowa Institute of Hydraulic Research, H. ROUSE Ed., John Wyley & Sons Publ., June, pp. 177–229.

Hunt, B. (1968). "Numerical Solution of an Integral Equation for Flow from a Circular Orifice." *Jl of Fluid Mech.*, Vol. 31, Part 2, pp. 361–177.

Hussain, A.K.M.F., and Reynolds, W.C. (1972). "The Mechanics of on Organized Wave in Turbulent Shear Flow. Part 2: Experimental Results." *Jl of Fluid Mech.*, Vol. 54, Part 2, pp. 241–261.

International Organization for Standardization (1979). "Units of Measurements." *ISO Standards Handbook*, No. 2, Switzerland.

Ippen, A.T., and Harleman, R.F. (1956). "Verification of Theory for Oblique Standing Waves." *Transactions*, ASCE, Vol. 121, pp. 678–694.

Japan Society of Mechanical Engineers (1988). "Visualized Flow: Fluid motion in Basic and Engineering Situations Revealed by Flow Visualization." *Pergamon*, Oxford, UK.

Jackson, D. (1995). "Osborne Reynolds: Scientist, Engineer and Pioneer." *Proceedings Mathematical and Physical Sciences*, Vol. 451, No. 1941, Osborne Reynolds Centenary Volume, pp. 49–86.

Jackson, D., and Launder, B. (2007). "Osborne Reynolds and the Publication of his Papers on Turbulent Flow." *Annual Rev. Fluid Mech.*, Vol. 39, pp. 19–35.

Jevons, W.S. (1858). "On Clouds; their Various Forms, and Producing Causes." *Sydney Magazine of Science and Art*, Vol. 1, No. 8, pp. 163–176.

Joukowski, N.E. (1890). "Modification of Kirchoff's Method for Determining the Two Dimensional Motion of a Fluid at a Prescribed Constant Velocity on a Given Streamline." *Proc. Math. Symp.*, Moscow, Russia, Vol. XV.

Karlsson, R.I., and Johansson, T.G. (1986). "LDV Measurements of Hihger Order Moments of Velocity Fluctuations in a Turbulent Boundary Layer." *Proc. 3rd Intl Symp. on Applications of Laser Anemometry to Fluid Mechanics*, Libon, Portugal. (also *Laser Anemometry in Fluid Mechanics III*: Selected Papers from the Third International Symposium on Applications of Laser Anemometry to Fluid Mechanics, 1988, R.J. Adrian, D.F.G. Durao, F. Durst, H. Mishina and J.H. Whitelaw Ed., Ladoan-Ist Publ., Chap. III, pp. 276–289.

Karman, T. von (1921). "Uber laminare und Zeitschrift für angewante Mathematik und Mechanik turbulente Reibung." *Zeitschrift für angewandte Mathematik und Mechanik (ZAMM)*, Vol. 1. (English translation: *NACA Tech. Memo. No. 1092*, 1946.)

Kazemipour, A.K., and Apelt, C.J. (1983). "Effects of Irregularity of Form on Energy Losses in Open Channel Flow." *Aust. Civil Engrg Trans.*, I.E.Aust., Vol. CE25, pp. 294–299.

Kelvin, Lord (1871). "The influence of Wind and Waves in Water Supposed Frictionless." *London, Edinburgh and Dublin Philosophical Magazine and Journal of Science*, Series 4, Vol. 42, pp. 368–374.

Knight, D.W., Demetriou, J.D., and Hamed, M.E. (1984). "Boundary Shear in Smooth Rectangular Channels." *Jl of Hyd. Engrg.*, ASCE, Vol. 110. No. 4, pp. 405–422.

Koch, C., and Chanson, H. (2005). "An Experimental Study of Tidal Bores and Positive Surges: Hydrodynamics and Turbulence of the Bore Front." *Report No. CH56/05*, Dept. of Civil Engineering, The University of Queensland, Brisbane, Australia, July, 170 pages (ISBN 1864998245).

Korn, G.A., and Korn, T.M. (1961). "Mathematical Handbook for Scientist and Engineers." *McGraw-Hill Book Comp.*, New York, USA.

Krogstad, P.A., Andersson, H.I., Bakken, O.M., and Ashrafian, A.A. (2005). An Experimental and Numerical Study of Channel Flow with Rough Walls." *Jl of Fluid Mech.*, Vol. 530, pp. 327–352.

Lagrange, J.L. (1781). "Mémoire sur la Théorie du Mouvement des Fluides." ('Memoir on the Theory of Fluid Motion.') in *Oeuvres de Lagrange*, Gauthier-Villars, Paris, France (printed in 1882) (in French).

Le Billon, P., and Waizenegger, A. (2007). "Peace in the Wake of Disaster? Secessionist Conflicts and the 2004 Indian Ocean Tsunami." *Trans. Institute of British Geographers*, Vol. 32, No. 3, pp. 411–427.

Lesieur, M. (1994). "La Turbulence." ('The Turbulence.') *Presses Universitaires de Grenoble*, Collection Grenoble Sciences, France, 262 pages (in French).

Liggett, J.A. (1993). "Critical Depth, Velocity Profiles and Averaging." *Jl of Irrig. and Drain. Engrg.*, ASCE, Vol. 119, No. 2, pp. 416–422.

Liggett, J.A. (1994). "Fluid Mechanics." *McGraw-Hill*, New York, USA.

Lynch, D.K. (1982). "Tidal Bores." *Scientific American*, Vol. 247, No. 4, Oct., pp. 134–143.

McDowell, S.E., and Rossby, H.T. (1978). *Science*, Vol. 202, p. 1085.

McKeon, B.J., Zagarola, M.V., and Smits, A.J. (2005). "A New Friction Factor Relationship for Fully-Developed Pipe Flow." *Jl of Fluid Mech.*, Vol. 538, pp. 429–443.

Macintosh, J.C. (1990). "Hydraulic Characteristics in Channels of Complex Cross-Section." *Ph.D. thesis*, Univ. of Queensland, Dept. of Civil Eng., Australia, Nov., 487 pages.

Macintosh, J.C., and Isaacs, L.T. (1992). "RPT – The Roving Preston Tube." *Proc. 11th AFMC Australasian Fluid Mechanics Conf.*, Hobart, Australia, Vol. II, Paper 8E-1, pp. 1049–1052.

Magnus, H.G. (1853). "The Drift of Shells." *Poggendorf's Annals*, Germany.

Mariotte, E. (1686). "Traité du Mouvement des Eaux et des Autres Corps Fluides." ('Treaty on the Motion of Waters and other Fluids.') Paris, France (in French) (Translated by J.T. Desaguliers, *Senex and Taylor*, London, UK, 1718).

Massey, B.S. (1989). "Mechanics of Fluids." *Van Nostrand Reinhold*, 6th edition, London, UK.

Meier, G.E.A. (2004). "Prandtl's Boundary Layer Concept and the Work in Göttingen. A Historical View on Prandtl' Scientific Life." *Proc. IUTAM Symp. One Hundred years of Boundary Layer Research*, Göttingen, Germany, G.E.A. Meier and K.R. Sreenivasan Ed., Springer, pp. 1–18.

Michell, J.H. (1890). "On the Theory of Free Streamlines." Phil. Trans. Roy. Soc., London, Part A, Vol. 181, pp.389–431.

Miller, A. (1971). "Meteorology." *Charles Merrill Publ.*, Colombus Oh, USA, 2nd ed., 154 pages.

Milne-Thomson, L.M. (1958). "Theoretical Dynamics." *Macmillan*, New York, London, 3rd edition, 414 pages.

Mises, R. von (1917). "Berechnung von Ausfluss und Uberfallzahlen." *Z. ver. Deuts. Ing.*, Vol. 61, p. 447 (in German).

Mises, R. von (1945). "Theory of Flight." *McGraw-Hill*, New York, USA, 629 pages.

Michioku, K., Maeno, S., Furusawa, T., and Haneda, M. (2005). "Discharge through a Permeable Rubble Mound Weir." *Jl of Hyd. Engrg.*, ASCE, Vol. 131, No. 1, pp. 1–10.

Montes, J.S. (1998). "Hydraulics of Open Channel Flow." *ASCE Press*, New-York, USA, 697 pages.

Morelli, C. (1971). "The International Gravity Standardization Net 1971 (I.G.S.N.71)." *Bureau Central de l'Association Internationale de Géodésie*, Paris, France.

Newman, J.N. (1977). "Marine Hydrodynamics." *MIT Press*, Massachusetts, USA.

Nezu, I., and Nakagawa, H. (1993). "Turbulence in Open-Channel Flows." *IAHR Monograph*, IAHR Fluid Mechanics Section, Balkema Publ., Rotterdam, The Netherlands, 281 pages.

Nikuradse, J. (1932). "Gesetzmässigkeit der turbulenten Strömung in glatten Rohren." ('Laws of Turbulent Pipe Flow in Smooth Pipes.') *VDI-Forschungsheft*, No. 356 (in German) (Translated in NACA TT F-10, 359).

Nikuradse, J. (1933). "Strömungsgesetze in rauhen Rohren." ('Laws of Turbulent Pipe Flow in Rough Pipes.') *VDI-Forschungsheft*, No. 361 (in German) (Translated in NACA Tech. Memo. No. 1292, 1950).

Nishioka, M., and Miyagi, T. (1987). "Measurements of Velocity Distributions in the Laminar Wake of a Flat Plate." Jl of Fluid Mechanics, Vol. 84, Partt 4, pp. 705–715.

Olivier, H. (1967). "Through and Overflow Rockfill Dams – New Design Techniques." *Proc. Instn. Civil Eng.*, March, 36, pp. 433–471. Discussion, 36, pp. 855–888.

Open University Course Team (1995). "Seawater: its Composition, properties and Behaviour." *Butterworth-Heinemann*, Oxford, UK, 2nd edition, 168 pages.

Osterlund, J.M. (1999). "Experimental Studies of Zero Pressure-Gradient Turbulent Boundary Layer Flow." *Ph.D. thesis*, Dept of Mechanics, Royal Institude of technology, Stockholm, Sweden. Databank: {http://www2.mech.kth.se/~jens/zpg/index.html}.

Patel, V.C. (1965). "Calibration of the Preston Tube and Limitations on its use in Pressure Gradients." *Jl of Fluid Mech.*, Vol. 23, Part 1, Sept., pp. 185–208.

Piquet, J. (1999). "Turbulent Flows. Models and Physics." *Springer*, Berlin, Germany, 761 pages.

Pohlhausen, K. (1921). "Zur näherungsweisen Integration der Differentialgleichung der laminaren Grenzschift." *Zeitschrift für angewandte Mathematik und Mechanik (ZAMM)*, Vol. 1, pp. 252–268.

Prandtl, L. (1904). "Über Flussigkeitsbewegung bei sehr kleiner Reibung." ('On Fluid Motion with Very Small Friction.') *Verh. III Intl. Math. Kongr.*, Heidelberg, Germany (in German) (also NACA Tech. Memo. No. 452, 1928).

Prandtl, L. (1925). "Über die ausgebildete Turbulenz." ('On Fully Developed Turbulence.') Z.A.M.M., Vol. 5, pp. 136–139 (in German).

Prasad, A., and Williamson, C.H.K. (1997). "A Method for the Reduction of Bluff Body Drag." *Jl of Wind Eng. & Industrial Aerodynamics*, Vol. 69–71, pp. 155–167.

Preston, J.H. (1954). "The Determination of Turbulent Skin Friction by Means of Pitot Tubes." *Jl Roy. Aeronaut. Soc., London*, Vol. 58, Feb., pp. 109–121.

Quintus Curcius (1984). "The History of Alexander." *Penguin*, New York, USA, translated by J. Yardley, 332 pages.

Rajaratnam, N. (1976). "Turbulent Jets." *Elsevier Scientific*, Development in Water Science, 5, New York, USA.

Raudkivi, A.J., and Callander, R.A. (1976). "Analysis of Groundwater Flow." *Edward Arnold Publisher*, London, UK.

Rayleigh, Lord (1883). "Investigation on the Character of the Equilibrium of an Incompressible Heavy Fluid of Variable Density." *Proc. London Mathematical Society*, Vol. 14, pp. 170–177.

Ré, R. (1946). "Etude du Lacher Instantané d'une Retenue d'Eau dans un Canal par la Méthode Graphique." ('Study of the Sudden Water Release from a Reservoir in a Channel by a Graphical Method.') *Jl La Houille Blanche*, Vol. 1, No. 3, May, pp. 181–187 & 5 plates (in French).

Reid, E.G. (1924). "Tests on Rotating Cylinders." *NACA Technical Note No. 209*, National Advisory Committee on Aeronautics, December, 47 pages.

Reynolds, O. (1883). "An Experimental Investigation of the Circumstances which Determine whether the Motion of Water shall be Direct or Sinuous, and the Laws of Resistance in Parallel Channels." *Phil. Trans. Roy. Soc. Lond.*, Vol. 174, pp. 935–982.

Richardson, L.F. (1922). "Weather Prediction by Numerical Process". London, UK.

Richardson, L.F. (1926). "Atmospheric Diffusion Shown on a Distance-Neighbour Graph." *Proc. Roy. Soc. Lond.*, Vol. A110, p. 709.

Richardson, P.L. (1993). "Tracking Ocean Eddies." *American Scientist*, Vol. 81, pp. 261–271.

Riley, J.P., and Skirrow, G. (1965). "Chemical Oceanography." *Academic Press*, London, UK, 3 volumes.

Rouse, H. (1938). "Fluid Mechanics for Hydraulic Engineers." *McGraw-Hill Publ.*, New York, USA (also Dover Publ., New York, USA, 1961, 422 pages).

Rouse, H. (1959). "Advanced Mechanics of Fluids." *John Wiley*, New York, USA, 444 pages.

Runge, C. (1908). "Uber eine Method die partielle Differentialgleichung $\Delta u$ = constant numerisch zu integrieren." *Zeitschrift der Mathematik und Physik*, Vol 56, pp. 225–232 (in German).

Sarrau (1884). "Cours de Mécanique." ('Lecture Notes in Mechanics.') *Ecole Polytechnique*, Paris, France (in French).

Sato, H., and Sakao, F. (1964). "An Experimental Investigation of the Instability of a Two-dimensional Jet at Low Reynolds Numbers." *Jl Fluid Mech.*, Vol. 20, Part 2, pp. 337–352.

Schetz, J.A. (1993). "Boundary Layer Analysis." *Prentice Hall*, Englewood Cliffs, USA.

Schlichting, H. (1960). "Boundary Layer Theory." *McGraw-Hill*, New York, USA, 4th edition.

Schlichting, H. (1979). "Boundary Layer Theory." *McGraw-Hill*, New York, USA, 7th edition.

Schlichting, H., and Gersten, K. (2000). "Boundary Layer Theory." *Springer Verlag*, Berlin, Germany, 8th edition, 707 pages.

Schnitter, N.J. (1994). "A History of Dams: the Useful Pyramids." *Balkema Publ.*, Rotterdam, The Netherlands.

Shuto, N. (1985). "The Nihonkai-Chubu Earthquake Tsunami on the North Akita Coast." *Coastal Eng. in Japan*, Vol. 20, pp. 250–264.

Sommer, A., and Mosley, W.H. (1972). "East Bengal Cyclone of November 1970. Epidemological Approach to Disaster Assessment." *Le Lancet*, Vol. 299, No. 7759, Sat. 13 May, pp. 1029–1036.

Spiegel (1968). "Mathematical Handbook of Formulas and Tables." *McGraw-Hill Inc.*, New York, USA.

Streeter, V.L. (1948). "Fluid Dynamics." *McGraw-Hill Publications in Aeronautical Science*, New York, USA.

Streeter, V.L., and Wylie, E.B. (1981). "Fluid Mechanics." *McGraw-Hill*, 1st SI Metric edition, Singapore.

Sutherland, W. (1893). "The Viscosity of Gases and Molecular Forces." *Phil. Mag.*, Ser. 5, Vol. 36, pp. 507–531.

Swanson, W.M. (1956). "An Experimental Investigation of the Magnus Effect." *Final Report*, OOR Project No. 1082, Calif. Inst. of Technology, Dec.

Swanson, W.M. (1961). "The Magnus Effect: a Summary of Investigations to Date." *Jl. Basic Engrg*, Trans. ASME, Series D, Vol. 83, pp. 461–470.

Tachie, M.F. (2001). "Open Channel Turbulent Boundary Layers and wall Jets on Rough Surfaces." *Ph.D. thesis*, Dept. of Mech. Eng., Univers. of Saskatchewan, Canada, 238 pages.

Tamburrino, A., and Gulliver, J.S. (2007). "Free-Surface Visualization of Streamwise Vortices in a Channel Flow." *Water Res. Res.*, Vol. 43, Paper W11410, doi:10.1029/2007WR005988.

Taylor, G.I. (1953). "Dispersion of Soluble Matter in Solvent Flowing Slowly Through a Tube." *Proc. Roy. Soc. Lond.*, Series A, Vol. 219, pp. 186–203.

Taylor, G.I. (1954). "The Dispersion of Matter in Turbulent Flow Through a Pipe." *Proc. Roy. Soc. Lond.*, Series A, Vol. 223, pp. 446–468.

Thom, A. (1934). "Effects of Discs on the Air Forces on a Rotating Cylinder." *Reports and Memoranda No. 1623*, Aircraft Research Coucil, UK.

Trevethan, M., Chanson, H., and Brown, R.J. (2006). "Two Series of Detailed Turbulence Measurements in a Small Subtropical Estuarine System." *Report No. CH58/06*, Div. of Civil Engineering, The University of Queensland, Brisbane, Australia, March, 153 pages.

Trevethan, M., Chanson, H., and Brown, R.J. (2007). "Turbulence and Turbulent Flux vents in a Small Sub-tropical Estuary." *Report No. CH65/07*, Hydraulic Model Report CH series, ivision of Civil Engineering, The University of Queensland, Brisbane, Australia, November, 67 pages.

Trevethan, M., Chanson, H., and Brown, R. (2008). "Turbulence Characteristics of a Small Subtropical Estuary during and after some Moderate Rainfall." *Estuarine Coastal and Shelf Science*, Vol. 79, No. 4, pp. 661–670 (DOI: 10.1016/j.ecss.2008.06.006).

Troskolanski, A.T. (1960). "Hydrometry: Theory and Practice of Hydraulic Measurements." *Pergamon Press*, Oxford, UK, 684 pages.

Tsukahara, T., Tillmark, N., and Alfredsson, P.H. (2010). "Flow Regimes in a Plane Couette Flow with System Rotation." *Jl. of Fluid Mech.*, Vol. 648, pp. 5–33 (doi:10.1017/S0022112009993880).

Vallentine, H.R. (1959). "Applied Hydrodynamics." *Butterworths*, London, UK, 1st edition.

Vallentine, H.R. (1969). "Applied Hydrodynamics." *Butterworths*, London, UK, SI edition.

Van DYKE, M. (1982). "An Album of Fluid Motion." *Parabolic Press*, Stanford, California, USA.

Virlogeux, M. (1993). "Wind Design and Analysis for the Normandy Bridge." *Structural Engineering in Natural Hazards Mitigation*, Proc. Structures Congress, Irvine CAL, USA, Vol. 1, pp. 478–483.

Webster, T.M. (2005). "The Dam Busters Raid: Success or Sideshow?" *Air Power History*, Vol. 52, Summer, pp. 12–25.

Xie, Q. (1998). "Turbulent Flows in Non-Uniform Open Channels: Experimental Measurements and Numerical Modelling." *Ph.D. thesis*, Dept. of Civil Eng., University Of Queensland, Australia, 339 pages.

Yu, Kwonkyu, and Yoon, Byungman (2005). "A Velocity-Profile Formula Valid in the Inner Region of Turbulent Flow." *Proc. 31th Biennial IAHR Congress*, Seoul, Korea, B.H. Jun, S.I. Lee, I.W. Seo and G.W. Choi Editors, pp. 378–385.

Zhu, Q., Wolfgang, M.J., Yue, D.K.P., and Triantafyllou, M.S. (2002). "Three-Dimensional Flow Structures and Vorticity Control in Fish-Like Swimming." *Jl of Fluid Mech.*, Vol. 468, pp. 1–28.

## INTERNET REFERENCES

| | |
|---|---|
| Wolfram Mathworld | {http://mathworld.wolfram.com/} |
| Structurae database | {http://www.structurae.de/en/index.php} |
| Structurae database – La Grande Arche | {http://www.structurae.de/en/structures/data/str00133.php} |
| Structurae database – Le Pont de Normandie | {http://www.structurae.de/en/structures/data/str00048.php} |
| IAHR Media Library | {http://www.iahrmedialibrary.net} |
| Research publications by Hubert Chanson | {http://espace.library.uq.edu.au/list.php?browse=author&author_id=193} |

## OPEN ACCESS REPOSITORIES

| | |
|---|---|
| UQeSpace open access repository | {http://espace.library.uq.edu.au/} |
| OAIster open acces catalogue | {http://www.oaister.org/} |
| Directory of Open Access Repositories | {http://www.opendoar.org/} |

## BIBLIOGRAPHY

### Books

Chanson, H. (2009). "Applied Hydrodynamics: An Introduction to Ideal and Real Fluid Flows." *CRC Press*, Taylor & Francis Group, Leiden, The Netherlands, 478 pages.

Library of Congress call number: TC171 .C54 2009
Comments: This book is the first version of the present book. It contains some colour plates. Based upon the lecture materials of the course Advanced Fluid Mechanics taught at the University of Queensland.

Vallentine, H.R. (1969). "Applied Hydrodynamics." *Butterworths*, London, UK, SI edition.

> Library of Congress call number: TC 171 .V3 1959 2
> Comments: This book is no longer published. However the book covers most of the program. It is well-written and the pedagogy is excellent. Professor H. Rupert Vallentine (1917–2010) worked at the Department of Civil Engineering, University of New South Wales, Australia.

Streeter, V.L. (1948). "Fluid Dynamics." *McGraw-Hill Publications in Aeronautical Science*, New York, USA.

> Library of Congress call number: QA 911 .S84 1948 1
> Comments: This book is no longer published. But the book covers most of the program. Good mathematical explanations. American approach. Professor Streeter worked at the Department of Civil Engineering, University of Michigan, USA.

Liggett, J.A. (1994). "Fluid Mechanics." *McGraw-Hill*, New York, USA.

> Library of Congress call number: QA901 .L54 1994
> Comments: This book is an advanced fluid mechanics textbook which an elegant presentation of the lecture material. Professor Liggett is an Emeritus Professor at Cornell University, USA and he was formerly the Editor of the Journal of Hydraulic Engineering.

Schlichting, H. (1979). "Boundary Layer Theory." *McGraw-Hill*, New York, USA, 7th edition.

> Library of Congress call number: TL574.B6 S283 1979
> Comments: This book is a basic reference in boundary layer theory. Although a 8th edition was published in 2000, many engineers and researchers continue to use the 1979 edition. Professor Schlichting was a student of Ludwig Prandtl and he became later Professor at the Technical University of Braunschweig, Germany.

Kennard, E.H. (1967). "Irrotational Flow of Frictionless Fluids, Mostly of Invariable Density." *Research and Development Report No. 2299*, Department of the Navy, David Taylor Model Basin, USA, 412 pages.
Comments: This report presents the fundamental of potential flow of frictionless fluids. Both two- and three-dimensional solutions are developed.

Kirchhoff, R.H. (1985). "Potential Flows. Computer Graphic Solutions." *Marcel Dekker Inc.*, New York, USA, 1985.

> Library of Congress call number: TA 357 .K57 1985 1
> Comments: This book presents graphic and computer solutions of several classical problems. Computer programs in Basic. Professor R.H. Kirchhoff is working at the Department of Mechanical Engineering, University of Massachussetts, USA.

Prandtl, L., and Tietjens, O.G. (1934). "Fundamentals of Hydro- and Aeromechanics." *Dover*, Engineering Societies Monographs, New York, USA.
Comments: Based upon the original lecture notes of Ludwig Prandtl.

Prandtl, L., and Tietjens, O.G. (1934). "Applied Hydro- and Aeromechanics." *Dover*, Engineering Societies Monographs, New York, USA.
Comments: Based upon the original lecture notes of Ludwig Prandtl.

Van Dyke, M. (1982). "An Album of Fluid Motion." *Parabolic Press*, Stanford, California, USA.

> Library of Congress call number: TA 357 .A53 1982 1
> Comments: The author assembled a collection of photographs of flow phenomena. Very good pictures. Professor Van Dyke was working at the Department of Mechanical Engineering, University of Stanford, California, USA.

## Multimedia digital materials

Homsy, G.M. (2000). "Multi-Media Fluid Mechanics: Multingual Version." *Cambridge University Press*, Cambridge, UK, CD-ROM.

> Library of Congress call number: TA 357.M85 2000
> Comments: One of the first successful multimedia documents in Fluid Mechanics. Simple and solid explanations. Good, simple video documentaries in Quicktime format.

Homsy, G.M. (2007). "Multi-Media Fluid Mechanics: Multingual Version." *Cambridge University Press*, Cambridge, UK, 2nd edition, CD-ROM.

> Library of Congress call number: TA 357.M852 2007

## Audiovisual materials

Appendix F (This book) presents a series of movies to illustrate some aspects of applied hydrodynamics and its applications into the real world.

Tasmania: Australia's awakening Island
> Videocassette VHS (4 mn 30 sec.)
> University of Queensland Library call number: GC63.A54T31991
> Comments: After underwater exploration in Bass Strait, Jean-Michel Cousteau leads the Cousteau expedition aboard the windship "Alcyon", a type of rotorship which utilises the Magnus effect for its propulsion, from its anchorage in Adventure Bay, Tasmania, up the Gordon river.

The Magnus effect
> Videocassette VHS (4 mn)
> University of Queensland Library call number: TA357.F64NO.11ETC
> Comments: Film on an experiment illustrating the Magnus effect.

"Les Fils de la Lune" (2005) by Philippe Lespinasse, Grand Angle production, France, 50 minutes.

> Comments: Documentary on tidal bores in France, UK, Brazil and China, and kayakists riding the bores. The documentary "Les Fils de la Lune" ('Sons of the Moon') was shown in Thalassa on channels FR3 and TV5 in November 2005 (over 120 Millions of subscribers worldwide).

"La Tribu du Mascaret" (2004) by Philippe Lespinasse, Grand Angle production, France, 30 minutes.
> Comments: Documentary on the tidal bore of the Dordogne River in France. The documentary "La Tribu du Mascaret" ('Surfing the Dordogne') was shown in Thalassa on channel France 3 on 9–10 Dec. 2004. The English translation was prepared by, and screened in Australia as "Surfing the Dordogne" by, SBS (Australia) on Sat. 7 Oct. 2006.

IAHR Media Library {http://www.iahrmedialibrary.net}
> Comments: Internet-based open access library of videos and photographs in hydraulic engineering and applied fluid mechanics.

## ABBREVIATIONS OF JOURNALS AND INSTITUTIONS

| | |
|---|---|
| *AFMC* | Australasian Fluid Mechanics Conference |
| *AGU* | American Geophysical Union (USA) |
| *AIAA Jl* | Journal of the American Institute of Aeronautics and Astronautics (USA) |
| *ANCOLD* | Australian Committee on Large Dams |
| *Ann. Chim. Phys.* | Annales de Chimie et Physique, Paris (France) |

| | |
|---|---|
| *ANSSR* | Academy of Sciences of the USSR, Moscow |
| *APHA* | American Public Health Association |
| *ARC* | Aeronautical Research Council (UK) |
| | Australian Research Council |
| *ARC RM* | Aeronautical Research Council Reports and Memoranda |
| *ARC CP* | Aeronautical Research Council Current Papers |
| *ASAE* | American Society of Agricultural Engineers |
| *ASCE* | American Society of Civil Engineers |
| *ASME* | American Society of Mechanical Engineers |
| *AVA* | Aerodynamische Versuchanstalt, Göttingen (Germany) |
| *BHRA* | British Hydromechanics Research Association (BHRA Fluid Engineering) |
| *BSI* | British Standards Instituion, London |
| *CIRIA* | Construction Industry Research and Information Association |
| *EDF* | Electricité] de France |
| *EPA* | Environmental protection Agency |
| *Ergeb. AVA Göttingen* | Ergebnisse Aerodynamische Versuchanstalt, Göttingen (Germany) |
| *Forsch. Ing. Wes.* | Forschung auf dem Gebiete des Ingenieur-Wesens (Germany) |
| *Forschunsheft* | Research supplement to Forsch. Ing. Wes. (Germany) |
| *Gid. Stroit.* | Gidrotekhnicheskoe Stroitel'stvo (Russia) |
| | (translated in Hydrotechnical Construction) |
| *IAHR* | International Association for Hydraulic Research |
| *IAWQ* | International Association for Water Quality |
| *ICOLD* | International Committee on Large Dams |
| *IEAust.* | Institution of Engineers, Australia |
| *IIHR* | Iowa Institute of Hydraulic Research, Iowa City (USA) |
| *Ing. Arch.* | Ingenieur-Archiv (Germany) |
| *JAS* | Journal of Aeronautical Sciences (USA) (replaced by JASS in 1959) |
| *JASS* | Journal of AeroSpace Sciences (USA) (replaced by AIAA Jl in 1963) |
| *Jl Fluid Mech.* | Journal of Fluid Mechanics (Cambridge, UK) |
| *Jl Roy. Aero. Soc.* | Journal of the Royal Aeronautical Society, London (UK) |
| *JSCE* | Japanese Society of Civil Engineers |
| *JSME* | Japanese Society of Mechanical Engineers |
| *Luftfahrt-Forsch.* | Luftfahrt-Forschung (Germany) |
| *NACA* | National Advisory Committee on Aeronautics (USA) |
| *NACA Rep.* | NACA Reports (USA) |
| *NACA TM* | NACA Technical Memoranda (USA) |
| *NACA TN* | NACA Technical Notes (USA) |
| *NASA* | National Aeronautics and Space Administration (USA) |
| *NBS* | National Bureau of Standards (USA) |
| *ONERA* | Office National d'Etudes et de Recherches Aérospatiales (France) |
| *Phil. Mag.* | Philosophical Magazine |
| *Phil. Trans. R. Soc. Lond.* | Philosophical Transactions of the Royal Society of London (UK) |
| *Proc. Cambridge Phil. Soc.* | Proceedings of the Cambridge Philosophical Society (UK) |
| *Proc. Instn. Civ. Engrs.* | Proceedings of the Institution of Civil Engineers (UK) |
| *Proc. Roy. Soc.* | Proceedings of the Royal Society, London (UK) |
| *Prog. Aero. Sci.* | Progress in Aerospace Sciences |
| *Proc. Cambridge Phil. Soc.* | Transactions of the Cambridge Philosophical Society (UK) |
| *SAF* | St Anthony Falls Hydraulic Laboratory, Minneapolis (USA) |
| *SHF* | Société Hydrotechnique de France |
| *SIA* | Société des Ingénieurs et Architectes (Switzerland) |

| | |
|---|---|
| *Trans. Soc. Nav. Arch. Mar. Eng.* | Transactions of the Society of Naval Architects and Marine Engineers |
| *USBR* | United States Bureau of Reclamation, Department of the Interior |
| *VDI Forsch.* | Verein Deutsche Ingenieure Forschungsheft (Germany) |
| *Wat. Res. Res.* | Water Resources Research Journal |
| *WES* | US Army Engineer Waterways Experiment Station |
| *Z.A.M.M.* | Zeitschrift für angewandete Mathematik und Mechanik (Germany) |
| *Z.A.M.P.* | Zeitschrift für angewandete Mathematik und Physik (Germany) |
| *Z. Ver. Deut. Ingr.* | Zeitschrift Verein Deutsche Ingenieure (Germany) |

## Common bibliographical abbreviations

| | |
|---|---|
| *Conf.* | Conference |
| *Cong.* | Congress |
| *DEng.* | Doctor of Engineering |
| *Intl.* | International |
| *Jl* | Journal |
| *Mitt.* | Mitteilungen |
| *Ph.D.* | Doctor of Philosophy |
| *Proc.* | Proceedings |
| *Symp.* | Symposium |
| *Trans.* | Transactions |

# Subject index

Subjects in *italic* refers to author names.

# Suggestion/Correction form

Thank you for your comments and suggestions regarding the book. They will be helpful to improve the book in the future. Further, if you find a mistake or an error, please record the error(s) on this page and address it to the author:

Professor Hubert Chanson
School of Civil Engineering, The University of Queensland, Brisbane QLD 4072, Australia
Fax.: (61 7) 33 65 45 99   Email: h.chanson@uq.edu.au   Url: http://www.uq.edu.au/~e2hchans/

Corrections and updates will be posted at
{http://www.uq.edu.au/~e2hchans/reprints/book15_2.htm}.

## Suggestion/Correction form (Applied Hydrodynamics: an Introduction)

### Contact

| Name: | |
|---|---|
| Address: | |
| Tel.: | |
| Fax: | |
| Email: | |

### Description of the suggestion, correction, error

| Part number: | |
|---|---|
| Page number: | |
| Line number: | |
| Figure number: | |
| Equation number: | |

**Proposed correction**

**Further comment**

For Product Safety Concerns and Information please contact our EU
representative GPSR@taylorandfrancis.com Taylor & Francis Verlag GmbH,
Kaufingerstraße 24, 80331 München, Germany

Printed and bound by CPI Group (UK) Ltd, Croydon, CR0 4YY
01/05/2025
01858561-0002